Biological and Medical Physics, Biomedical Engineering

This series is intended to be comprehensive, covering a broad range of topics important to the study of the physical, chemical and biological sciences. Its goal is to provide scientists and engineers with monographs- both, authored and contributed volumes to address the growing need for information. The fields of biological and medical physics and biomedical engineering are broad, multidisciplinary and dynamic. They lie at the crossroads of frontier research in physics, biology, chemistry, and medicine.

Books in the series emphasize established and emergent areas of science including molecular, membrane, and mathematical biophysics; photosynthetic energy harvesting and conversion; information processing; physical principles of genetics; sensory communications; automata networks, neural networks, and cellular automata. Equally important is coverage of applied aspects of biological and medical physics and biomedical engineering such as molecular electronic components and devices, biosensors, medicine, imaging, physical principles of renewable energy production, advanced prostheses, and environmental control and engineering.

Nandakumar Kalarikkal · B. C. Bhadrapriya ·
Bosely Anne Bose · Parasuraman Padmanabhan ·
Sabu Thomas · Murukeshan Vadakke Matham
Editors

Multimodal Biomedical Imaging Techniques

Springer

Editors
Nandakumar Kalarikkal
Mahatma Gandhi University
Kottayam, Kerala, India

Bosely Anne Bose
Mahatma Gandhi University
Kottayam, Kerala, India

Sabu Thomas
Vice Chancellor, Mahatma Gandhi University
Kottayam, Kerala, India

B. C. Bhadrapriya
Mahatma Gandhi University
Kottayam, Kerala, India

Parasuraman Padmanabhan
Lee Kong Chian School of Medicine
Nanyang Technological University
Singapore, Singapore

Murukeshan Vadakke Matham
School of Mechanical and Aerospace Engineering
Nanyang Technological University
Singapore, Singapore

ISSN 1618-7210 ISSN 2197-5647 (electronic)
Biological and Medical Physics, Biomedical Engineering
ISBN 978-981-96-1123-2 ISBN 978-981-96-1124-9 (eBook)
https://doi.org/10.1007/978-981-96-1124-9

This Springer imprint is published by the registered company Springer Nature Singapore Pte Ltd.
The registered company address is: 152 Beach Road, #21-01/04 Gateway East, Singapore 189721, Singapore

Preface

In the present world scenario, biomedical imaging can easily be considered one of the modern medical techniques to diagnose, treat, and monitor diseases. Being able to visualise the internal structures and functions of the body without invasive procedures opens new avenues in research as well as in clinical practice. This book, *Multimodal Biomedical Imaging Techniques*, provides a comprehensive view of what is present and going to emerge in the field of biomedical imaging from its historical backgrounds to the state-of-the-art technologies that are revolutionizing health care. It dives deep into such technological innovations, methodologies, and applications that are continuously making biomedical imaging an essential tool in medical practice today.

Each chapter of the book is carefully crafted to give a comprehensive look at specific imaging modalities, materials, and methodologies that shape the field. The area of biomedical imaging requires collaborative input from physics, biology, chemistry, nanotechnology, and computer science in order to expand the paradigm of medical diagnostics and treatment.

Chapter 1, Biomedical Imaging: State of the Art, sets the stage by overviewing the main biomedical imaging technologies, from X-rays to MRI, CT scanning, ultrasound, and nuclear medicine. By detailing the development of these techniques—from their origins to the current state—the chapter showcases how advancements, including AI, molecular imaging, and more, significantly influence the diagnostic capacities of these techniques, particularly in guiding personalized treatment plans and enhancing patient recovery.

In Chap. 2, In Vivo Fluorescence Imaging and its Role in Multimodal Bioimaging, emphasis is laid on in vivo bioimaging in real-time visualization of biological processes without unnecessary invasion. The chapter discusses the progress towards the development of novel imaging probes, such as quantum dots and near-infrared fluorescent probes, that push the limit of resolution and specificity to produce better images than are possible with conventional modalities. These probes circumvent two main disadvantages of traditional imaging—attenuation of light and autofluorescence—for the realization of more productive diagnostics.

Chapter 3 focuses on hybrid imaging systems that contain hybrid magnetic resonance imaging/fluorescence molecular tomography hybrid imaging. This has emerged as one of the newest developments, with MRI along with fluorescence molecular tomography being used for an in-depth study of molecular and cellular processes. The chapter focuses on the merits and demerits of the technique, along with its applications toward understanding diseases, such as cancer and neurodegenerative diseases.

Chapter 4 on Rare Earth-Based Nanostructures for Medical Bio-Imaging sheds light on the employment of rare earth elements in applications of bioimaging, such as multimodal nanostructure composed of superior photostability and biocompatibility, therefore, ultimately becoming crucial components of improving medical imaging quality. Synthesis and functionalization of nanomaterials guide its application as a multimodal imaging agent.

Synthesis routes of ultrasmall renal clearable nanoparticles for medical imaging applications are introduced in Chap. 5. Such ultrasmall nanoparticles less than 10 nm are cleared by the kidneys, which reduces long-term toxicity. Different synthesis techniques, along with the importance of size control, surface modification, and biocompatibility for medical imaging purposes, are reviewed in this chapter.

Chapter 6, MRI Contrast Agents Based on Iron Oxide Nanoparticles, shifts focus to the iron oxide nanoparticles to function as safer substitutes in place of gadolinium-based contrast agents. The chapter discusses the morphology, biocompatibility, and biodegradability of these nanoparticles to reflect their future promise with minimal toxicity to be good MRI contrast agents.

Chapter 7, Recent Advances in Biomedical Imaging for Cancer Diagnosis and Therapy, discusses the constant breakthroughs in biomedical imaging, primarily in diagnostics and treatment of cancer. This chapter will discuss how AI, machine learning, and nanotechnology advance innovations in medical imaging frontiers that detect cancer very early on and allow drugs to target cancer cells better.

Chapter 8 covers applications in bioimaging of metal and metal oxide nanoparticles: exploiting nanoparticle technology in advancing tissue imaging and the detection of disease. The metal and metal oxide nanohybrids are very fascinating due to their unique properties, making them adaptable for gene delivery, theranostics, and cell tracking. The chapter also discusses the potential application of metal and metal oxide nanoparticles for the visualization of a specific region of disease.

In Chap. 9, Advances in Graphene-Based Nanostructures for Multimodal Bioimaging, the spotlight has been placed on graphene nanostructures. The versatility of graphene and its derivatives is of immense value for various imaging modalities, ranging from MRI to photoacoustic imaging. This chapter provides a comprehensive review of ways in which graphene is being used to improve the resolution of imaging and functional versatility.

Tailoring Nanoscale Metal-Organic Frameworks for Medical Imaging Applications, Chap. 10, discusses in good detail the promise of MOFs to provide diagnostic imaging in various applications. The nanoscale structures are tailored to provide high-resolution images with improved contrast, which offers targeted imaging solutions that revolutionize techniques for diagnosis.

The stimuli-responsive nanomaterials discussed in Chap. 11 open exciting new possibilities for bioimaging applications. Tunable contrast and specificity of diagnostic imaging with pH, temperature, or light-responsive materials shall further improve the quality of imaging.

Chapter 12, Potential of MRI in Clinical Medicine, covers the broader applications of MRI-from neurodegenerative disorders to diseases of the cardiovascular system. The imaging detail makes MRI such an important tool not only for detailed diagnosis but also as an aid to treatment planning in so many different medical specialities.

It goes further to explore the horizon of neuroscience in the study of activity in the neurons and connectivity between the different parts of the brain with the help of fMRI. In Chap. 13, Progress and Prospects in Functional MRI (fMRI) Studies of Neural Networks, it seems that integration into other neuroimaging techniques comes with a promise of new comprehensive views of brain functions and opens avenues for personalized treatments.

Chapter 14, Biomedical Imaging: Scope for Future Studies and Applications, can be seen as the final beacon towards the future. Innovative views toward this promising area of study are thereby presented. The scope for future studies with applications relating to the inclusion of AI, nanotechnology, and molecular imaging offers a window into what is to be expected and, as such, pushes the envelope for healthcare and medical research.

The purpose of this book is to be an invaluable reference for graduates and postgraduates, research scholars (primarily in the field of materials science, biomedical imaging, and nanoscience and nanotechnology), professionals and students in the field of biomedical field, radiologic technologists, and those who want to nurture their scientific temper, with the aim of gaining insight into the basic principles and emerging technologies in biomedical imaging.

Kottayam, Kerala, India — Nandakumar Kalarikkal
Kottayam, Kerala, India — B. C. Bhadrapriya
Kottayam, Kerala, India — Bosely Anne Bose
Singapore, Singapore — Parasuraman Padmanabhan
Kottayam, Kerala, India — Sabu Thomas
Singapore, Singapore — Murukeshan Vadakke Matham

Contents

About the Editors

Prof. Nandakumar Kalarikkal is the Director of International Centre for Ultrafast Studies at Mahatma Gandhi University, Kerala, India. He was the former Director and Joint Director of the International and Inter University Centre for Nanoscience and Nanotechnology and former Director and Chair of School of Pure and Applied Physics and Director-in-Charge of School of Nanoscience and Nanotechnology, Mahatma Gandhi University, India. With more than 30 years of research and academic experience, his research group is specialized in syntheses, characterization, and applications of various nanomaterials. The different topics of current research works include nanomultiferroics, nanosemiconductors, nanophosphors, nanocomposites, nanomedicine, nanosensors, LASER-matter interactions, catalysis, water splitting, phase transitions, and ion beam irradiation effects. He is a recipient of research fellowships and associateships from prestigious government organizations such as the Department of Science and Technology and Council of Scientific and Industrial Research of Government of India. He is CNRS Professor, Claud-Bernard Lyon University, Lyon, France; Professor@Lorraine and Visiting Professor, Institute of Jean Lamour, Nancy, France; Visiting Professor, Stockholm University, Sweden; Visiting Professor, Jožef Stefan Institute, Ljubjana, Slovenia; Guest Scientist, Leibniz-Institut fur Polymerforschung Dresden, Germany; Visiting Fellow, Jawaharlal Nehru Centre for Advanced Scientific Research, Bangalore, India, etc. He has published over 300 research articles in international refereed journals, edited 35 books, and served as Principal Investigator (PI) and Co-Principal Investigator (Co-PI) of 35 national and international projects. He also holds 8 patents and has supervised/co-supervised 32 Ph.D. and 30 Master's students.

B. C. Bhadrapriya is a research scholar at Mahatma Gandhi University in Kottayam, Kerala, India. She holds a Bachelor's degree in Physics from Kerala University and a Master's and M.Phil. degrees in Physics from Mahatma Gandhi University, Kerala, India. Her research focuses on developing magnetic nanoparticles and magnetoelectric nanocomposites for energy and sensing applications. She has published 4 research articles in peer-reviewed international journals and has authored 5 book chapters to her credit.

Bosely Anne Bose is a research scholar at Mahatma Gandhi University, Kottayam, Kerala, India. She holds a Bachelor's, Master's, and M.Phil. degrees in Physics from Mahatma Gandhi University, Kottayam, Kerala, India. Her current research focuses on developing carbon-based nanostructures for environmental remediation applications. She published 3 research articles in peer-reviewed international journals and has authored 5 book chapters to her credit.

Dr. Parasuraman Padmanabhan is the Head of Operations at the Cognitive Neuroimaging Centre (CoNiC) at Lee Kong Chian School of Medicine, NTU, Singapore. He brings extensive expertise in establishing various in vivo imaging modalities, including Optical, MRI, PET, SPECT, and CT, for monitoring cellular events in mice, rats, and non-human primates. His primary research focuses on developing innovative imaging assays to study cellular signal transduction networks in living animals. With over 30 years of research and teaching experience, Dr. Parasuraman has contributed to several prestigious institutions, including Cornell, Stanford University, and A*STAR's SBIC Research Institute.

Prof. Sabu Thomas is the Former Vice Chancellor of Mahatma Gandhi University, Kottayam, Kerala, India. He currently serves as the Director and Professor of the International and Interuniversity Centre for Nanoscience and Nanotechnology at Mahatma Gandhi University, Kerala, and is the Chairman of the Trivandrum Engineering Science and Technology Research Park (TrEST Research Park). He has guided over 120 Ph.D. students. He has received several national and international awards for best researcher and is a member of numerous scientific bodies and committees. So far, 3 universities have honored him with honorary doctorates, which include the University of South Brittany, France; the University of Lorraine, France; and the Siberian Federal University, Russia. He has authored more than 1,400 research publications and 205 edited books with renowned publishers like the Royal Society of Chemistry, Elsevier Wiley, etc. He has a notable h-index of 132 and has accumulated over 82,100+ citations.

Prof. Murukeshan Vadakke Matham is the Director of the Center for Optical and Laser Engineering (COLE) and Deputy Director of The Photonics Institute. He leads a research group specializing in Biomedical Optics, Nanoscale Optics, Laser-based Micro and Nanoscale Fabrication, and Applied Optics for Metrology. With over 25 years of professional experience, including 3 years of postdoctoral work and 22 years of teaching, Dr. Murukeshan has an extensive academic profile. He has published over 200 international journal papers and over 170 international conference proceedings/presentations, co-authored a book, and holds 22 patents and 12 innovation disclosures. To date, he has supervised 26 Ph.D. students and 4 Master's students and is currently supervising 4 Ph.D. students. Dr. Murukeshan is also Associate Editor for Optical Engineering and the International Journal of Optomechatronics (IJO). He is a Fellow of SPIE, a Fellow of the Institute of Physics (UK), and a Distinguished Fellow of the Optical Society of India.

Chapter 1
Biomedical Imaging: State of the Art

Sayantan Ghosh, Parasuraman Padmanabhan, Nandakumar Kalarikkal, Murukeshan Vadakke Matham, and Balázs Gulyás

Abstract Biomedical imaging involves using non-invasive physical devices to create anatomical visualizations crucial for clinical analysis and medical diagnostics. This technology aids in early disease diagnosis, better understanding of pathological conditions, and precise treatment planning. The field encompasses various modalities that capture images of organs, bones, and tissues, providing essential information for timely medical interventions. Historically, biomedical imaging began with Roentgen's discovery of X-rays in 1895, followed by advancements like CT in the 1970s and MRI in the 1980s. Each of these modalities, including advanced technologies, such as ultrasound and nuclear medicine techniques like PET and SPECT, has its unique mechanisms and benefits. Understanding these principles is key to recognizing the capabilities and limitations of each imaging technique. Recent developments in AI, molecular imaging, and wearable devices, along with breakthroughs in quantum imaging, are pushing the boundaries of what is possible. These technologies play a vital role in diagnostics, guiding interventions, monitoring treatments, and enabling personalized medicine across various medical fields. Ongoing research

S. Ghosh
Department of Integrative Biology, Vellore Institute of Technology, Vellore 632014, India

M. V. Matham
School of Mechanical and Aerospace Engineering, Centre for Optical and Laser Engineering (COLE), Nanyang Technological University, 50 Nanyang Avenue, Singapore 639798, Singapore

P. Padmanabhan (✉) · M. V. Matham · B. Gulyás
Imaging Probe Development Platform, Nanyang Technological University, 59 Nanyang Drive, Singapore 636921, Singapore
e-mail: ppadmanabhan@ntu.edu.sg

P. Padmanabhan · B. Gulyás
Cognitive Neuroimaging Centre, Nanyang Technological University, 59 Nanyang Drive, Singapore 636921, Singapore

B. Gulyás
Department of Clinical Neuroscience, Karolinska Institute, 17176 Stockholm, Sweden

N. Kalarikkal
School of Pure and Applied Physics, Mahatma Gandhi University, Kottayam 686560, Kerala, India

N. Kalarikkal et al. (eds.), *Multimodal Biomedical Imaging Techniques*,
Biological and Medical Physics, Biomedical Engineering,
https://doi.org/10.1007/978-981-96-1124-9_1

and innovations promise to address existing challenges and propel the field forward, emphasizing the importance of interdisciplinary collaboration to fully realize the potential of biomedical imaging.

1.1 Introduction

Biomedical imaging involves the use of physical devices to non-invasively create anatomical visualizations for clinical analysis and medical diagnostics. This helps in early diagnosis, better understanding of diseases, and precise treatment planning. Biomedical imaging encompasses various modalities using different approaches to capture images ranging from organs, bones, and even tissues, providing crucial information for timely medical interventions [1, 2] as discussed in the following sections.

Biomedical imaging stands at the forefront of modern healthcare, revolutionizing diagnostics and treatment strategies through non-invasive technologies. These instruments, ranging from X-rays to advanced MRI scanners, play a pivotal role in early disease detection and therapeutic planning. By capturing detailed images of organs, bones, and tissues, they provide clinicians with invaluable insights into anatomical structures and pathological conditions.

The evolution of biomedical imaging traces back to Wilhelm Roentgen's groundbreaking discovery of X-rays in 1895, which opened new avenues for medical diagnostics. Subsequent milestones, such as the advent of computed tomography (CT) in the 1970s and magnetic resonance imaging (MRI) in the 1980s, expanded the capabilities of medical imaging, enabling deeper anatomical visualization and tissue characterization.

Each imaging modality, whether ultrasound, PET, SPECT, or molecular imaging techniques, operates on distinct principles tailored to specific clinical needs. Ultrasound, for instance, utilizes sound waves to create real-time images, ideal for examining fetal development or soft tissue abnormalities. In contrast, PET and SPECT employ radioactive tracers to visualize metabolic processes and cellular functions, essential for oncological and neurological assessments.

Recent advancements in artificial intelligence (AI) have further augmented the diagnostic precision of these imaging technologies. AI algorithms can analyze vast datasets from imaging studies, assisting radiologists in detecting subtle anomalies and predicting disease progression with unprecedented accuracy. Moreover, the integration of wearable devices and molecular imaging techniques promises personalized approaches to treatment, optimizing therapeutic outcomes based on individual patient profiles.

Looking forward, quantum imaging holds promise for achieving even higher resolution and sensitivity in biomedical imaging. By harnessing quantum phenomena, researchers aim to surpass current limitations, potentially revolutionizing how diseases are diagnosed and monitored in clinical practice. Continued interdisciplinary collaboration between physicists, engineers, and medical professionals will be crucial in harnessing these technologies to their fullest potential.

In conclusion, biomedical imaging continues to evolve rapidly, driven by technological innovation and the pursuit of enhanced patient care. As these advancements unfold, they promise to reshape medical diagnostics, treatment strategies, and personalized medicine, ushering in a new era of precision healthcare.

1.2 Historical Perspective

Biomedical imaging started with the discovery of X-rays by Roentgen in 1895, enabling non-intrusive visualization of bones and other internal structures and bringing about the field of radiology. Computed Tomography (CT) was invented in the 1970s, paving the way for the generation of three-dimensional images. Magnetic resonance imaging (MRI) emerged in the 1980s, providing soft tissue contrast without dependence on ionizing radiation. Ultrasound became vital in obstetrics and cardiology due to its real-time imaging capabilities. The advent of nuclear medicine techniques like positron emission tomography (PET) and single-photon emission computed tomography (SPECT) allowed for functional imaging. These historical milestones laid the foundation for the sophisticated imaging technologies used today [3, 4].

1.3 Basic Principles of Biomedical Imaging

Understanding the principles of various modalities is instrumental in grasping the precise workings of each modality, which helps in appreciating the advantages and limitations of each.

Modalities considered

We consider the primary modalities that are mostly used in the industry along with the ones that are emerging in the market. Here, we discuss the modalities, mechanism of imaging, and the process of image acquisition and reconstruction.

1.3.1 X-ray Imaging

This modality utilizes X-rays to produce images generated by the selective permeability of electromagnetic (EM) radiation by various body parts such as bones and soft tissues. Dense tissues like bone absorb more X-rays and appear white, while softer tissues absorb less and appear darker [3].

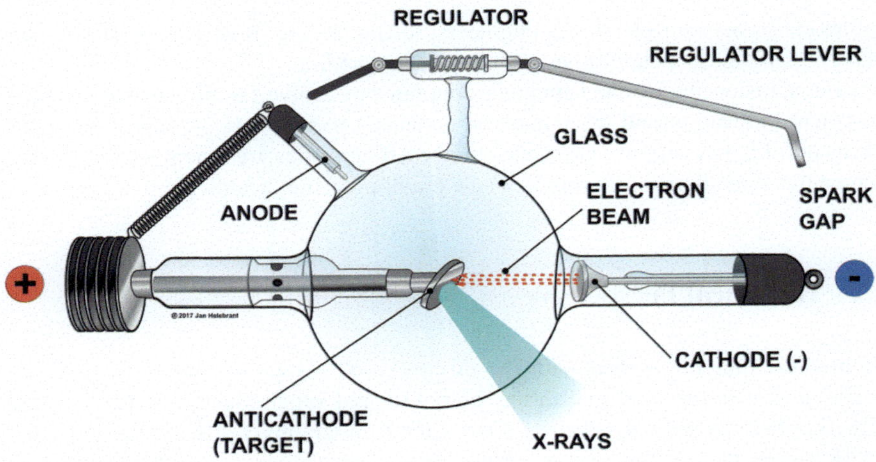

Fig. 1.1 Crookes X-Ray tube. (Licensed to Jhelebrant, CC0, via Wikimedia Commons)

Figure 1.1 demonstrates an X-ray tube developed by Crookes that used a cathode ray tube to generate the EM radiation. Images are directly captured on film.

1.3.2 Computed Tomography (CT)

CT combines X-ray measurements taken from different angles to produce cross-sectional images, providing detailed views of internal organs. CT scanners rotate around the patient, capturing multiple X-ray images from different angles. These images are then reconstructed by a computer to create detailed cross-sectional views of the body [1, 3, 4].

Figure 1.2 shows a CT scan device with a subject placed inside. In CT, multiple X-ray images are processed using sophisticated algorithms to reconstruct detailed cross-sectional views.

Also, cone beam CT is a variation of traditional CT imaging that uses a cone-shaped X-ray beam to produce detailed 3D images. It is particularly useful in dental and maxillofacial imaging, orthopedics, and interventional radiology. Its ability to produce high-resolution images with lower radiation doses makes it a preferred choice for certain applications (Scarfe et al., 2006).

1.3.3 Magnetic Resonance Imaging (MRI)

MRI relies on the principles of nuclear magnetic resonance. In this modality, protons in the body align with a strong magnetic field and are disturbed by radiofrequency

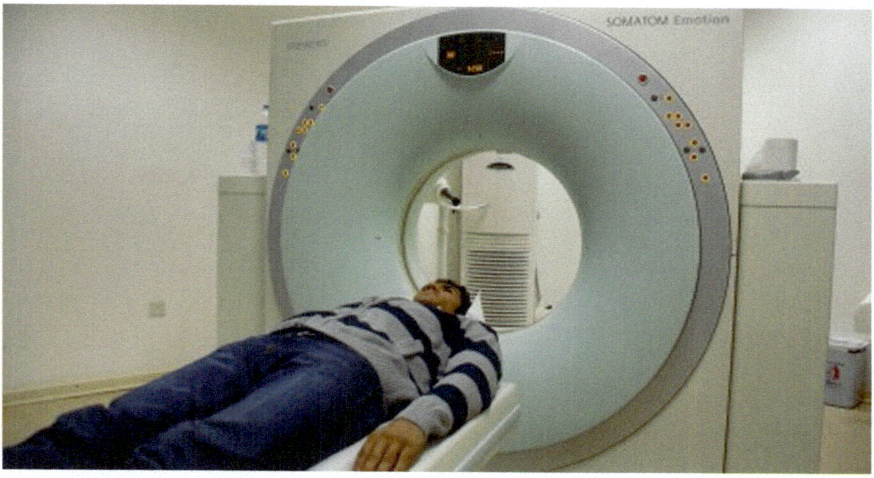

Fig. 1.2 CT scan device (Licensed to Ramin. essamanas, CC BY-SA 4.0, via Wikimedia Commons)

pulses. As the protons realign with the magnetic field, they emit signals that are detected and converted into images [5].

Figure 1.3 shows an MRI scanner. Signals emitted by protons are captured by coils and processed by computers to create images. Various pulse sequences can be used to emphasize different tissue properties.

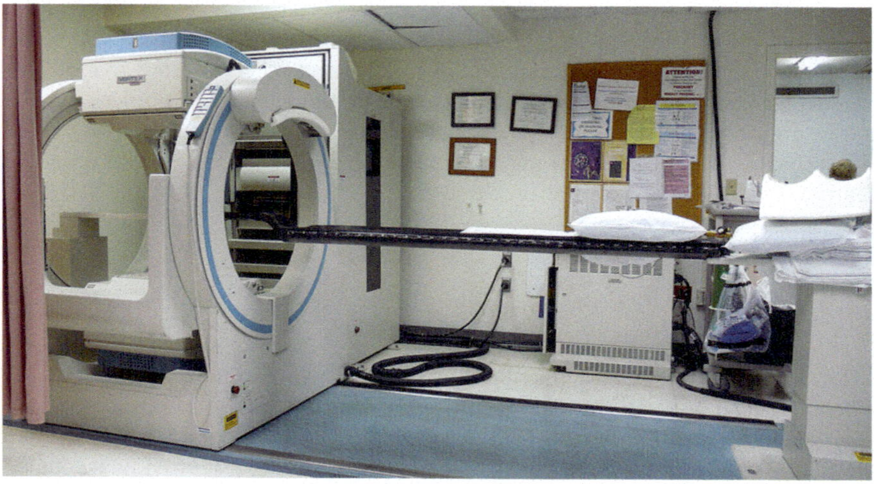

Fig. 1.3 MRI scanner (Licensed to liz west from Boxborough, MA, CC BY 2.0, via Wikimedia Commons)

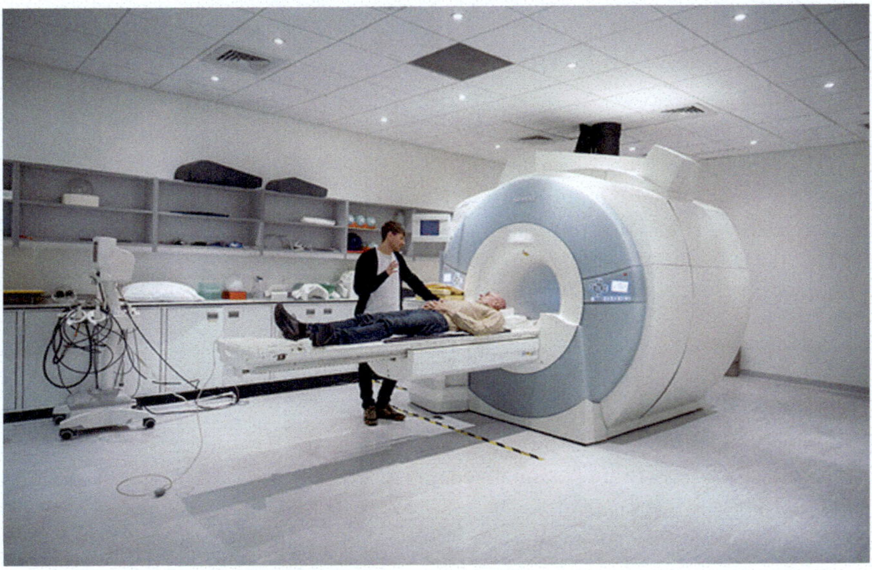

Fig. 1.4 (Licensed to: Thomas Angus, Imperial College London, CC BY-SA 4.0, via Wikimedia Commons)

1.3.4 Functional-MRI (fMRI)

Functional MRI (fMRI) measures brain activity by detecting changes in blood flow. It is based on the principle that cerebral blood flow and neuronal activation are coupled. When an area of the brain is more active, it receives more blood. This imaging technique is critical in understanding brain functions and disorders, such as in pre-surgical planning for epilepsy and brain tumors [6].

Figure 1.4 shows functional magnetic resonance imaging setup at the Imperial Centre for Psychedelic Research.

1.3.5 Diffusion Tensor Imaging (DTI)

DTI is an MRI-based neuroimaging method that maps the diffusion process of molecules, mainly water, in biological tissues. It is especially useful for imaging white matter tracts in the brain. DTI helps in diagnosing and monitoring diseases like multiple sclerosis, stroke, and traumatic brain injuries by providing detailed images of the brain's connectivity [7].

Figure 1.5 demonstrates a schematic diagram of DTI housed within an MRI setup.

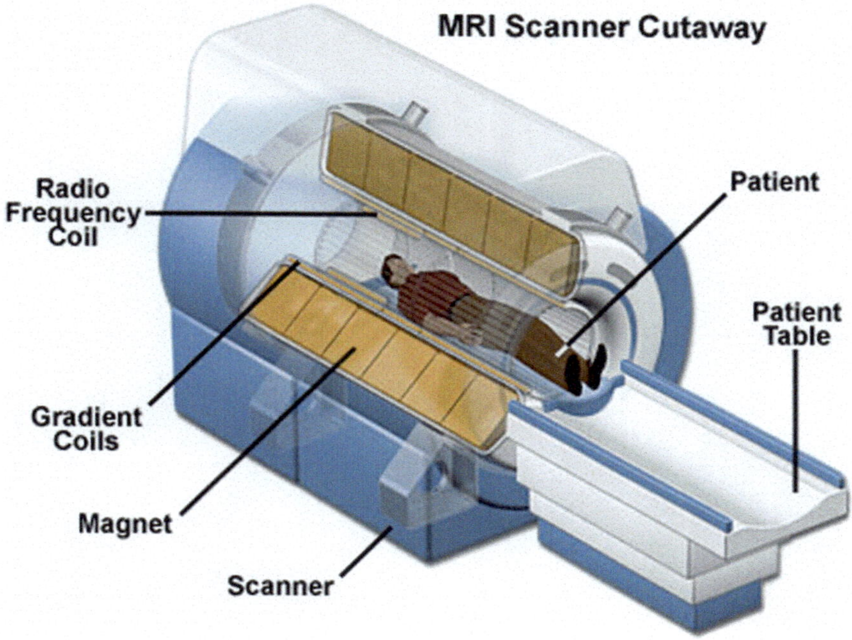

Fig. 1.5 Diffusion tensor imaging (DTI) with an MRI (Licenced to: JVizcarra, CC BY-SA 3.0 Unported, via Quantum Day)

1.3.6 Magnetic Resonance Spectroscopy (MRS)

MRS is a non-invasive diagnostic tool for measuring biochemical changes in the brain, especially useful in the diagnosis of metabolic disorders, brain tumors, and neurodegenerative diseases. By measuring the concentration of specific metabolites, MRS provides a chemical profile of tissues [8].

Figure 1.6 illustrates the workflow for magnetic resonance spectroscopy (MRS) and the resulting spectrum that highlights various molecular levels within a specific region of interest. The labeled spectrum indicates the concentrations of key molecules measured by MRS, which are critical for assessing brain health. Specifically, MRS provides quantifiable insights into the integrity of neuronal and glial cells through molecules such as N-acetylaspartate (NAA) and myo-inositol (mI), intercellular communication via glutamate (Glu), glutamine (Gln), and gamma-aminobutyric acid (GABA), energy metabolism through creatine (Cr), phosphocreatine (PCr), and lactate, and cell membrane turnover with phosphocholine (PC) and glycerophosphocholine (GPC).

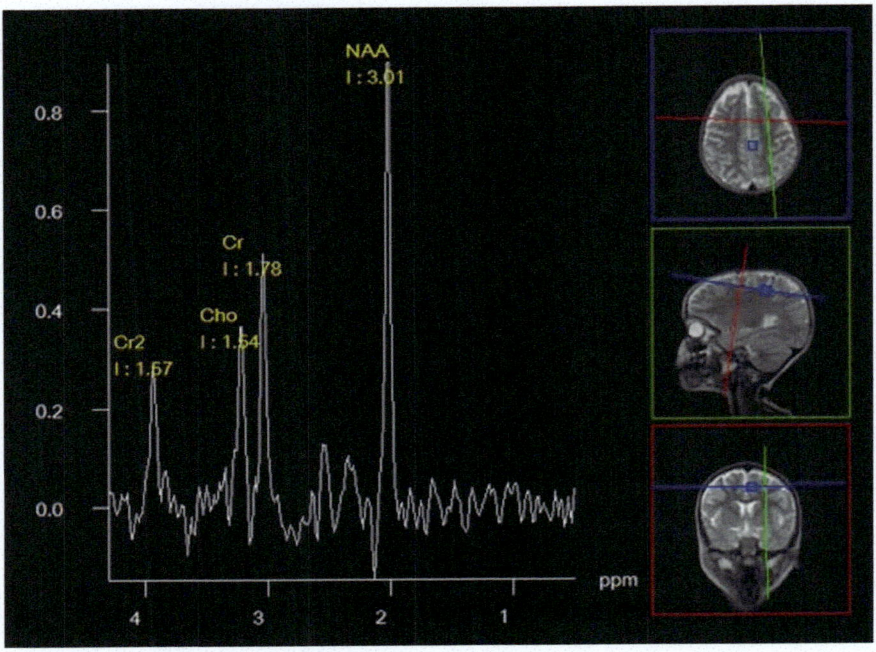

Fig. 1.6 Diagram for MRS (Licenced to: Nevit Dilmen Creative Commons Attribution ShareAlike 3.0 (Generic))

1.3.7 Ultrasound Imaging

Ultrasound devices emit high-frequency sound waves that travel through the body and bounce back when they encounter different tissues. The returning echoes are used to construct real-time images of the internal structures [9].

Figure 1.7 shows an ultrasound scanner. Sound waves are emitted and received by a transducer. The echoes are processed in real-time to generate moving images, which can be analyzed immediately.

Another subset, Doppler ultrasound, measures blood flow and velocity, providing vital information for diagnosing conditions related to blood flow, such as blood clots, poor circulation, and heart valve defects. This technique uses the Doppler effect to assess the movement of blood through vessels [10].

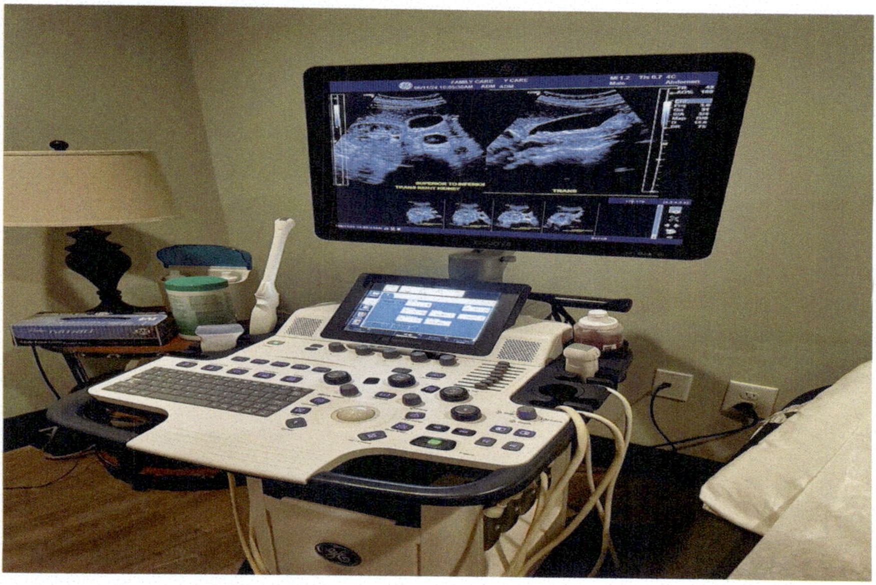

Fig. 1.7 A modern ultrasound scanner (Licensed to Harrison Keely, CC BY 4.0, via Wikimedia Commons)

1.3.8 Nuclear Medicine Imaging

In PET and SPECT, radioactive tracers are injected into the body and accumulate in specific tissues. These tracers emit gamma rays, which are detected by the scanner to create images showing physiological processes [11, 12].

Figure 1.8 shows a nuclear medicine cabinet. Gamma rays emitted by radioactive tracers are detected by gamma cameras. PET and SPECT use different methods to create three-dimensional images from these signals.

1.3.9 Elastography

Elastography is an imaging modality that maps the elastic properties of soft tissue. It is useful for detecting liver fibrosis, tumors, and other pathologies that alter tissue stiffness. By providing a visual representation of tissue stiffness, elastography aids in early diagnosis and treatment planning [13].

Figure 1.9 displays a vaginal tactile imager (VTI) that captures mechanical characteristics of the vaginal tissues crucial for diagnosing and managing conditions related to pelvic floor disorders. Measuring tissue responses to applied pressure, VTI provides insights into tissue integrity, stiffness, and functional properties, which are key aspects of elastography.

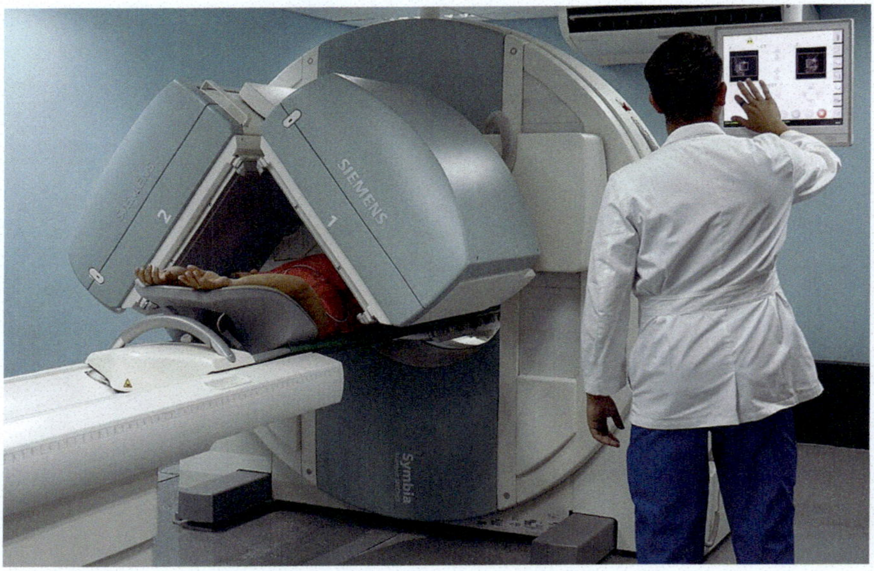

Fig. 1.8 Nuclear medicine (Licensed to IAEA Imagebank, CC BY 2.0, via Wikimedia Commons)

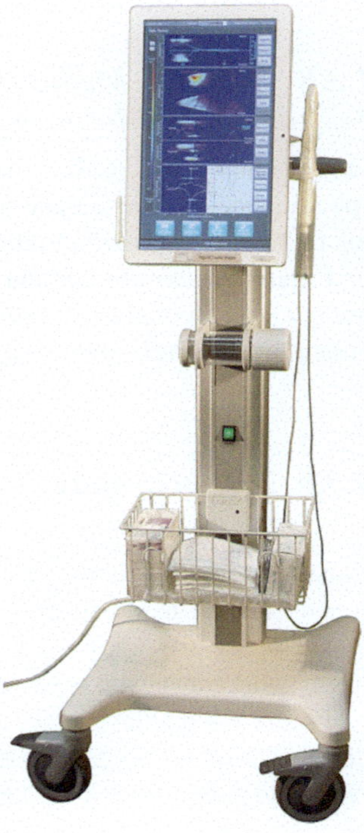

Fig. 1.9 Vaginal Tactile Imager (Licensed to: Egorov123, CC BY-SA 3.0, via Wikimedia Commons)

1.3.10 Positron Emission Tomography (PET)

PET imaging is crucial for oncology, cardiology, and neurology. By using radioactive tracers, PET provides detailed images of metabolic activity and is invaluable in cancer detection, monitoring heart diseases, and studying brain disorders. It provides functional information that complements anatomical imaging [14].

Figure 1.10 shows a PET facility. However, nowadays it is more common to find a combination of PET/CT setups as explained in the next section. The trend towards PET/CT is driven by its improved diagnostic accuracy, efficiency, and the ability to provide comprehensive data in a single session. This integration has become a standard in modern imaging facilities, reflecting the ongoing advancements in medical imaging technology.

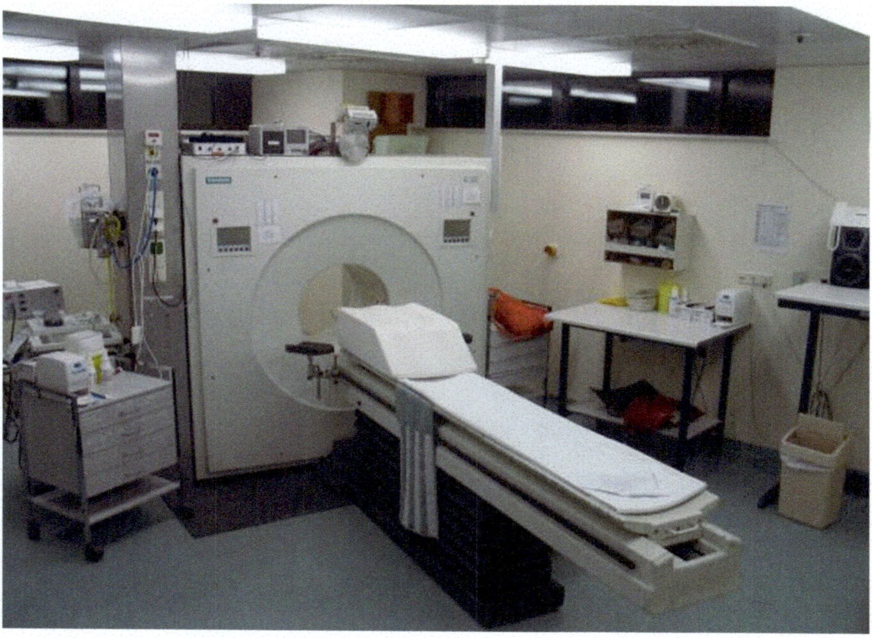

Fig. 1.10 ECAT exact HR + PET scanner (Licensed to: Jens Maus, Public domain, via Wikimedia Commons)

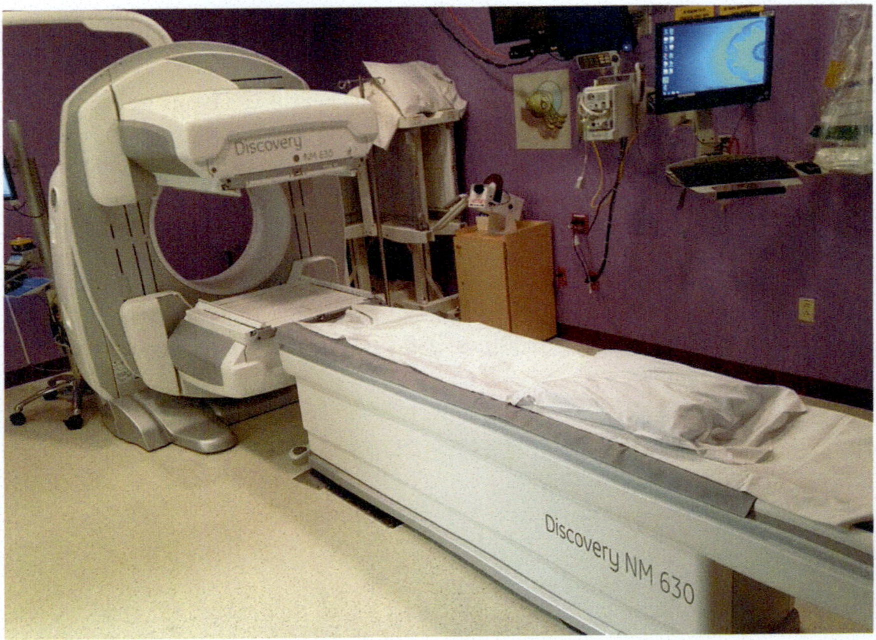

Fig. 1.11 (Licenced to: Nightryder84, CC BY-SA 3.0, via Wikimedia Commons)

1.3.11 Single Photon Emission Computed Tomography (SPECT)

SPECT imaging uses gamma rays to create 3D images and is used for diagnosing and monitoring various conditions, including heart disease, bone disorders, and brain abnormalities. SPECT provides information on blood flow and cellular activity, aiding in the diagnosis and treatment of diseases [15].

Figure 1.11 shows a single photon emission computed tomography (SPECT/CT) device that is an advanced imaging system that combines the functional imaging capabilities of SPECT with the anatomical imaging capabilities of computed tomography (CT).

1.3.12 Fluorescence Imaging

Fluorescence imaging uses fluorescent markers to visualize cellular and molecular processes in real-time. It is widely used in research and clinical diagnostics, particularly in oncology, where it helps in detecting cancerous tissues [16].

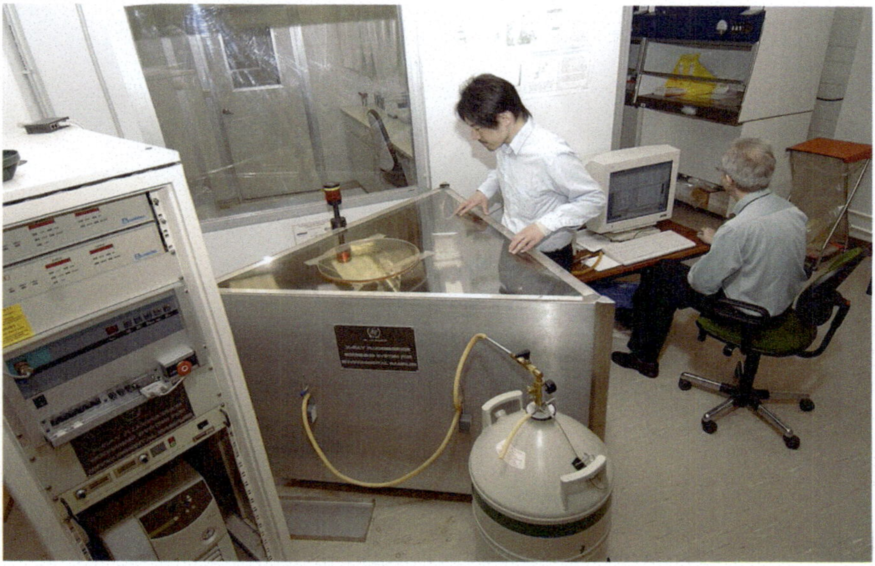

Fig. 1.12 Screening measurements on swipe samples by X-Ray Fluorescence Spectrometry. (Licensed to: IAEA Imagebank, CC BY-SA 2.0, via Wikimedia Commons)

Figure 1.12 shows a lab technician performing screening operations on X-ray fluorescence spectrometry at Clean Laboratory, Seibersdorf, Austria.

1.3.13 Optical Coherence Tomography (OCT)

OCT is a non-invasive imaging technique that provides high-resolution cross-sectional images of biological tissues. It is widely used in ophthalmology for retinal imaging, as well as in cardiology and dermatology for examining the structure of tissues [17].

Figure 1.13 illustrates OCT's operation, which closely resembles ultrasound but employs light waves instead. OCT enables images to achieve a resolution ten times greater than ultrasound due to its use of light. However, a notable drawback is that light can be reflected or completely absorbed by various tissues, restricting OCT's applicability across different organs.

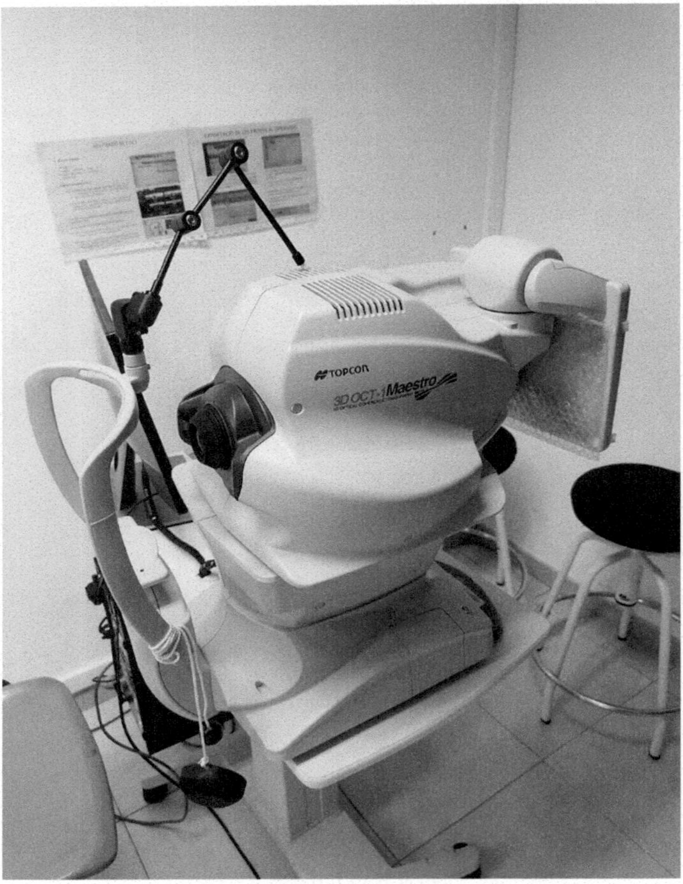

Fig. 1.13 Tomografía de Coherencia Óptica (OCT) device. (Licensed to: Sergio soto-Rondan, CC BY-SA 4.0, via Wikimedia Commons)

1.4 Hybrid Imaging Techniques

Combining different imaging modalities can provide more comprehensive diagnostic information.

1.4.1 PET-CT

PET-CT combines metabolic information from PET with anatomical information from CT. This hybrid imaging technique is particularly useful in oncology for accurate tumor localization and staging, providing both functional and structural details in a single scan [18]. The integration of PET and CT technologies allows for the simultaneous acquisition of metabolic and anatomical information, leading to more accurate diagnosis and treatment planning.

1.4.2 PET-MRI

PET-MRI combines the metabolic imaging capabilities of PET with the high-resolution soft tissue imaging of MRI. This combination is beneficial for detailed imaging of soft tissues, such as in neurological and oncological applications, enhancing both functional and anatomical detail [19].

1.5 Recent Advances and Emerging Technologies

1.5.1 Artificial Intelligence and Machine Learning in Imaging

AI and machine learning are transforming medical imaging by enabling automated image analysis, predictive analytics, and enhanced image reconstruction. These technologies help improve diagnostic accuracy and efficiency, potentially reducing human error and speeding up the diagnostic process [20].

1.5.2 Automated Image Analysis

AI algorithms can analyze large volumes of imaging data, identifying patterns and anomalies that may be missed by the human eye. This capability is particularly useful in radiology, where AI can assist in detecting diseases such as cancer, fractures, and other conditions [21].

1.5.3 Predictive Analytics

Machine learning models can predict patient outcomes based on imaging data, helping clinicians make informed decisions about treatment plans. Predictive analytics can also identify patients at risk for certain conditions, allowing for earlier intervention [22, 23].

1.5.4 Enhanced Image Reconstruction

AI-enhanced image reconstruction techniques improve image quality and reduce noise, allowing for clearer and more accurate diagnostic images. These techniques are particularly beneficial in low-dose imaging scenarios, such as CT and PET scans, where minimizing radiation exposure is crucial [24, 25].

1.5.5 Molecular Imaging

Molecular imaging techniques provide detailed pictures of what is happening inside the body at the molecular and cellular levels. This field is rapidly advancing with new imaging agents and techniques that offer unprecedented insights into biological processes.

1.5.6 Target Imaging Agents

Target imaging agents are designed to bind to specific cellular targets, such as receptors or enzymes, allowing for the visualization of specific molecular processes. These agents are crucial for the early detection of diseases, such as cancer, and for monitoring the effectiveness of treatments [26].

1.5.7 Theranostics

Theranostics combines therapy and diagnostics in a single platform, using molecular imaging to guide targeted therapies. This approach allows for personalized treatment plans that are tailored to the specific molecular characteristics of a patient's disease [27].

1.6 Applications in Medicine

1.6.1 Diagnostic Imaging

Biomedical imaging is integral to diagnostic medicine, offering non-invasive methods to visualize internal body structures, thereby facilitating accurate diagnosis and treatment planning. Various modalities cater to different diagnostic needs, enhancing our understanding and management of diseases. Biomedical imaging plays a pivotal role in diagnostic medicine, offering non-invasive methods to visualize internal body structures. These techniques are essential for accurate diagnosis and treatment planning across various medical specialties. Modalities such as MRI, CT, PET, ultrasound, and others cater to diverse diagnostic needs, providing detailed anatomical and functional information. MRI, for example, uses magnetic fields and radio waves to create detailed images of soft tissues like organs and muscles, making it invaluable in neurology and orthopedics. CT scans utilize X-rays to produce cross-sectional images used extensively in oncology for tumor detection and staging. PET scans employ radioactive tracers to detect metabolic activity, aiding in cancer diagnosis and monitoring treatment responses. Ultrasound, based on sound waves, offers real-time imaging useful in obstetrics and cardiology for monitoring fetal development and heart function. Each modality brings unique advantages to clinical practice, enhancing our ability to detect diseases early and tailor treatments for optimal patient outcomes.

1.6.2 Oncology

In oncology, imaging techniques like MRI, CT, PET, and ultrasound play critical roles in detecting, staging, and monitoring cancer. Advanced imaging allows for precise tumor localization and assessment of metastasis, aiding in effective treatment planning and monitoring therapeutic response [28]. In oncology, imaging techniques are indispensable for detecting and managing cancers throughout the disease continuum. MRI provides detailed images of soft tissue structures, helping oncologists visualize tumors in the brain, spine, and musculoskeletal system with exceptional clarity. CT scans offer high-resolution images that are crucial for identifying tumor locations and assessing their size and spread. PET imaging, by highlighting areas of increased metabolic activity, enables precise tumor staging and monitoring of treatment responses. Ultrasound, although primarily used for screening and guiding biopsies, also plays a role in detecting tumors in superficial organs like the breast and thyroid. These imaging modalities not only aid in accurate diagnosis but also guide therapeutic decisions, such as surgical planning, radiation therapy targeting, and monitoring of chemotherapy effectiveness. Advances in imaging technology continue to improve our ability to detect cancers at earlier stages, personalize treatment strategies, and monitor patients' responses to therapy, ultimately improving survival rates and quality of life for cancer patients worldwide.

1.6.3 Cardiology

Cardiovascular imaging includes modalities such as echocardiography, cardiac MRI, and CT angiography. These techniques provide detailed images of the heart and blood vessels, helping in the diagnosis and management of conditions like coronary artery disease, heart failure, and congenital heart defects [29]. Cardiovascular imaging encompasses a range of modalities that provide essential insights into heart structure and function. Echocardiography, using sound waves to create real-time images of the heart, is pivotal in diagnosing structural abnormalities like valve defects and assessing cardiac function and blood flow. Cardiac MRI offers detailed images of the heart's chambers, valves, and blood vessels, crucial for diagnosing conditions such as cardiomyopathy and congenital heart defects. CT angiography provides 3D images of blood vessels, aiding in the assessment of coronary artery disease and guiding interventions like stent placement. These imaging techniques not only diagnose cardiovascular diseases but also monitor disease progression and evaluate the effectiveness of treatments such as medication and surgical interventions. By providing detailed anatomical and functional information, cardiovascular imaging plays a critical role in improving patient outcomes, reducing complications, and enhancing quality of life for individuals with heart conditions.

1.6.4 Neurology

Neuroimaging modalities, including MRI, fMRI, and PET, are essential in diagnosing and managing neurological disorders such as stroke, Alzheimer's disease, and multiple sclerosis. These techniques provide detailed images of brain structure and function, aiding in early diagnosis and monitoring disease progression (Gorelick et al., 2011). Neuroimaging modalities are essential tools in diagnosing and managing neurological disorders affecting the brain and nervous system. MRI, with its ability to create detailed images of brain structures and detect abnormalities such as tumors and strokes, is invaluable in neurology. Functional MRI (fMRI) goes further by mapping brain activity during tasks, aiding in the understanding of neurological functions and disorders like epilepsy and Alzheimer's disease. PET scans, using radioactive tracers, highlight areas of abnormal brain activity, aiding in the diagnosis and monitoring of conditions like Parkinson's disease and dementia. These imaging techniques not only facilitate early diagnosis but also guide treatment decisions, monitor disease progression, and assess the efficacy of therapies. Neuroimaging continues to evolve with advancements in technology and data analysis, offering new insights into brain function and paving the way for more targeted and personalized approaches to neurological care.

1.7 Image-Guided Intervention

1.7.1 Interventional Radiology

Interventional radiology uses imaging guidance to perform minimally invasive procedures. Techniques like fluoroscopy, CT, and ultrasound guide interventions such as angioplasty, stent placement, and tumor ablation, improving precision and reducing recovery times [30]. Interventional radiology utilizes imaging techniques such as fluoroscopy, CT, and ultrasound to guide minimally invasive procedures. Fluoroscopy provides real-time X-ray images used in procedures like angioplasty to open blocked blood vessels and stent placement to reinforce vessel walls. CT guidance offers precise localization of tumors for procedures such as biopsy and ablation, where cancerous tissue is destroyed using heat or cold. Ultrasound, known for its versatility and safety, guides procedures like draining fluid from abscesses or cysts and inserting catheters into blood vessels. These imaging-guided interventions minimize the need for traditional surgery, reducing recovery times, complications, and hospital stays for patients. By combining imaging with therapeutic procedures, interventional radiology enhances treatment precision while preserving healthy tissue, improving patient outcomes across a wide range of medical conditions.

1.7.2 Surgical Navigation

Advanced imaging is crucial in surgical navigation, providing real-time guidance during complex surgeries. Techniques like intraoperative MRI and CT improve the accuracy of procedures such as tumor resection and spinal surgery, enhancing patient outcomes [31–34]. Advanced imaging technologies play a crucial role in surgical navigation, providing real-time guidance and enhancing precision during complex procedures. Intraoperative MRI and CT scans allow surgeons to visualize tumors and critical structures with high resolution during brain surgeries, ensuring complete tumor removal while minimizing damage to healthy brain tissue. Image-guided navigation systems use preoperative imaging data to create 3D maps of surgical targets, guiding surgeons with pinpoint accuracy in procedures such as spinal fusion and joint replacements. Optical coherence tomography (OCT), offering micron-level resolution, aids in delicate surgeries like retinal procedures, enabling surgeons to visualize tissue layers and make precise interventions. These technologies not only improve surgical outcomes but also reduce risks, shorten recovery times, and enhance patient safety, making them indispensable in modern surgical practice.

1.7.3 Radiation Therapy

Imaging techniques are pivotal in radiation therapy planning, ensuring precise targeting of cancerous tissues while sparing healthy ones. CT and MRI are used to delineate tumor boundaries and plan the optimal delivery of radiation doses, improving treatment efficacy and reducing side effects [35, 36]. Imaging plays a pivotal role in radiation therapy, ensuring precise targeting of cancerous tissues while sparing healthy organs. CT and MRI scans provide detailed images used to plan radiation treatments by delineating tumor boundaries and identifying nearby critical structures. Advanced imaging techniques like PET-CT combine metabolic and anatomical information to create comprehensive treatment plans that maximize tumor control while minimizing side effects. Image-guided radiation therapy (IGRT) uses real-time imaging during treatment sessions to adjust for patient positioning and tumor motion, ensuring accurate radiation delivery. These technologies improve treatment outcomes by allowing higher radiation doses to be delivered precisely to cancerous tissues, reducing the likelihood of recurrence and improving overall survival rates. Ongoing advancements in imaging technology continue to refine radiation therapy techniques, offering new opportunities to enhance treatment efficacy and quality of life for cancer patients.

1.8 Monitoring and Management

1.8.1 Oncological Monitoring

Imaging modalities like PET, MRI, and CT are essential for monitoring the treatment response in oncology. These techniques enable the assessment of tumor shrinkage, metastasis, and recurrence, guiding adjustments in therapy [37]. PET scans detect changes in metabolic activity within tumors, providing early indications of treatment effectiveness or disease recurrence. MRI offers detailed anatomical images used to measure tumor size, assess tumor shrinkage, and identify new lesions that may require additional treatment. CT scans provide 3D images that help oncologists monitor changes in tumor volume and evaluate the spread of cancer to distant organs. These imaging techniques enable oncologists to tailor treatment strategies based on individual patient responses, optimizing therapeutic outcomes and improving long-term survival rates. By providing objective data on tumor response and disease status, oncological imaging supports informed clinical decisions and enhances patient care throughout the cancer journey.

1.8.2 Cardiovascular Management

Imaging plays a crucial role in managing cardiovascular diseases, from diagnosis to monitoring treatment efficacy. Echocardiography, cardiac MRI, and CT angiography provide detailed images of heart structure and function, aiding in the management of heart diseases [38]. Echocardiography, cardiac MRI, and CT angiography provide detailed images of heart structure and function, aiding in the diagnosis and management of conditions such as coronary artery disease and heart failure. These imaging modalities assess cardiac function, evaluate blood flow dynamics, and detect anatomical abnormalities that may require surgical intervention or medical therapy. By guiding treatment decisions and assessing treatment effectiveness, cardiovascular imaging improves patient outcomes, reduces complications, and enhances quality of life for individuals with heart disease.

1.8.3 Chronic Disease Management

For chronic diseases like diabetes and chronic obstructive pulmonary disease (COPD), imaging techniques such as CT, MRI, and ultrasound help in monitoring disease progression and complications. These modalities provide detailed insights into organ structure and function, aiding in effective management [39]. These modalities detect disease progression, monitor complications, and guide treatment decisions to optimize patient care and improve outcomes. In diabetes management, imaging helps assess vascular complications such as diabetic retinopathy and nephropathy, guiding timely interventions to prevent vision loss and kidney damage. In COPD, imaging evaluates lung function, detects emphysema and bronchial changes, and monitors disease progression to adjust therapeutic strategies and improve respiratory function. By integrating imaging with clinical assessments, chronic disease management becomes more personalized and effective, enhancing patient outcomes and quality of life.

1.9 Personalized Disease Management

1.9.1 Precision Oncology

Precision oncology leverages detailed genetic and molecular profiling, guided by advanced imaging, to tailor treatments to individual patients. Imaging techniques like PET and MRI help in understanding the molecular characteristics of tumors, enabling personalized treatment plans [40]. Precision oncology integrates advanced imaging with genetic and molecular profiling to tailor treatment strategies to individual patient characteristics. Imaging techniques such as PET and MRI play crucial roles in identifying tumor characteristics, assessing treatment responses, and monitoring disease progression. PET scans detect changes in metabolic activity within tumors, guiding decisions on targeted therapies and treatment adjustments. MRI offers detailed anatomical images used to plan surgery and radiation therapy, ensuring precise tumor targeting and minimizing damage to healthy tissues. By combining imaging with genomic data, precision oncology optimizes treatment outcomes, improves survival rates, and reduces treatment-related side effects for cancer patients.

1.9.2 Genetic Disorders

Imaging plays a vital role in diagnosing and managing genetic disorders. Techniques like MRI and ultrasound can detect structural abnormalities in organs, aiding in the diagnosis of conditions like cystic fibrosis and muscular dystrophy [41]. Techniques like MRI and ultrasound provide detailed images used to identify genetic conditions such as cystic fibrosis, muscular dystrophy, and congenital heart defects. These modalities enable early diagnosis, guide treatment decisions, and monitor disease progression to improve patient outcomes and quality of life. In genetic disorders affecting the brain, neuroimaging techniques like MRI and CT scans visualize brain anatomy and abnormalities, aiding in the diagnosis of conditions like tuberous sclerosis and neurofibromatosis. By integrating imaging with genetic testing, clinicians gain comprehensive insights into disease mechanisms, facilitating personalized care strategies and enhancing patient well-being.

1.9.3 Pharmacogenomics

Pharmacogenomics involves tailoring drug treatments based on individual genetic profiles. Imaging techniques help in assessing the efficacy of these personalized treatments by monitoring the physiological and biochemical effects of drugs [42]. Imaging techniques such as PET and MRI monitor physiological responses to medications, evaluating drug distribution, metabolism, and interactions with target tissues. These modalities provide quantitative data on drug efficacy and toxicity, guiding adjustments in treatment regimens to optimize therapeutic outcomes. In oncology, pharmacogenomic imaging assesses tumor responses to chemotherapy and targeted therapies, identifying patients likely to benefit from specific treatments and minimizing adverse effects. By integrating imaging with genomic data, pharmacogenomics enhances personalized medicine approaches, tailoring treatments to individual patient characteristics and improving clinical outcomes across diverse medical specialties.

1.10 Challenges and Future Directions

1.10.1 Technical Challenges

Imaging technologies face several technical challenges that need addressing to improve their effectiveness and applicability.

1.10.1.1 Resolution and Sensitivity

Improving the resolution and sensitivity of imaging modalities is essential for detecting minute anatomical and functional changes. This requires advancements in hardware and software, including better detectors and image processing algorithms [43]. Advances in detector technology, image processing algorithms, and hardware design are essential for achieving higher spatial and temporal resolutions, enhancing diagnostic capabilities across modalities such as MRI, CT, PET, and ultrasound.

1.10.1.2 Contrast and Specificity

Enhancing contrast and specificity in imaging helps in distinguishing between different tissue types and disease states. Developing novel contrast agents and imaging techniques can improve the specificity of diagnostic imaging [44]. Research into novel contrast agents, imaging protocols, and spectroscopic techniques promises to enhance tissue characterization and disease detection across diverse medical conditions.

1.10.1.3 Artefact Reduction

Artifacts in imaging can obscure important diagnostic information. Techniques to reduce artifacts, such as advanced reconstruction algorithms and motion correction methods, are essential for improving image quality [42, 45].

1.10.1.4 Radiation Exposure

Reducing radiation exposure while maintaining image quality is a significant challenge in modalities like CT and PET. Innovations in imaging technology, such as low-dose protocols and AI-based reconstruction methods, are crucial for minimizing radiation risks [46].

1.10.2 Practical Challenges

1.10.2.1 Clinical Challenges

Integrating advanced imaging techniques into clinical practice involves overcoming various practical challenges, including cost, accessibility, and training. Ensuring that clinicians are well-versed in using and interpreting these technologies is essential for their effective application [41, 47].

1.10.2.2 Integration into Clinical Workflows

Seamlessly integrating advanced imaging technologies into clinical workflows can enhance diagnostic efficiency and patient care. This requires robust IT infrastructure, interoperability between different imaging systems, and streamlined processes [48].

1.10.2.3 Cost and Accessibility

High costs and limited accessibility of advanced imaging technologies can restrict their widespread use. Efforts to develop cost-effective solutions and improve access in underserved regions are crucial for maximizing the benefits of these technologies [49].

1.10.2.4 Interdisciplinary Collaboration

Collaboration between various disciplines, including radiology, oncology, cardiology, and bioengineering, is vital for advancing imaging technologies. Interdisciplinary research and training programs can foster innovation and improve clinical outcomes [50].

1.11 Future Trends and Potential Advancements

1.11.1 Artificial Intelligence and Machine Learning

The future of biomedical imaging is poised to be significantly influenced by AI and machine learning. These technologies will enhance image analysis, predictive modeling, and personalized treatment planning, making diagnostic processes more accurate and efficient [51].

1.11.2 Artificial Intelligence and Machine Learning

AI and machine learning (ML) are set to revolutionize biomedical imaging by enhancing image analysis, predictive modeling, and personalized treatment planning. These technologies will make diagnostic processes more accurate and efficient by enabling automated interpretation of images, identifying patterns that may not be visible to the human eye, and integrating imaging data with other clinical information for comprehensive diagnostics [52].

1.11.3 Enhanced Image Analysis

AI algorithms can automatically detect and classify abnormalities in medical images, such as tumors in radiographs, with high accuracy. Deep learning models have shown promise in identifying specific conditions in radiology, pathology, and dermatology, often matching or exceeding the performance of human experts [53].

1.11.4 Predictive Modeling

ML can predict disease progression and patient outcomes by analyzing historical imaging data combined with other patient information. This predictive power can help clinicians anticipate complications and make informed decisions about treatment options [54].

1.11.5 Personalized Treatment Planning

By integrating imaging data with genetic and molecular information, AI can help design personalized treatment plans that are tailored to the specific characteristics of a patient's disease. This approach is particularly beneficial in oncology, where precision medicine is crucial for effective treatment [55].

1.11.6 Molecular and Functional Imaging

Advances in molecular and functional imaging are providing deeper insights into the biological processes underlying diseases. These techniques are crucial for early detection, treatment planning, and monitoring response to therapy [56].

1.11.7 Fluorescence Imaging

This technique uses fluorescent markers to visualize cellular and molecular processes in real-time. It is widely used in research and clinical diagnostics, particularly for detecting cancerous tissues [57].

1.11.8 Optical Coherence Tomography (OCT)

OCT provides high-resolution cross-sectional images of biological tissues, widely used in ophthalmology for retinal imaging and in cardiology and dermatology for examining tissue structure [58].

1.11.9 Theranostics

Theranostics combines therapeutic and diagnostic capabilities into a single platform, allowing for personalized treatment approaches that can be adjusted based on real-time feedback from imaging. This approach is particularly effective in oncology for targeting specific molecular markers and monitoring treatment response [59].

1.11.10 Targeted Imaging Agents

These agents are designed to bind to specific cellular targets, enabling visualization of specific molecular processes and allowing for targeted therapies [60].

1.11.11 Theranostic Agents

These combine diagnostic imaging with therapeutic capabilities, such as delivering chemotherapy directly to tumor cells while simultaneously imaging their distribution and effects [61].

1.11.12 Integration of Multi-Modal Imaging

The integration of different imaging modalities, such as combining MRI and PET scans, can provide comprehensive insights by capturing multiple aspects of tissue properties. Hybrid imaging techniques like PET-CT and PET-MRI are already improving the diagnosis and management of complex diseases by providing both anatomical and functional information [62].

1.11.13 PET-CT

This hybrid modality combines the metabolic imaging capabilities of PET with the high-resolution anatomical imaging of CT. It is widely used in oncology for accurate tumor localization, staging, and monitoring treatment response [63].

1.11.14 PET-MRI

PET-MRI combines the metabolic and functional insights from PET with the superior soft tissue contrast of MRI. This integration is particularly useful in neuroimaging and oncology, where detailed visualization of soft tissues is crucial [64].

1.12 Conclusion

The journey from the discovery of X-rays to the development of sophisticated imaging modalities like MRI, CT, and PET highlights the remarkable progress in biomedical imaging. Understanding the fundamental mechanisms of each imaging modality is crucial for appreciating their applications and advancements [64–67]. Innovations such as functional MRI and dual-energy CT represent the pinnacle of current imaging capabilities [68].

The integration of AI, the development of molecular imaging, the advent of wearable and portable devices, and breakthroughs in quantum imaging and nanotechnology are pushing the boundaries of what is possible in biomedical [69]. Imaging technologies are vital in diagnostics, guiding interventions, monitoring treatment, and enabling personalized medicine across various medical fields, including oncology, cardiology, and neurology [70]. While significant challenges exist, ongoing research and technological innovations promise to address these issues and lead to further advancements in the field [71–73].

As biomedical imaging continues to evolve, it will play an increasingly critical role in advancing medical knowledge, improving patient care, and shaping the future of healthcare [51, 52, 74]. Collaboration between scientists, engineers, clinicians, and industry will be key to driving innovation and overcoming the challenges faced by the field. By embracing new technologies and fostering interdisciplinary cooperation, the potential of biomedical imaging to revolutionize medicine will be fully realized [75, 76].

References

1. M. Chappell, *Principles of Medical Imaging for Engineers: From Signals to Images* (2019), pp. 1–169
2. N. Smith, A. Webb, *Introduction to Medical Imaging : Physics, Engineering and Clinical Applications* (2011), p. 286
3. M. Mahesh, The essential physics of medical imaging, third edition. Med. Phys. **40** (2013) [cited 2024 Aug 21]. Available from: https://pubmed.ncbi.nlm.nih.gov/28524933/
4. E. Seeram, *Computed Tomography : Physical Principles, Patient Care, Clinical Applications, and Quality Control* (2023), p. 515
5. S. Waite, A. Grigorian, R.G. Alexander, S.L. Macknik, M. Carrasco, D.J. Heeger et al., Analysis of perceptual expertise in radiology—current knowledge and a new perspective. Front. Hum. Neurosci. **13** (2019) [cited 2024 Aug 21]. Available from: /pmc/articles/PMC6603246/
6. R.B. Buxton, The physics of functional magnetic resonance imaging (fMRI). Rep. Prog. Phys. **76**, 096601 (2013) [cited 2024 Aug 21]. Available from: https://doi.org/10.1088/0034-4885/76/9/096601
7. D. Le Bihan, M. Iima, Diffusion magnetic resonance imaging: what water tells us about biological tissues. PLoS Biol **13**, e1002203 (2015) [cited 2024 Aug 21]. Available from: https://doi.org/10.1371/journal.pbio.1002203
8. H. Hall, S. Cuellar-Baena, C. Dahlberg, R.I. Zandt, V. Denisov, D. Kirik, Magnetic resonance spectroscopic methods for the assessment of metabolic functions in the diseased brain. Curr. Top. Behav. Neurosci. **11**, 169–198 (2011) [cited 2024 Aug 21]. Available from: https://doi.org/10.1007/7854_2011_166
9. F.E. Bell, R. Haddad, *The Basics of Ultrasound Physics* (2022), pp. 11–57.
10. A.E. Roher, Z. Garami, S.L. Tyas, C.L. Maarouf, T.A. Kokjohn, M. Belohlavek et al., Transcranial doppler ultrasound blood flow velocity and pulsatility index as systemic indicators for Alzheimer's disease. Alzheimer's Dement. **7**, 445–455 (2011)
11. G. Crişan, N.S. Moldovean-cioroianu, D.G. Timaru, G. Andrieş, C. Căinap, V. Chiş, Radiopharmaceuticals for PET and SPECT imaging: a literature review over the last decade. Int. J. Mol. Sci. **23**, 5023 (2022) [cited 2024 Aug 21]. Available from: https://www.mdpi.com/1422-0067/23/9/5023/htm
12. A. Cuocolo, M. Petretta, PET and SPECT specialty grand challenge. when knowledge travels at the speed of light, photons take to the field. Front. Nucl. Med. **1** (2021)
13. X.W. Cui, K.N. Li, A.J. Yi, B. Wang, Q. Wei, G.G. Wu et al., Ultrasound elastography. Endosc. Ultrasound **11**, 252–274 (2022) [cited 2024 Aug 21]. Available from: https://journals.lww.com/eusjournal/fulltext/2022/11040/ultrasound_elastography.2.aspx
14. S.L. Kitson, V. Cuccurullo, A. Ciarmiello, D. Salvo, L. Mansi, Clinical applications of positron emission tomography (pet) imaging in medicine: oncology, brain diseases and cardiology.
15. M.M. Khalil, Elements of gamma camera and SPECT systems. Basic Sci. Nucl. Med., 155–178 (2010) [cited 2024 Aug 21]. Available from: https://doi.org/10.1007/978-3-540-85962-8_10
16. G. Rabut, J. Ellenberg, Automatic real-time three-dimensional cell tracking by fluorescence microscopy. J. Microsc. **216**, 131–137 (2004) [cited 2024 Aug 21]. Available from: https://doi.org/10.1111/j.0022-2720.2004.01404.x
17. S. Aumann, S. Donner, J. Fischer, F. Müller, Optical coherence tomography (OCT): principle and technical realization, in *High Resolution Imaging in Microscopy and Ophthalmology* (2019), pp. 59–85 [cited 2024 Aug 21]. Available from: https://doi.org/10.1007/978-3-030-16638-0_3
18. A. Zhu, D. Lee, H. Shim, Metabolic PET imaging in cancer detection and therapy response. Semin. Oncol. **38**, 55 (2011) [cited 2024 Aug 22]. Available from: /pmc/articles/PMC3075495/
19. Y. Zhu, X. Zhu, MRI-driven PET image optimization for neurological applications. Front. Neurosci. **13** (2019) [cited 2024 Aug 22]. Available from: /pmc/articles/PMC6684790/
20. M. Khalifa, M. Albadawy, AI in diagnostic imaging: revolutionising accuracy and efficiency. Comput. Methods Programs Biomed. Update. **5**, 100146 (2024)

21. L. Pinto-Coelho, How artificial intelligence is shaping medical imaging technology: a survey of innovations and applications. Bioengineering **10** (2023) [cited 2024 Aug 22]. Available from: /pmc/articles/PMC10740686/
22. D. Dixon, H. Sattar, N. Moros, S.R. Kesireddy, H. Ahsan, M. Lakkimsetti et al., Unveiling the influence of AI predictive analytics on patient outcomes: a comprehensive narrative review. Cureus **16** (2024) [cited 2024 Aug 22]. Available from: /pmc/articles/PMC11161909/
23. C.C. Aggarwal, Neural networks and deep learning. Neural Netw. Deep Learn. (2018)
24. S. Kaplan, Y.M. Zhu, Full-dose PET image estimation from low-dose PET image using deep learning: a pilot study. J. Digit. Imaging **32**, 773–778 (2019)
25. J. Seah, Z. Brady, K. Ewert, M. Law, Artificial intelligence in medical imaging: implications for patient radiation safety. Br. J. Radiol. **94** (2021) [cited 2024 Aug 22]. Available from: /pmc/articles/PMC9328044/
26. E.E. Kim, Targeted molecular imaging. Korean J. Radiol. **4**, 201 (2003) [cited 2024 Aug 22]. Available from: /pmc/articles/PMC2698097/
27. B.J. Burkett, D.J. Bartlett, P.W. McGarrah, A.R. Lewis, D.R. Johnson, K. Berberoğlu et al., A review of theranostics: perspectives on emerging approaches and clinical advancements. Radiol. Imaging Cancer **5** (2023) [cited 2024 Aug 22]. Available from: /pmc/articles/PMC10413300/
28. C.D. Collins, PET/CT in oncology: for which tumours is it the reference standard? Cancer Imaging **7**, S77 (2007) [cited 2024 Aug 22]. Available from: /pmc/articles/PMC2727966/
29. Q. Counseller, Y. Aboelkassem, Recent technologies in cardiac imaging. Front. Med. Technol. **4** (2022) [cited 2024 Aug 22]. Available from: /pmc/articles/PMC9872125/
30. R. Mashoufi, R. Mashoufi, Interventional radiology for disease management: a narrative review. Cureus **15** (2023) [cited 2024 Aug 22]. Available from: /pmc/articles/PMC10711338/
31. M. Sharma, S.A. Uddin, G. Hanna, B. Ugiliweneza, T.T. Kim, J.P. Johnson et al., Trends and long-term health care utilization of computer-assisted neuronavigation in spine fusions: an exact matched analysis of national administrative database. World Neurosurg. **166**, e850–e858 (2022)
32. T. Tjardes, S. Shafizadeh, D. Rixen, T. Paffrath, B. Bouillon, E.S. Steinhausen et al., Image-guided spine surgery: state of the art and future directions. Eur. Spine J. **19**, 25–45 (2010)
33. J.Z. Mao, J.O. Agyei, A. Khan, R.M. Hess, P.K. Jowdy, J.P. Mullin et al., Technologic evolution of navigation and robotics in spine surgery: a historical perspective. World Neurosurg. **145**, 159–167 (2021)
34. J.P. Wilson, L. Fontenot, C. Stewart, D. Kumbhare, B. Guthikonda, S. Hoang, Image-guided navigation in spine surgery: from historical developments to future perspectives. J. Clin. Med. **13**, 2036 (2024) [cited 2024 Aug 22]. Available from: /pmc/articles/PMC11012660/
35. O. Bin-Alamer, H. Abou-Al-Shaar, Z.C. Gersey, S. Huq, J.A. Kallos, D.J. McCarthy et al., Intraoperative imaging and optical visualization techniques for brain tumor resection: a narrative review. Cancers (Basel) **15** (2023) [cited 2024 Aug 22]. Available from: /pmc/articles/PMC10571802/
36. V. Grégoire, M. Guckenberger, K. Haustermans, J.J.W. Lagendijk, C. Ménard, R. Pötter et al., Image guidance in radiation therapy for better cure of cancer. Mol. Oncol. **14**, 1470 (2020) [cited 2024 Aug 22]. Available from: /pmc/articles/PMC7332209/
37. S. Stieb, K. Kiser, L. van Dijk, N.R. Livingstone, H. Elhalawani, B. Elgohari et al., Imaging for response assessment in radiation oncology: current and emerging techniques. Hematol. Oncol. Clin. North Am. **34**, 293 (2020) [cited 2024 Aug 22]. Available from: /pmc/articles/PMC7253297/
38. P.J. Slomka, D. Dey, A. Sitek, M. Motwani, D.S. Berman, G. Germano, Cardiac imaging: working towards fully-automated machine analysis & interpretation. Expert Rev. Med. Devices **14**, 197 (2017) [cited 2024 Aug 22]. Available from: /pmc/articles/PMC5450918/
39. N. Sverzellati, F. Molinari, T. Pirronti, L. Bonomo, P. Spagnolo, M. Zompatori, New insights on COPD imaging via CT and MRI. Int. J. Chron. Obstruct. Pulmon. Dis. **2**, 301 (2007) [cited 2024 Aug 22]. Available from: /pmc/articles/PMC2695207/
40. E.R. Malone, M. Oliva, P.J.B. Sabatini, T.L. Stockley, L.L. Siu, Molecular profiling for precision cancer therapies. Genome Med. **12** (2020)

41. S. Hussain, I. Mubeen, N. Ullah, S.S.U.D. Shah, B.A. Khan, M. Zahoor et al., Modern diagnostic imaging technique applications and risk factors in the medical field: a review. Biomed. Res. Int. (2022) [cited 2024 Aug 22]. Available from: /pmc/articles/PMC9192206/
42. T.P. Aneesh, S. Sekhar, A. Jose, L. Chandran, S.M. Zachariah, Pharmacogenomics: the right drug to the right person. J. Clin. Med. Res. **1**, 191 (2009) [cited 2024 Aug 22]. Available from: /pmc/articles/PMC3299179/
43. S.R. Cherry, The 2006 Henry N. Wagner Lecture: of mice and men (and positrons)—advances in PET imaging technology. J. Nucl. Med. **47**(11), 1735–1745 (November 2006)
44. T.F. Massoud, S.S. Gambhir, Molecular imaging in living subjects: seeing fundamental biological processes in a new light. Genes Dev. **17**(5), 545–580 (2003). https://doi.org/10.1101/gad.1047403
45. B. Zahneisen, B. Keating, A. Singh, M. Herbst, T. Ernst, Reverse retrospective motion correction. Magn. Reson. Med. **75**, 2341–2349 (2016)
46. S.S. Dudhe, G. Mishra, P. Parihar, D. Nimodia, A. Kumari, Radiation dose optimization in radiology: a comprehensive review of safeguarding patients and preserving image fidelity. Cureus **16** (2024) [cited 2024 Aug 22]. Available from: /pmc/articles/PMC11191847/
47. M. Laal, Innovation process in medical imaging. Procedia Soc. Behav. Sci. **81**, 60–64 (2013)
48. S.V. Bhagat, D. Kanyal, Navigating the future: the transformative impact of artificial intelligence on hospital management—a comprehensive review. Cureus **16** (2024) [cited 2024 Aug 23]. Available from: /pmc/articles/PMC10955674/
49. N. Aderinto, D. Olatunji, M. Abdulbasit, M. Edun, The essential role of neuroimaging in diagnosing and managing cerebrovascular disease in Africa: a review. Ann. Med. **55**, 2251490 (2023) [cited 2024 Aug 23]. Available from: /pmc/articles/PMC10496522/
50. P. Nagpal, Collaborations between cardiology and radiology on the development of clinical programs (cardiac imaging). J. Am. Heart Assoc. Cardiovasc. Cerebrovasc. Dis. **8** (2019) [cited 2024 Aug 23]. Available from: /pmc/articles/PMC6951048/
51. L. Pinto-Coelho, How artificial intelligence is shaping medical imaging technology: a survey of innovations and applications. Bioengineering **10** (2023) [cited 2024 Aug 23]. Available from: /pmc/articles/PMC10740686/
52. N. Ghaffar Nia, E. Kaplanoglu, A. Nasab, Evaluation of artificial intelligence techniques in disease diagnosis and prediction. Discover Artif. Intell. **3** (2023)
53. A. Hosny, C. Parmar, J. Quackenbush, L.H. Schwartz, H.J.W.L. Aerts, Artificial intelligence in radiology. Nat. Rev. Cancer **18**, 500 (2018) [cited 2024 Aug 23]. Available from: /pmc/articles/PMC6268174/
54. D. Dixon, H. Sattar, N. Moros, S.R. Kesireddy, H. Ahsan, M. Lakkimsetti et al., Unveiling the influence of AI predictive analytics on patient outcomes: a comprehensive narrative review. Cureus **16** (2024) [cited 2024 Aug 23]. Available from: /pmc/articles/PMC11161909/
55. K.B. Johnson, W.Q. Wei, D. Weeraratne, M.E. Frisse, K. Misulis, K. Rhee et al., Precision medicine, AI, and the future of personalized health care. Clin. Transl. Sci. **14**, 86 (2021) [cited 2024 Aug 23]. Available from: /pmc/articles/PMC7877825/
56. A. Pulumati, A. Pulumati, B.S. Dwarakanath, A. Verma, R.V.L. Papineni, Technological advancements in cancer diagnostics: improvements and limitations. Cancer Rep. **6** (2023) [cited 2024 Aug 23]. Available from: /pmc/articles/PMC9940009/
57. Y. Woo, S. Chaurasiya, M. O'Leary, E. Han, Y. Fong, Fluorescent imaging for cancer therapy and cancer gene therapy. Mol. Ther. Oncolytics **23**, 231 (2021) [cited 2024 Aug 23]. Available from: /pmc/articles/PMC8531657/
58. M. Zeppieri, S. Marsili, E.S. Enaholo, A.O. Shuaibu, N. Uwagboe, C. Salati et al., Optical coherence tomography (OCT): a brief look at the uses and technological evolution of ophthalmology. Medicina (B Aires) **59** (2023) [cited 2024 Aug 23]. Available from: /pmc/articles/PMC10744394/
59. F. Chen, E.B. Ehlerding, W. Cai, Theranostic nanoparticles. J. Nucl. Med. **55**, 1919 (2014) [cited 2024 Aug 23]. Available from: /pmc/articles/PMC4255955/
60. J. Yang, A. Griffin, Z. Qiang, J. Ren, Organelle-targeted therapies: a comprehensive review on system design for enabling precision oncology. Signal Transduct. Target. Ther. **7**, 1–27 (2022) [cited 2024 Aug 23]. Available from: https://www.nature.com/articles/s41392-022-01243-0

61. S. Hapuarachchige, D. Artemov, Theranostic pretargeting drug delivery and imaging platforms in cancer precision medicine. Front. Oncol. **10**, 1131 (2020) [cited 2024 Aug 23]. Available from: /pmc/articles/PMC7387661/
62. B.J. Pichler, H.F. Wehrl, A. Kolb, M.S. Judenhofer, PET/MRI: the next generation of multi-modality imaging? Semin. Nucl. Med. **38**, 199 (2008) [cited 2024 Aug 23]. Available from: /pmc/articles/PMC2762705/
63. E. Even-Sapir, Z. Keidar, R. Bar-Shalom, Hybrid imaging (SPECT/CT and PET/CT)—improving the diagnostic accuracy of functional/metabolic and anatomic imaging. Semin. Nucl. Med. **39**, 264–275 (2009)
64. Q. Yan, X. Yan, X. Yang, S. Li, J. Song, The use of PET/MRI in radiotherapy. Insights Imaging **15,** 1–15 (2024) [cited 2024 Aug 23]. Available from: https://doi.org/10.1186/s13244-024-01627-6
65. R.M. Lebel, Performance characterization of a novel deep learning-based MR image reconstruction pipeline (2020) [cited 2024 Aug 23]. Available from: http://arxiv.org/abs/2008.06559
66. E. Bercovich, M.C. Javitt, Medical imaging: from roentgen to the digital revolution, and beyond. Rambam Maimonides Med. J. **9**, e0034 (2018) [cited 2024 Aug 23]. Available from: /pmc/articles/PMC6186003/
67. T. Beyer, D.L. Bailey, U.J. Birk, I. Buvat, C. Catana, Z. Cheng et al., Medical physics and imaging–a timely perspective. Front. Phys. **9**, 634693 (2021) [cited 2024 Aug 23]. Available from: www.frontiersin.org
68. L.S. Alizadeh, T.J. Vogl, S.S. Waldeck, D. Overhoff, T. D'Angelo, S.S. Martin et al., Dual-energy CT in cardiothoracic imaging: current developments. Diagnostics **13** (2023) [cited 2024 Aug 23]. Available from: /pmc/articles/PMC10297493/
69. E.H. Herskovits, Artificial intelligence in molecular imaging. Ann. Transl. Med. **9**, 824–824 (2021) [cited 2024 Aug 23]. Available from: /pmc/articles/PMC8246206/
70. European Society of Radiology (ESR), Medical imaging in personalised medicine: a white paper of the research committee of the European Society of Radiology (ESR). Insights Imaging **6**, 141 (2015) [cited 2024 Aug 23]. Available from: /pmc/articles/PMC4376812/
71. L. Floridi, J. Cowls, A unified framework of five principles for AI in society. Harv. Data Sci. Rev. (2019) [cited 2024 Aug 23]. Available from: https://hdsr.mitpress.mit.edu/pub/l0jsh9d1
72. S.B. Junaid, A.A. Imam, A.O. Balogun, L.C. De Silva, Y.A. Surakat, G. Kumar et al., Recent advancements in emerging technologies for healthcare management systems: a survey. Healthcare **10** (2022) [cited 2024 Aug 23]. Available from: /pmc/articles/PMC9601636/
73. A.W.K. Yeung, A. Torkamani, A.J. Butte, B.S. Glicksberg, B. Schuller, B. Rodriguez et al., The promise of digital healthcare technologies. Front. Public Health **11**, 1196596 (2023) [cited 2024 Aug 23]. Available from: /pmc/articles/PMC10562722/
74. S.A. Alowais, S.S. Alghamdi, N. Alsuhebany, T. Alqahtani, A.I. Alshaya, S.N. Almohareb et al., Revolutionizing healthcare: the role of artificial intelligence in clinical practice. BMC Med. Educ. **23**, 1–15 (2023). Available from: https://doi.org/10.1186/s12909-023-04698-z
75. S. Flessa, C. Huebner, Innovations in health care—a conceptual framework. Int. J. Environ. Res. Public Health **18** (2021) [cited 2024 Aug 23].Available from: /pmc/articles/PMC8508443/
76. S.E. Hernandez, D.A. Conrad, M.S. Marcus-Smith, P. Reed, C. Watts, Patient-centered innovation in health care organizations: a conceptual framework and case study application. Health Care Manage. Rev. **38**, 166–175 (2013)

Chapter 2
In Vivo Fluorescence Imaging and Its Role in Multimodal Bioimaging

Soorya G. Nath and E. I. Anila

Abstract In vivo bioimaging involves the non-invasive imaging of internal cells or organs within the human body, offering real-time insights into dynamic biological processes. Its pivotal role in clinical contexts is highlighted by its potential for early disease detection, particularly in critical conditions such as cancer. Despite the existence of established imaging techniques like ultrasound, magnetic resonance imaging, and fluorescence imaging, there is an ongoing exploration for innovative imaging probes that can offer enhanced visibility, aiming to deliver high-resolution images. The advent of quantum dots, near-infrared (NIR) fluorescent probes, persistent luminescent nanoprobes, and upconversion luminescent probes for in vivo imaging techniques introduces exciting improvements, ushering in a new era of enhanced bioimaging outcomes. The advent of these nanoprobes offers numerous advantages over traditional fluorescent probes, especially in their capacity to overcome challenges related to light attenuation within biological tissues and autofluorescence. The chapter incorporates sections that elucidate the significance of persistent luminescence and upconversion luminescence specifically in the context of in vivo imaging. Furthermore, the chapter explores the role of various fluorescence imaging techniques in the context of multimodal bioimaging.

2.1 Introduction to in Vivo Fluorescence Imaging

Biomedical imaging has revolutionized our ability to visualize intricate biological processes within living organisms, playing a pivotal role in diagnosing various diseases by revealing the internal structures of organs and cells. Detecting malignant

S. G. Nath · E. I. Anila
Optoelectronic and Nanomaterials' Research Lab, Department of Physics, Union Christian College, Aluva 683102, Kerala, India

E. I. Anila (✉)
Department of Physics and Electronics, CHRIST (Deemed to be University), Bengaluru 560029, Karnataka, India
e-mail: anilaei@gmail.com; anila.ei@christuniversity.in

N. Kalarikkal et al. (eds.), *Multimodal Biomedical Imaging Techniques*,
Biological and Medical Physics, Biomedical Engineering,
https://doi.org/10.1007/978-981-96-1124-9_2

cells early, particularly in cancer therapy, poses a significant challenge. Established imaging techniques like CT, MRI, PET, and SPECT are essential in the biological imaging industry. However, they often rely on contrast agents to enhance image accuracy and sensitivity, yet these agents have limitations, including toxicity and short circulation periods.

In vivo imaging faces challenges due to animal tissues' natural opaqueness, leading to the necessity for contrast agents. While gadolinium (III) based agents are commonly used in MRI, concerns about their toxicity due to potential free Gd leakage have prompted exploration of alternatives like highly paramagnetic iron oxide nanoparticles. The rapid advancements in the bioimaging industry have driven the development of multimodal imaging techniques by combining various imaging modes, such as PET/CT and PET/MRI, to enhance diagnostic capabilities.

Capturing highly detailed images of microscopic biological specimens necessitates greater resolution than what standard microscopy typically offers. For instance, in the examination of transparent cells primarily composed of water, conventional visible light-based microscopy fails to unveil the intricate structures within the cell. Due to the majority of light wavelengths between 400 and 700 nm effortlessly passing through, this method makes it unfeasible to observe sub-microscopic biological elements like intracellular structures or proteins. Fluorescence imaging represents a non-invasive method that uses fluorescence to visualize biological processes occurring within a living organism.

Fluorescence refers to the process where an atom or molecule absorbs a photon, elevating its energy level to an excited state, subsequently releasing a photon of lower energy in the visible region as it returns to a lower energy state. Figure 2.1 is the Jablonski diagram used to illustrate fluorescence. S_0 and S_1 represent the singlet electronic states whereas V_0, V_1, V_2, etc. denote vibrational energy levels. When a photon interacts with a molecule possessing the suitable energy, it can be absorbed, prompting an electron to jump to an excited state like S_1 as depicted in the diagram. Once in an excited state, the electron rapidly dissipates its energy through vibrations (heat), a process known as vibrational relaxation, returning to the lowest level of the first excited state (S_1). Subsequently, the electron may descend to one of the sub-levels within the ground state (S_0), emitting visible light known as fluorescence. This emission occurs within a nanosecond time frame (10^{-9}–10^{-8} s) after the initial photon absorption.

Fluorescence imaging is significant due to its relatively safe nature, relying on fluorescent probes that produce optical signals when activated. These probes commonly involve organic fluorophores and luminescent proteins. Presently, there is extensive interest in developing ultra-small fluorescent nanoparticles for precise in vivo targeting. These nanoparticles possess unique traits like heightened specificity, excellent resolution, and the safe emission of optical radiations, thereby reducing the toxicity of the fluorophores. To further enhance the non-toxic and targeting efficiency of these probes, surface functionalization using diverse proteins is explored. Nano-sized probes like quantum dots (QD) and lanthanide-activated nanoparticles are currently under extensive investigation for their potential clinical application in targeting and treating cancer.

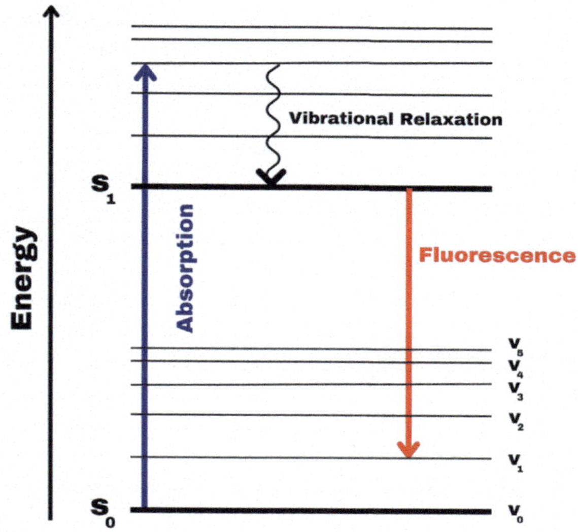

Fig. 2.1 Jablonski diagram of fluorescence

Numerous studies have explored the utilization of fluorescent nanoparticles for in vivo imaging, employing various excitation wavelengths such as UV (ultraviolet), visible, NIR (near-infrared), and x-ray wavelengths [1–4]. These nanoparticles encompass diverse materials, including organic dyes, inorganic semiconductor quantum dots, and carbon dots. However, UV and visible light excitation encounter significant limitations such as photon scattering, absorption, and tissue autofluorescence. Tissue autofluorescence refers to the natural emission of colour by certain cell molecules upon exposure to UV/visible light, impacting the performance of fluorescent probes. This intrinsic fluorescence of cells affects the overall efficacy of the probes, as detailed in earlier reports [5]. Notably, NIR-excitable fluorescent probes significantly circumvent these issues related to autofluorescence and scattering, offering promising prospects for more effective imaging methodologies. Image resolution properties can be improved by combining fluorescence with other imaging techniques.

2.2 In Vivo Fluorescence Imaging Using Quantum Dots

Numerous types of fluorescent probes have been created for scientific use, with some common ones being cyanines or carbocyanine dyes. Among these, indocyanine green (ICG) is a significant one, approved by the FDA (Food and Drug Administration) for clinical trials. There are also other organic probes like methylene blue (MB), BIODPY (difluoroboron dipyrromethene), nicotinamide adenine dinucleotide (NADH), 5-aminolevulinic acid (5-ALA), rhodamine, flavin adenine dinucleotide (FAD), and porphyrins. These organic dyes are often used in imaging techniques but face issues like low durability under light and emitting a wide range of light. To address these limitations, new fluorescent probes such as quantum dots have been developed. Quantum dots have greater durability and emit light based on their size, improving upon the drawbacks of conventional organic dyes.

Scientists have been researching fluorescent probes based on semiconductor quantum dots (QDs) for in vivo imaging within living organisms. These QDs have remarkable properties dependent on their size, such as long-lasting stability and brightness. They emit light ranging from visible to near-infrared (NIR) and have high quantum efficiency.

To make QDs work better within cells and target tumours, researchers modify them by attaching different proteins. This not only improves their ability to bind to cells but also helps the body accept and remove them effectively. Many researchers have studied using QDs for biological imaging. For example, B. Ballou and colleagues showed that QDs encapsulated in polymers could be used for long-term imaging in live mice, proving that QDs remain stable even after being injected [6].

In other studies, water-soluble QDs were used to clearly visualize blood vessels in live mice using two-photon imaging [7]. In another experiment, bio-conjugated and polymer-coated CdSe-ZnS QDs were used as fluorescent probes to image human prostate cancer cells developing in mice [8]. Researchers have even explored making self-illuminating QDs for live imaging, as seen in work by X. Sun and others [9]. Moreover, rare-earth doped semiconductor QDs have been used for high-quantum efficiency of luminescence with a narrow range. More in-depth information about similar studies using QDs can be found in a review article by X. Michalet and colleagues [10].

Despite numerous advantageous properties exhibited by quantum dots (QDs), such as a tunable narrow emission range achieved through particle size manipulation, high photostability, enhanced fluorescence emission, and increased quantum efficiency, a significant drawback is their cytotoxicity, particularly in QDs containing toxic Cd ions. Accidental release of these ions can result in adverse effects when employed for in vivo imaging purposes. Nevertheless, the material holds immense promise owing to advancements in self-illuminating and near-infrared (NIR) emitting QDs. Therefore, the creation of non-toxic quantum dots with persistent luminescence (PersL) has the potential to make substantial contributions to the fluorescence imaging industry.

2.3 Near Infrared (NIR) Fluorescent Probes for Bioimaging

One of the major limitations associated with in vivo fluorescence imaging is lower penetration depth. This is because most of the animal tissues are usually opaque to the electromagnetic spectrum in the visible region that light cannot penetrate deep in to the tissues due to high photon scattering/absorption in the tissues which leads to poor resolution images. Autofluorescence is another reason for low contrast images. In order to get best imaging results, the signal intensity must be high compared to the background noises or in other words, signal to background ratio should be high. Fluorescent probes in the NIR region can overcome most of the above-mentioned limitations.

The near-infrared (NIR) portion of the electromagnetic spectrum is segmented into three distinct regions often termed as the first, second, and third biological windows. The first biological window, designated as NIR-I, encompasses wavelengths ranging from 700 to 950 nm. The subsequent region, known as NIR-II, spans wavelengths from 1000 to 1350 nm, while the third biological window, termed NIR-III, covers wavelengths from 1550 to 1870 nm [11].

Of these regions, the NIR-I region, often referred to as the 'biological transparent window', garners considerable attention in current research due to its relatively higher depth of penetration in biological tissues. However, recent advancements in material science and chemistry have spurred the development of new probes specifically tailored for imaging within the NIR-II and NIR-III regions, expanding the scope of in vivo imaging possibilities.

The first ever NIR fluorophore that came into clinical use is indocyanine green (ICG) and it has an emission wavelength of ~800 nm [12]. This was followed by the use of methylene blue (MB) with emission at ~700 nm. But both these dyes are charged, and this causes non-specific binding to body proteins which in turn results in increased noise. Thus, the research focussed on the development of new probes in the NIR region with low noise levels and can specifically bind to the target tissues. Many such NIR emitting fluorescent probes were developed so far and this includes QDs, lanthanide doped nanoparticles, carbon nano tubes, semiconducting polymers, etc. [13]. NIR fluorophores which do not require external excitation can greatly avoid autofluorescence effect by the tissues. Such self-illuminating fluorophores based on NIR-I emitting QDs usually exhibit in vivo fluorescence emission through energy transfer mechanisms such as bioluminescence, chemiluminescence and fluorescence resonance energy transfer [13]. For example, on the introduction of persistent luminescence (PersL) agents along with NIR fluorophores can transfer energy through fluorescence resonance energy transfer mechanism thus enabling self-illumination during in vivo imaging. PersL is a phenomenon in which the optical energy is released slowly and it can last from minutes to hours even after cutting off the excitation. The application of PersL for in vivo imaging was tested in a work by Maldiney et al. by preparing nano particles of $Ca_2Si_5N_8$:Eu^{2+}, Tm^{3+} from bulk material [14]. Here, the material exhibited NIR PersL at 610 nm and in vivo imaging experiments were performed in mice by injecting the particles after a pre-excitation using UV light for 5 min. Mikael Palner and his co-workers report a semiconducting conjugating polymer nanoparticle that exhibits both NIR emission as well as PersL which last for about an hour which was also utilized for in vivo imaging in mice [15]. NIR dye coated polymer nanoparticles were synthesized in this work and these particles were used for in vivo imaging in mice the particles were excited by white light before injection. The work demonstrated that the dye coating in the polymer nanoparticles enabled the storage of more excitation energy in nanoparticles. In another work, the authors developed a nanoprobe based on chromium activated zinc gallate which exhibited PersL upon excitation with UV as well as red LED light. The material was also used for in vivo tumour targeting in mice after PEG capping [16]. During in vivo experiments the NIR PersL signal was detectable even 2 h after ending the UV excitation. 15 h after the initial imaging experiment using UV excitation, orange/red LED was used to irradiate the mice and intense NIR PersL signal was obtained even form deep tissues and this indicated that initial UV excitation was not compulsory for long-lasting imaging. Mn^{2+} doped nanophosphor showing long lasting luminescence was used for in vivo imaging in yet another work [17]. Here, the material exhibiting afterglow emission at ~690 nm was used for in vivo imaging in mice and the material was excited by UV light for 5 min before injecting in to the animal. The biodistribution and toxicity effects of zinc gallium oxide doped with Cr^{3+} ions were tested in mice in a work by Xia Sun and his co-workers and the work demonstrated the risk effect of the PEG functionalized $Zn_{1.1}Ga_{1.8}Sn_{0.1}O_4$: Cr^{3+} nanoparticles when used for long term PersL bioimaging [18]. The Fig. 2.2 shows the in vivo and ex vivo persistence luminescence imaging performed in Balb/c mice. For in vivo imaging, the mouse was injected with the sample (10 mg/kg) and in situ irradiated with an LED with wavelength 655 nm for 5 min. NIR PersL signal was obtained one day after injection from most parts of the body and predominantly it occurred in the reticuloendothelial system (RES). The signal intensity slowly decreased with time but it existed even 60 days after injection which shows that the sample was distributed almost in all organs through blood circulation and it stays in the body for long time (Fig. 2.2a). To study the biodistribution of the nanoparticles for long term, the major organs from the body were collected at different times post injection and they are irradiated by LED for 3 min and then imaged. One day after injection, strong PersL was observed from liver, spleen and lungs and no signals were obtained from heart and kidney (Fig. 2.2b). The signals from liver and lungs decreased with time (Fig. 2.2c) but existed up to 60 days after injection proving the long term biodistribution of the material. More details about NIR fluorescent probes and PersL can be found from published review articles [13, 19].

NIR fluorescent probes with the highest quantum yields are listed in Table 2.1, and the other NIR probes and their details are available from the published review article [13].

In vivo imaging in the third biological window (NIR-III region) became more prominent after it was first reported in 2011 by Kamimura et al. in which they used 1550 nm NIR radiation from rare-earth doped nanophosphors for in vivo imaging in mouse [20]. In this work PEG coated Y_2O_3: Er, Yb nanoparticles were used for in vivo imaging in NIR-III region. During imaging experiments in mice, the nanoparticles were excited at 980 nm using a diode laser, and the NIR emission at 1550 nm was detected using InGaAs CCD camera. More details about such works in this window can be found from the review article by Hemmer et al. [11].

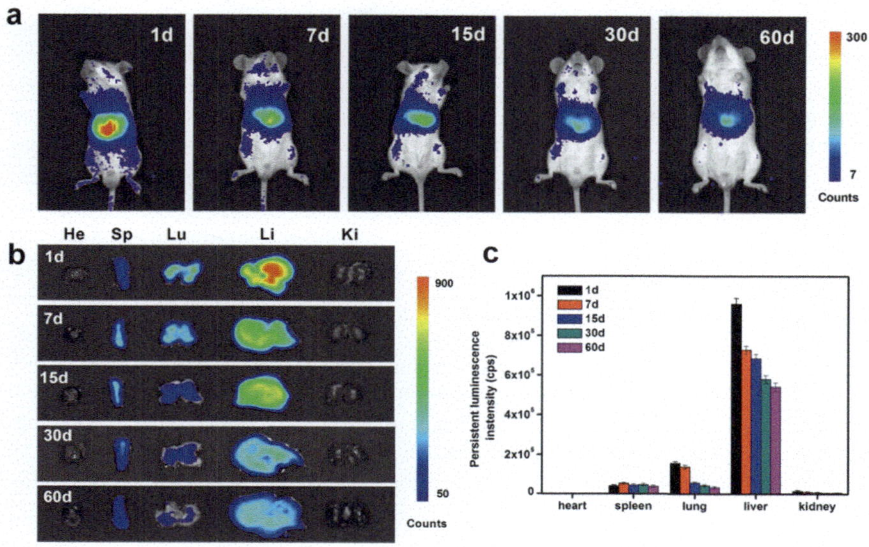

Fig. 2.2 In vivo and ex vivo persistent luminescence imaging. **a** In vivo persistent luminescence images of Balb/c mice with the intravenous injection of ZGO (10 mg/kg) at different time points, **b** ex vivo persistent luminescence imaging of the heart (He), liver (Li), spleen (Sp), lungs (Lu) and kidneys (Ki) collected from Balb/c mice injected with ZGO at different time points, and **c** persistent luminescence intensities of ZGO from isolated organs at different time points. The data represent the mean ± standard deviation (n = 5). Reprinted from Ref. [18]

2.4 Upconversion Nanoparticles as an Efficient Bioimaging Probe

The majority of materials utilized in the bioimaging industry respond to higher energy excitation and emit lower energy photons, a characteristic known as Stokes shift. However, some materials display a unique property called anti-Stokes luminescence, which encompasses phenomena like two-photon luminescence, second-harmonic generation, and upconversion luminescence (UCL). Both the two-photon absorption process and second-harmonic generation necessitate coherent excitation sources like lasers. Specifically, the two-photon process requires expensive pulsed lasers, typically femtosecond lasers with high power, for efficient excitation of material. Conversely, UCL nanophosphors can be effectively excited by low-power sources such as standard xenon lamps or continuous-wave lasers.

Before moving to the in vivo imaging applications using UCL materials, a brief introduction about the UCL mechanisms need to be discussed. UCL nanophosphors convert lower energy photons to higher energy photons, or in other words, these materials can convert NIR wavelengths to UV/visible wavelengths thus exhibiting anti-Stokes process. Upconversion process can occur via various mechanisms namely excited state absorption (ESA), energy transfer upconversion (ETU) and photon

Table 2.1 List of NIR fluorescent probes with highest quantum yield

NIR fluorescent probe	In vivo experiment performed	Quantum yield	References
Boron dipyrromethene (BODIPY)	For detection of β-amyloid (Aβ) plaques responsible for Alzheimer disease	~100%	[21]
Boron dipyrromethene (BODIPY) conjugated with hydrophilic poly (ethylene glycol) (MPEG)	For NIR imaging in H22 cancer bearing mouse	~100%	[22]
Rare-earth nanoparticles (E.g.: SrF_2: Nd, LaF_3: Nd)	Autofluorescence free deep tissue imaging in mouse; First-time in vivo imaging in second biological window using Nd doped nanoparticles	50–90%	[23, 24]
Non-sulfonated cyanine dyes (CyX Series) and hybrid inorganic (silica) particles	First-in-human clinical trial in patients with metastatic melanoma	28%	[25, 26]
Quantum dots	In vivo NIR imaging in mice with and without tumour	~8–17% and higher	[27–29]
Zwitterionic cyanine dyes	Uptake of NIR fluorophore in mice and rats	15.10%	[30]

avalanche (PA). These are illustrated in the Fig. 2.3 in which Fig. 2.3a represents the ESA process where the electron moves to an excited metastable state E_I through the absorption of a pump photon (hv_1) and then to an excited state E_{II} using another pump photon (hv_2). The Upconversion emission occurs when the electron returns to the ground state (G) with maximum energy equal to $hv_1 + hv_2$. This kind of ESA process occurs in the case of singly doped material.

The ETU mechanism is the most efficient upconversion process, which occurs in doubly doped materials (i.e., it occurs in 2 neighbouring lanthanide ions) in which one ion act as sensitizer (S) and the other act as activator (A). There are five different types of ETU process namely ET followed by ESA, successive ET, cross relaxation (CR), cooperative sensitization and cooperative luminescence of which ET following ESA is depicted in Fig. 2.3b. Here the S ion absorbs the pump photon to get excited to metastable level E_I and this energy is then transferred to A ion which is used for exciting the A ion to its metastable level E_I and this is followed by ESA process to excite the A ion to E_{II}. More ETU schemes can be found from previous reports [31]. PA process involves the combined effects of ESA and ETU processes and in this process the pump energy is resonant with the metastable state and excited state only and a simple energy level diagram representing PA mechanism is shown in Fig. 2.3c. Suppose ion 1 is in its ground state and ion 2 absorbs a photon through non-resonant absorption and get excited to its metastable level E_I and it is further excited to E_{II} through ESA process (through absorption of pump photon). Now, ion

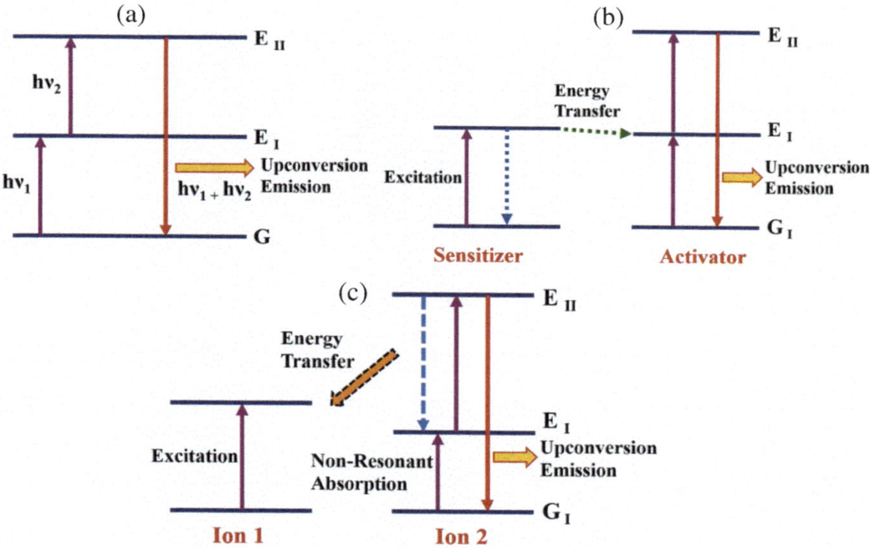

Fig. 2.3 UCL mechanisms. **a** ESA process, **b** Energy transfer followed by ESA process in ETU mechanism and **c** PA process

2 can lose its energy through many processes including fluorescence and this de-excites the second ion to E_I and during which this energy is transferred to first ion (through cross relaxation) thus exciting ion 1 to its metastable level (E_I) thus making the 2 ions in their metastable levels. Thus, both these ions can now act as sensitizer ions which can again produce four ions in metastable state and the process can continue thus forming the avalanche effect leading to upconversion luminescence. This process occurs only when the excitation intensity is above a threshold value and hence it is the least occurring upconversion process. The detailed processes can be found from published reports [31–33]. Recent studies demonstrate on the use of UCL materials for in vivo applications and the predominant UCL materials include lanthanide doped materials. One of the widely studied materials in this context is $NaRF_4$, where R represents rare earth elements and Yttrium (Y). $NaYF_4$ upconversion nanophosphor doped with Tm and Yb was used for in vitro imaging of pancreatic cancer cells and in vivo imaging in mouse [34]. Another work reported the synthesis of amine functionalized UCL material ($NaYF_4$: Yb, Er) denoted by UCNPs-NH_2 and the same conjugated with folic acid (FA) denoted as UCNPs-FA through simple hydrothermal microemulsion method [35].

The presence of amine group enabled the conjugation of the material with folic acid (FA) and both the materials (with and without FA conjugation) were injected to a nude mouse bearing HeLa tumour (in the right hind leg) for demonstrating the capability of the material for targeted in vivo imaging (Fig. 2.4). This was followed by the imaging experiment in which excitation was at 980 nm and the UC luminescence was observed at 600–700 nm range. UC luminescence signals were obtained from

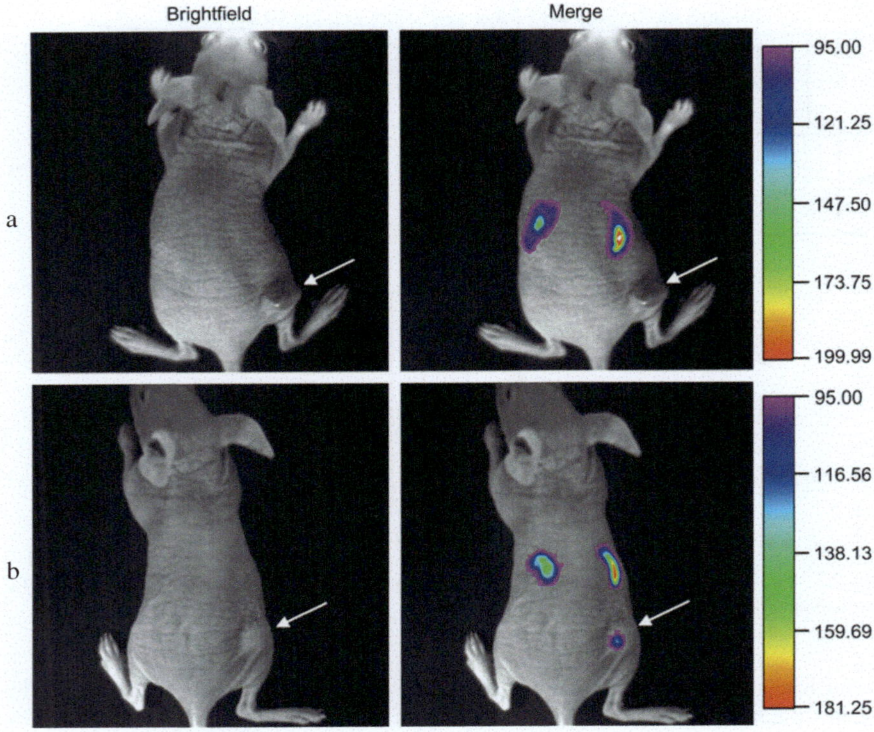

Fig. 2.4 In vivo upconversion luminescence imaging of subcutaneous HeLa tumour-bearing athymic nude mice (right hind leg, pointed by white arrows) after intravenous injection of amine functionalized UC nanoparticles (**a**) and FA conjugated UC nanoparticles (**b**). All images were acquired under the same instrumental conditions (power density z 120 mW/cm2 on the surface of mouse). Reprinted with permission from Ref. [35]. Copyright 2009 Elsevier Ltd.

the tumour when injected with FA conjugated material and no signals were obtained for the UCNPs-NH_2 injected tumour (Fig. 2.4a, b). Liqin Xiong et al. reported the use of peptide modified UCL nanoparticles for tumour targeting in small animals [36].

Qian Liu and his co-workers successfully performed in vivo imaging using UCL material in a black mouse (which usually exhibit high absorption) and they reported a penetration depth of ~2 nm [37]. UC nanophosphors were used by several researchers for imaging the blood vessels or vascular imaging in small animals [38, 39]. There are many other imaging studies using upconversion nanophosphors which can be found from published review articles [31, 40].

UC nanophosphors exhibiting PersL phenomenon are now a days widely investigated due to the peculiar properties of UC materials. Xiaochen Qiu et al. in their work reported the synthesis of a hybrid nanocluster comprising both upconverting nanoparticles and PersL nanomaterial and they also report the use of the material for in vivo lymphatic imaging in nude mouse [41]. UC and PersL nanoparticles were developed in a work by P. Zhao et al. for in vivo imaging guided photothermal therapy [42]. Zhanjun Li and his co-workers integrated UC and PersL nanoparticles to develop NIR activated upconversion persistent luminescent (UCPersL) material [43]. $Zn_3Ga_2GeO_8$: Cr^{3+}, Yb^{3+}, Er^{3+} UCPersL nanoparticles exhibiting long-lasting NIR-UC PersL were developed in yet another work and the material was used for in vivo imaging in mice [44]. Also, in this work both excitation and emission falls in the NIR region (biological transparent region) and the emission signal lasted up to 10 h during the in vivo experiment.

2.5 Persistent Luminescence

Persistent luminescence (PersL) is an optical phenomenon where we obtain light emission even after the removal of source of excitation. Here the excitation source can be UV radiation, visible light, X rays, plasma beam, electron beam, etc. It is also referred to as long-lasting phosphorescence, though the luminescence mechanism is different in both cases. Phosphorescence can be understood using Jablonski diagram (Fig. 2.5a) in which both singlet and triplet states are involved. Consider an electron getting excited from singlet state S_0 to S_2 from where it decays to S_1 by radiation less decay known as internal conversion (IC) as S_2 is more unstable compared to S_1 with lifetimes ~10^{-11} s and 10^{-8} s, respectively. From S_1, the electron can return to the singlet ground state through two paths; one is direct transition known as fluorescence. Another path is through intersystem crossing where the electron changes its spin multiplicity from excited singlet state to excited triplet state. This electron undergoes radiative transition to singlet ground state known as phosphorescence. This is a slow and forbidden transition and its timescale is 10^{-3} to 100 s.

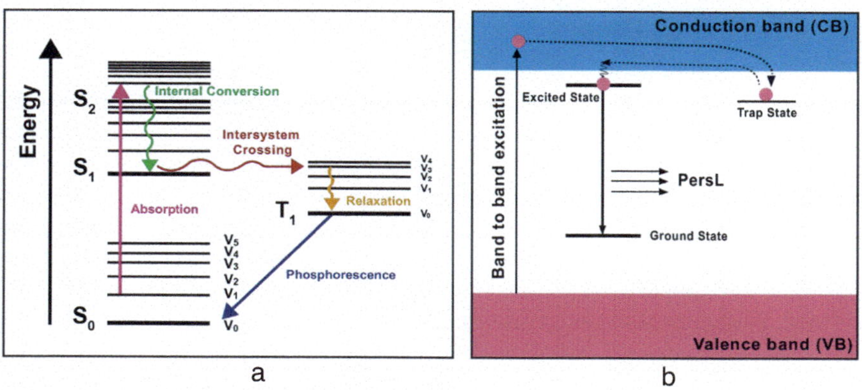

Fig. 2.5 **a** Partial Jablonski diagram for phosphorescence, **b** A schematic diagram for the illustration of persistent luminescence (PersL) mechanism

In the case of persistent phosphors, there exist trap centres along with luminescence emission centres. The long decay time in persistent luminescence is due to the involvement of traps in the process. The trap centres are shallow lying at an energy gap of less than 1 eV below conduction band or above valance band in the case of electrons and holes, respectively. The electron or hole trapping centres are generated by intrinsic defects such as vacancies, F centres, intentionally added co-dopants, etc. Excitation corresponds to the band gap energy, which means electrons are excited to the conduction band. These electrons fall into the shallow traps due to Coulombic force or by quantum tunnelling process through forbidden band, from where they will be detrapped again to conduction band as the energy levels of trapping centres are equivalent to room temperature thermal energy. They are further migrated to the excited state of luminescence centre from where they undergo radiative transition to ground state. The whole process takes several minutes or hours leading to persistent luminescence (Fig. 2.5b). In the work reported by N Basavaraju et al., defects due to inversion cites act as trapping centres giving rise to red PersL in $MgGa_2O_4$: Cr^{3+}phosphor [45]. Luminescence life time in PersL is three orders of magnitude higher than that of phosphorescence.

PersL materials have a variety of applications. The safety signage in aeroplanes are persistent phosphors. They are used in toys, road markings, watches, wall decorations, etc. They can be used for other applications like solar energy storage, photocatalysis, temperature sensors, dosimeters, etc. depending on the depth of trap centres.

PersL materials are suitable for in vitro and in vivo bioimaging applications. Compared to normal fluorescent probes, better resolution can be obtained in imaging using PersL materials since imaging is done after excitation source is switched off. Moreover, autofluorescence is inhibited due to the lack of exciting source.

2.6 Fluorescence Imaging Using X-Ray Triggered Fluorescent Probes

The mechanism behind PersL phenomena involves slow release of electrons from the traps to generate long lasting luminescence. Typically, materials like sulphides and oxides with a bandgap energy ranging between 3 and 7 eV necessitate UV or visible light for inducing luminescence. There are certain wide band gap materials which can be charged by X-rays to produce PersL. The creation of defects in lanthanides (Ln^{3+}) upon charging by x-ray radiation was discussed by Li et al. and the work also proposed the mechanism behind PersL in large bandgap materials doped with Ln^{3+} ions [46]. While the use of short-wavelength UV light for excitation can potentially harm tissues by inducing cell damage, materials activated by low-energy x-rays emitting deep UV persistent radiation have promising applications in disinfection and sterilization [47].

The applications of NIR emitting nanomaterials for in vivo imaging applications were discussed in the previous sections but all of them were excited in the visible/NIR regions. There are few works reporting the x-ray activated PersL materials for in vivo imaging. In such a work, the authors developed x-ray activated Ln doped nanoparticles which exhibit NIR persistent luminescence and they studied the material for its in vivo imaging applications [4]. They performed in vivo imaging studies in several areas of a living mouse such as blood vessels, tumour, ureter, etc. In another work, Cr doped zinc gallate nanophosphors (X-PLNPs) with persistent emission in the NIR region and activated by x-rays was used for in vivo imaging after modification with poly ethylene glycol (PEG). The in vivo imaging was performed through oral administration as well as intravenous injection (Figs. 2.6 and 2.7) of nanoparticles in a mouse [48]. After intravenous injection of the nanoparticles and in situ x-ray irradiation, whole body imaging was performed. Significant NIR PersL signal was obtained in liver after excitation (Fig. 2.6) and the signal decayed slowly but it was again obtained after recharging using x-ray. This shows that the material can be successfully recharged constantly.

The in vivo imaging experiment was also performed after oral administration of the nanoparticles followed by x-ray excitation. In this case, the NIR PersL signals were obtained from the stomach and intestines (Fig. 2.7) and it existed up to 50 min. Here also, the luminescence was obtained after repeated excitation. This work demonstrated the use of x-ray activated NIR PersL material for in vivo deep penetrating and renewable imaging.

2.7 Role of Various Fluorescence Imaging Techniques in Multimodal Bioimaging

Multimodal imaging involves the merging of multiple imaging techniques for the creation of a new imaging mode and this produces better diagnosis results. For instance, combination of PET and CT imaging modes to get PET/CT technique and this mode combine the advantages of both modalities. Similarly, scan time is reduced in PET/MRI imaging system with respect to single MRI imaging. This means that production of high-resolution images is made possible through the development of such multimodal imaging techniques. Although fluorescence imaging can give high resolution images, 3D images of anatomical structures could not be normally produced in this imaging mode. Therefore, combining various fluorescence imaging techniques with other molecular imaging modes can give both high resolution images as well as the 3D structures of various organs. For example, through the combination MRI, SPECT and fluorescence imaging techniques, concerned organs or tissues can be visualized at much better resolution. In this section, the combination of various imaging modalities with different fluorescence imaging techniques is discussed in detail.

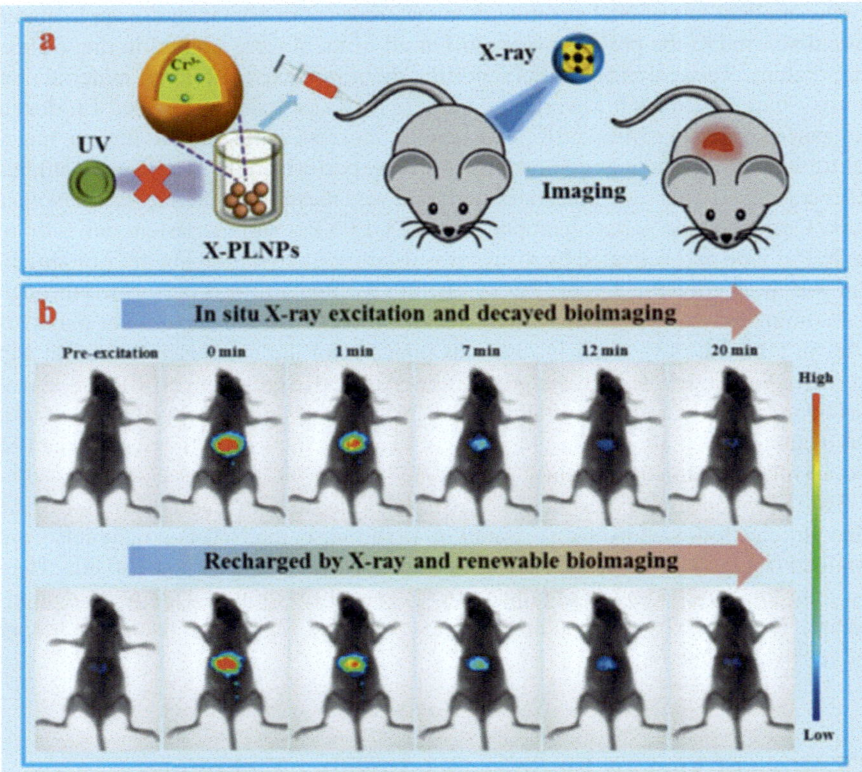

Fig. 2.6 **a** A schematic representation of in vivo whole-body persistent bioimaging of mouse after intravenous injection with PEG-modified X-PLNPs by in situ x-ray excitation, **b** in vivo renewable and decayed whole-body bioimaging of intravenously injected mouse after ceasing x-ray excitation (5 min, 45 kVp). Reprinted with permission from Ref. [48] Copyright © 2017, American Chemical Society

2.7.1 MRI/SPECT/Fluorescence Imaging

There are a number of molecular imaging modes as discussed before, and each mode has their own strengths as well as associated limitations. MRI imaging has many advantages like it uses non-ionizing radiation, allows soft tissue imaging, high penetration depth, etc. The major drawbacks of optical imaging are its lower penetration depth and the inability to describe 3D structures. It has many advantages too that it allows real-time imaging and provides images with very high spatial and temporal resolution. It also has an added advantage of determining the biodistribution of the particles in organs or tissues, which is an ability that other imaging modes lack. SPECT imaging uses radioactive tracers which emit gamma rays and it is cheaper than PET imaging and the tracers have longer half-life. The inherent disadvantages

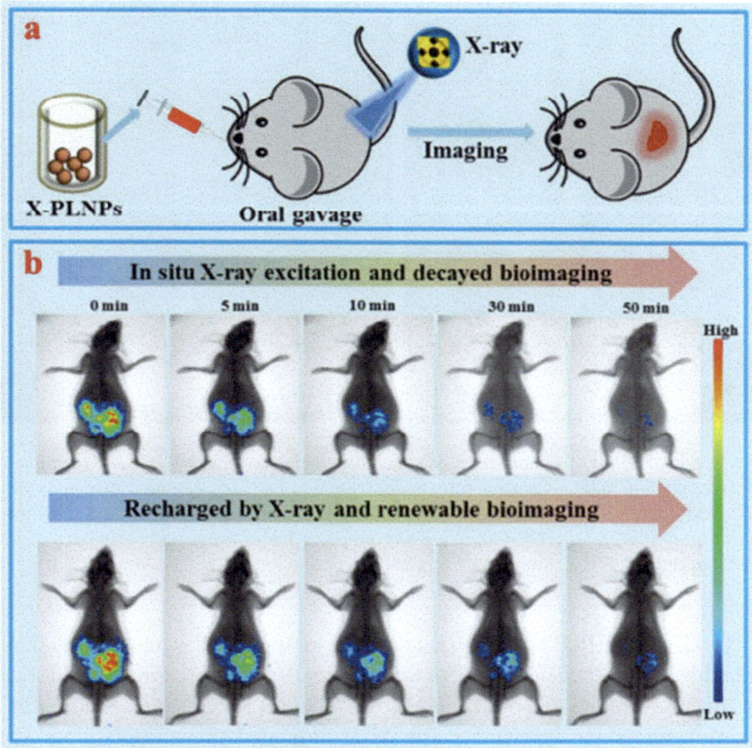

Fig. 2.7 **a** A schematic diagram of in vivo bioimaging of mouse with oral administration of PEG-modified X-PLNPs after in situ x-ray excitation, **b** rechargeable in vivo NIR persistent bioimaging of mouse at different time via oral gavage of PEG-modified X-PLNPs after stoppage of x-ray excitation (5 min, 45 kVp). Reprinted with permission from Ref. [48] Copyright © 2017, American Chemical Society

of this mode are lower spatial resolution and longer imaging time. If all these three modes are combined to produce a single MRI/SPECT/fluorescence imaging mode, the limitations of individual modes can be compensated and the advantages of each mode can be obtained. In one such work by Saatchi and co-workers, they fused MRI, SPECT and fluorescence imaging modes to produce a trimodal imaging technique [49]. In this work, the authors synthesized a trimodal imaging probe based on high molecular weight polyglycerol (HPG). HPG was labelled with ^{111}In, Gd and Alexa dye (fluorescence probe). ^{111}In is a radioactive tracer used in SPECT imaging mode and Gd is as a contrast agent in MRI imaging. HPG was also modified by DOTA (1,4,7,10-tetraazacyclododecane-1,4,7,10-tetraacetic acid) for binding with the metal ions (^{111}In and Gd). Biodistribution studies of radiolabelled HPG were performed on a female mice injected with tumour cells and the activity at 24 h and 72 h post injection

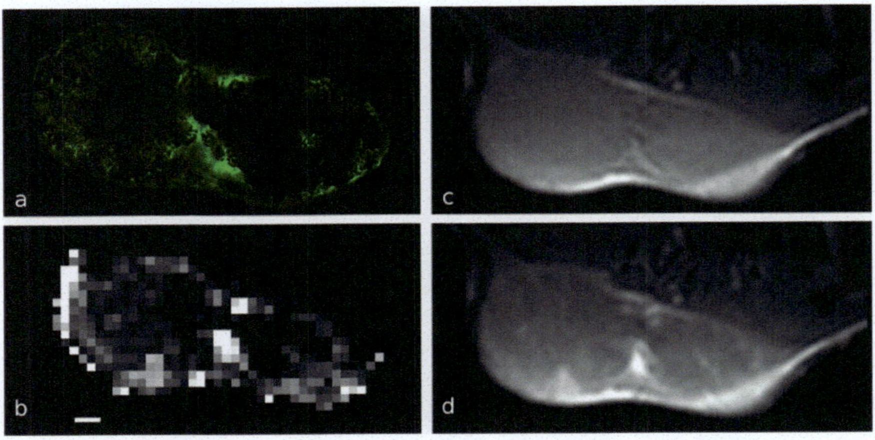

Fig. 2.8 **a** Image of HPGDF-Gd fluorescence of a tumour cryosection. **b** HPGDF-Gd concentration from calibrated change in R1 relaxation rate (range from 0 to 1 s^{-1}) of the same slice and orientation in vivo. Corresponding T1-weighted RARE MR images are shown **c** prior to and **d** 40 min post-injection of 600 mg/kg of HPGDF-Gd. Scale bar indicates 1 mm. Reprinted with permission from Ref. [49]. Copyright © 2012, American Chemical Society

of the ^{111}In-HPGD (HPGD means ligand functionalized HPG) and a biological half-life of 32.6 h was obtained. To check the suitability for MRI imaging and fluorescence imaging, tumour uptake of HPGDF-Gd (dye labelled HPGD) was studied in mice. T1 relaxation rate measurements using MRI indicated that tumour uptake of the probe increased with time. The results were also verified ex vivo by fluorescence microscopy of the tumour (Fig. 2.8).

Similarly, the design of a trimodal probe for MRI/SPECT/fluorescence imaging was reported in another work by Madru et al. where they used nanoparticles of superparamagnetic iron oxide labelled with ^{99m}Tc and a fluorescent dye (Alexa Fluor 647) as the imaging probe (^{99m}Tc-AF-SPIONs) [50]. The iron oxide nanoparticles were functionalized using poly ethylene glycol (PEG) and primary amino group. The in vivo imaging studies were performed by injecting the probe in the hind paw of Wistar rats and the probe was found to be accumulated in the sentinel lymph node (SLN). MRI imaging identified the accumulation of the probe in SLN and also the T2 relaxation rate was found to be shortened. SPECT images also detected the accumulated probes in SLN as well as a small accumulation in the inguinal lymph node and renal lymph node also after 5 h of injection. Fluorescence imaging also recognized probe accumulation in SLN and the results confirm the suitability to use the material as a trimodal imaging probe for clinical applications.

2.7.2 MRI/CT/Upconversion Luminescence (UCL) Imaging

CT imaging provides the 3D structural details of various parts of the body based on x-ray absorption characteristics. Even though, this is one of the widely used imaging technique, it has an inherent drawback which is its low sensitivity to soft tissues and MRI imaging can overcome this limitation due to its higher sensitivity to soft tissues. Therefore, combining both these imaging modes can give better tissue imaging results. Both these imaging modalities (CT and MRI) face a common drawback which is their inability to provide cellular details during the imaging process. Here comes the importance of fluorescence imaging which can provide the cellular details with high resolution. Upconversion nanophosphors which convert low energy excitation to high-energy emission can provide better imaging results. This is because it mainly emits in the NIR region which gives higher penetration depth and avoids autofluorescence effect. Therefore, combining these three imaging modalities can bring efficient diagnosis results.

Zhou and his co-workers synthesized upconversion nanophosphors based on $NaLuF_4$ and they developed a triple modal imaging probe for CT/MRI/UCL imaging using the material [51]. In this work, $NaLuF_4$ based UCL nanophosphors co-doped with Gd, Yb, and Er/Tm (denoted as Lu-UCNPs) were successfully prepared using microemulsion method and the suitability of this material for using as a trimodal imaging agent was proved. The material's capability as T_1 weighted contrast agent in MRI was verified both using in-vitro and in vivo experiments. In vivo lymphatic MRI imaging in mice involved the injection of the material into the left paw, followed by image capture after 30 min (Fig. 2.9). The increased uptake of the contrast agent (CnA) in the lymph nodes suggested the suitability of using the material as T_1-positive CnA in MRI. In vivo CT imaging was conducted using Lu-UCNPs, administered similarly into the left paw, with images recorded 30 min post-injection. The 3D CT images provided the exact position of the lymph node, thus proving the ability of the material as CT contrast agent (Fig. 2.10).

The location of lymph node was also determined using in vivo UCL imaging in mice. The position of the sentinel lymph node was exactly located 30 min after injection of 80 μL of the material. UCL signal at 800 nm was obtained at the injection site as well as from the lymph node (Fig. 2.11a). In-situ images were also recorded after removing the skin and fatty tissues which again provided UCL signal from the lymph node (Fig. 2.11b). Ex vivo images after the removal of the lymph node from the body was also recorded (Fig. 2.11c).

This work was the first-time report of using Lu-UCNPs for MRI/CT/UCL trimodal lymphatic imaging of mice. The authors also proved the water-stability, long-term stability, and the less toxicity of the material.

Another work reported the synthesis of Fe_3O_4@$NaLuF_4$: Yb, Er/Tm nanoparticles with excellent upconversion characteristics for using as a trimodal imaging agent in MRI/CT/UCL imaging [52]. Here, Fe_3O_4 was the core and it exhibited the magnetic properties essential for MRI imaging and $NaLuF_4$: Yb, Er/Tm with excellent UCL properties was the shell and it enhanced the contrast in CT imaging. Xing

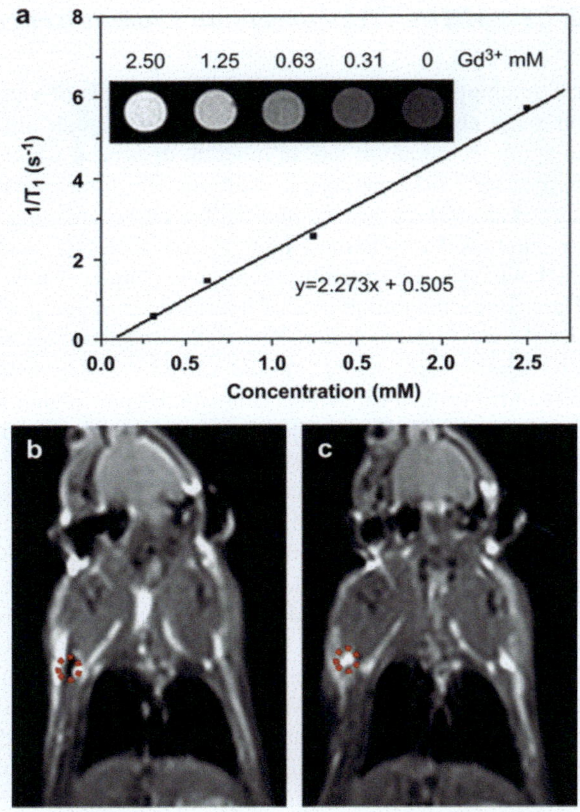

Fig. 2.9 **a** Relaxation rate R_1 ($1/T_1$) vs various molar concentrations of Gd^{3+} from Lu-UCNPs (2.50, 1.25, 0.63 and 0.31 mM of Gd^{3+}) using a 3 T MRI scanner. Inset: T_1-weighted MR images of Lu-UCNPs. Transversal cross-section images of lymph node, 0 min (**b**) and 30 min (**c**) after intradermal injection of Lu-UCNPs (80 mL, 3 mg/mL). Reprinted with permission from Ref. [51]. Copyright 2012 Elsevier Ltd.

et al. developed a PEG coated triple modal imaging agent NaY/GdF_4: Yb, Er, Tm @SiO_2-Au with UCL properties in which Gd acted as MRI contrast agent and Au nanoparticles enhanced the CT contrast [53]. There are similar published reports which combined the UCL properties with MRI and CT imaging modes for better imaging results [54, 55].

2.7.3 MRI/PersL

The importance of PersL nanoprobes such as very less noise levels, deep tissue penetration property, easier handling, etc. make them an ideal for multimodal imaging applications. The long afterglow emission and high image resolution properties exhibited by such materials can be extremely useful when combined with other imaging modes such as MRI, CT, etc. The bimodal imaging combining PersL with MRI was first reported in a work by Abdukayum et al. in which they developed a PersL

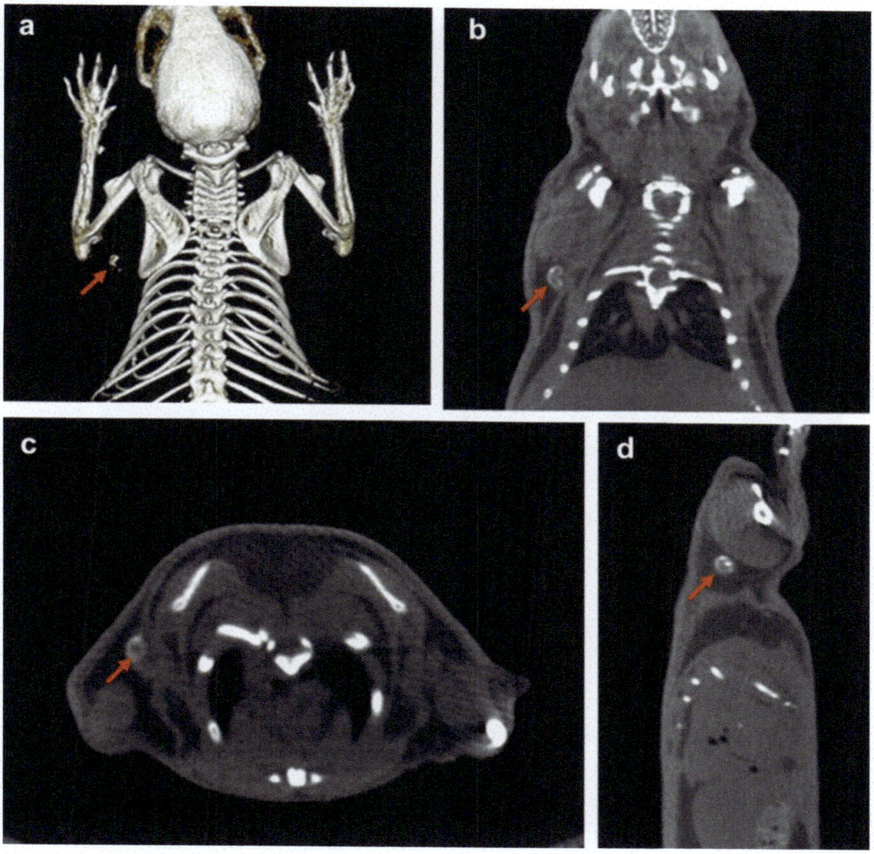

Fig. 2.10 In vivo three-dimensional CT volume-rendered (**a**) and maximum intensity projections of coronal (**b**); transversal (**c**) and sagittal (**d**) images of the lymph node were obtained 30 min after intradermal injection of Lu-UCNPs solution (80 mL, 3 mg/mL) in left paws. The red arrows indicate the locations of the lymph nodes. Reprinted with permission from Ref. [51]. Copyright 2012 Elsevier Ltd.

nanophosphor ($Zn_{2.94}Ga_{1.96}Ge_2O_{10}$: Cr^{3+}, Pr^{3+}) surface modified by diethylenetriamine pentaacetic acid (DTPA) and 3-amino propyltriethoxysilane (APTES) [56]. Gd^{3+} ions were introduced into this material for MRI contrast enhancement. The sample solution was prepared in PBS and 300 μL of this solution was injected to a nude mouse for in vivo imaging. The sample solution was pre-excited at 254 nm 10 min before the injection and images were recorded for more than 6 h without any further excitation (Fig. 2.12). The PersL signals obtained mainly originated form liver, and this was an example of long-term in vivo imaging.

The material was also used for T_1 – weighted MRI imaging in mouse by injecting 300 μL of the sample solution. The liver could not be located in the MRI images taken before the injection but the same was clear 15 min after injection (Fig. 2.13).

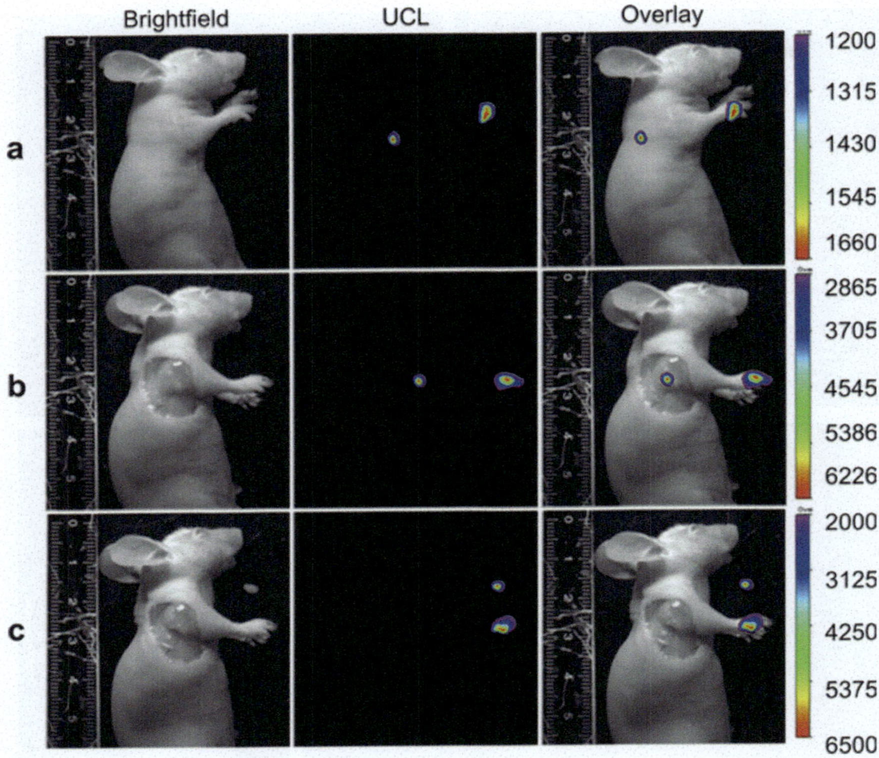

Fig. 2.11 In vivo (**a**), in situ (**b**), and ex vivo (**c**) lymphatic images were detected at lymph node of the nude mice after injection of Lu-UCNPs (80 mL, 3 mg/mL). UCL signals were collected at 800 ± 12 nm. The power density is 150 mW/cm^2 on the surface of the mouse. Reprinted with permission from Ref. [51]. Copyright 2012 Elsevier Ltd.

Similarly, in a work by Wang and his co-workers, amino functionalized PersL nanoparticles of $ZnGa_2O_4:Cr^{3+}$ were synthesized and linked to hyaluronic acid functionalized Gd_2O_3 for using as a bimodal imaging agent in MRI/PersL imaging [57]. Another work by Maldiney et al. reported the synthesis of $ZnGa_2O_4$: Cr^{3+}, Gd^{3+} nanoparticles for bimodal MRI/PersL imaging [58]. The nanoparticles were suspended in 5% glucose solution for using in in vivo imaging experiments. After pre-excitation by UV light (254 nm), the solution was injected into the tail vein of a healthy mice (BALB/c) and PersL signals were obtained from liver after 5 min of injection and the retention of the nanoparticles in liver and spleen was confirmed from the images taken after 45 min of injection. T_2-weighted MRI images were also recorded before and after injection, and comparison of these images confirmed a change in contrast within the liver.

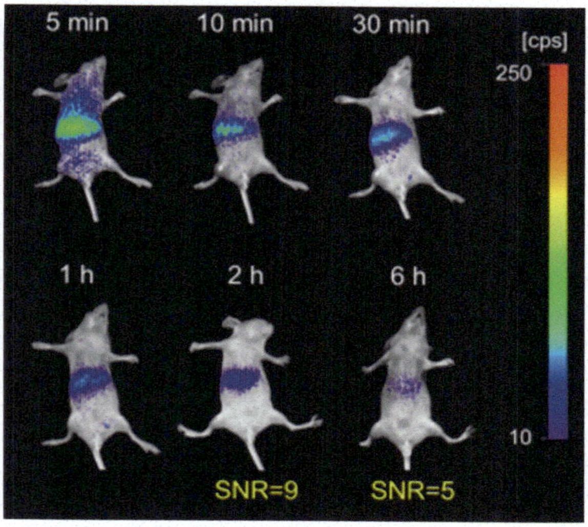

Fig. 2.12 In vivo NIR luminescence images of a normal mouse after intravenous injection of Gd (III)-PLNPs (0.3 mg, 10 min irradiation with a 254 nm UV lamp before injection but without irradiation during imaging). Reprinted with permission from Ref. [56]. Copyright © 2014, American Chemical Society

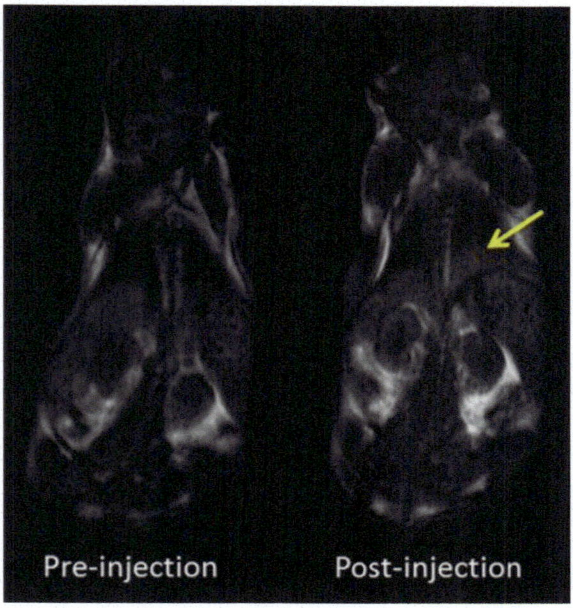

Fig. 2.13 In vivo T1-weighted MR images of the mouse before and after intravenous injection of Gd (III)-PLNPs (0.3 mg); the yellow arrow indicates the liver. Reprinted with permission from Ref. [56]. Copyright © 2014, American Chemical Society

2.7.4 CT/PersL

The advantages PersL nanoparticles and CT imaging in medical diagnosis was already discussed in previous sections. Synthesis of a bimodal imaging probe useful for both CT and PersL can give 3D images with very high resolution. Such a bimodal imaging probe useful for both PersL and CT imaging techniques was developed by Jinlei et al. using $Bi_2Ga_4O_9$: Cr (BGOC) nanoparticles [59]. Here, Cr^{3+} ions are the luminescence centres and Bi^{3+} ions act as CT contrast agent. The biocompatibility of the nanoparticles was enhanced by capping them with NH_2 groups, and folic acid (FA) was coated for specific binding.

The in vivo experiments were performed on a Kunming mouse bearing H22 tumour. The BGOC—NH_2 nanoparticles and BGOC—FA nanoparticles were injected into the tail vein of the mouse to perform PersL imaging. The nanoparticles were pre-excited by UV light (254 nm) for 5 min before injection. The first strong PersL signal was obtained from the liver and LED as well as x-ray reactivation was used when the PersL signal diminished (Fig. 2.14). Stronger signals appeared from the tumour when x-rays were used for reactivation due to the higher tissue penetration property of x-rays. The work demonstrated that long-term tumour targeted imaging was possible with FA caped BGOC nanoparticles when x-rays were used for re-activation.

To compare the performance of the synthesized probes as CT contrast agent, in-vitro experiments were performed using the probe as well as with iohexol (clinical CT contrast agent) and the results were compared. The Hounsfield unit (HU) characterises the CT signal intensity and the HU values of the nanoprobes were found to be larger than that of iohexol (Fig. 2.15a, b). The BGOC-FA nanoparticles were also used for in vivo CT imaging in tumour-bearing mouse. The nanoprobes were injected to the mouse and CT images prove that the CT signal was present even after 1 h of injection (Fig. 2.15c–e).

Similarly, integration of PersL and CT imaging was also performed in another published work where the authors developed $Zn_{2.94}Ga_{1.96}Ge_2O_{10}$: Cr^{3+}, Pr^{3+}@TaO_x@SiO_2 as a bimodal imaging probe [60]. Here, the core was $Zn_{2.94}Ga_{1.96}Ge_2O_{10}$: Cr^{3+}, Pr^{3+} and it exhibited NIR PersL and the SiO_2 shell was fabricated on TaO_x layer for ensuring the biocompatibility and also for the surface functionalization using tumour targeting peptides (PEG and cyclic NGR peptides). The work demonstrated that the NGR modified probes have tumour targeting capability and the material could be used as a bimodal imaging probe.

2.7.5 *MRI/CT/PersL*

Having discussed the role of PersL nanoparticles in various dual-modal imaging techniques such as MRI/PersL and CT/PersL, we are in a position to discuss the role of PersL nanoparticles in tri-modal imaging platforms. Liu and his co-workers developed such a nanoprobe with NIR PersL capability for triple modal imaging i.e., for MRI/CT/PersL imaging [61]. Here, they developed $GdAlO_3$: Mn^{4+}, Ge^{4+}@Au (GAMG@Au) core–shell nanoparticles in which the $GdAlO_3$: Mn^{4+}, Ge^{4+} nanoparticles presented PersL around 700 nm and the presence of Gd- and Au-aided MRI and CT imaging modes, respectively.

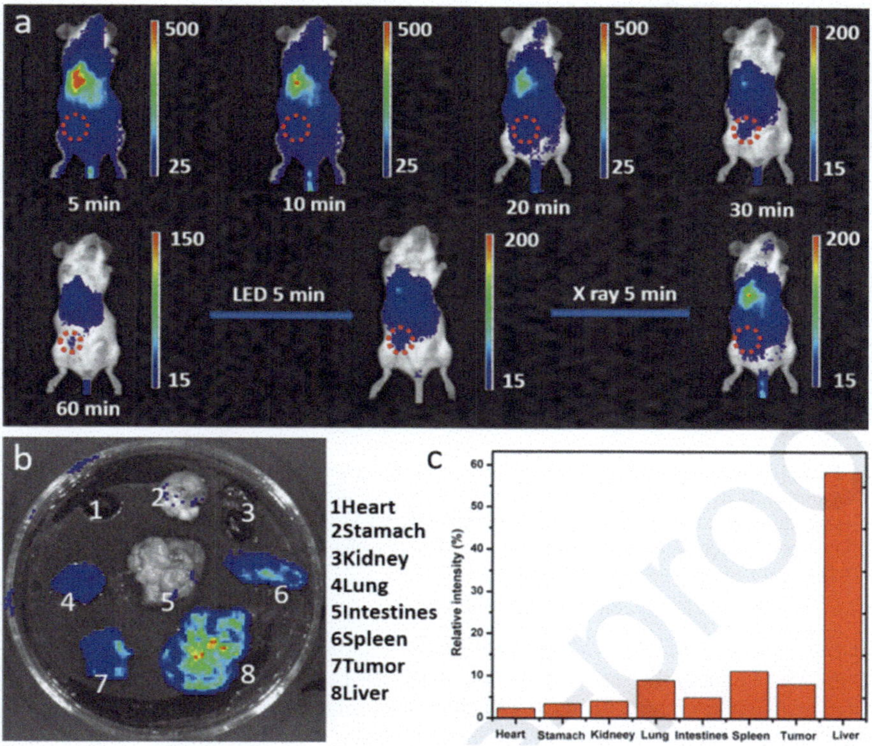

Fig. 2.14 **a** Specific tumour persistent luminescence imaging of H22 tumour-bearing mouse using BGOC-FA nanoprobes. **b** NIR luminescence images of organs from a H22 tumour-bearing mouse at 24 h post-intravenous injection of BGOC-FA nanoprobes (0.2 mg): (1) heart, (2) stomach, (3) kidney, (4) lung, (5) intestines, (6) spleen, (7) tumour, (8) liver. **(c)** Distribution of BGOC-FA nanoprobes in the isolated organs of H22 tumour-bearing mouse. Reprinted with permission from Ref. [59]. Copyright 2021 Elsevier Ltd.

In vivo PersL imaging was performed in healthy mouse as well as in a nude mouse bearing MCF—7 tumour and the experiments were performed after modifying GAMG@Au using folic acid—PEG—thiol (FA-PEG-GAMG@Au) for specific binding. The nanoprobes dispersed in PBS solution were injected to the mouse after a pre-excitation at 254 nm for 10 min. The probes distributed throughout the body of the healthy mouse and PersL signal was recorded for about 8 h without further excitation and the strongest signal was observed from the liver (Fig. 2.16a). PEG-thiol functionalized GAMG@Au (PEG-GAMG@Au) was used a control in the experiment.

The injected FA-PEG-GAMG@Au spread throughout the tumour-bearing mouse's body and the PersL signal appeared at the site of tumour after 1 h of injection and then it disappeared gradually (Fig. 2.16b). The intensity of the PersL signal of the PEG-FA modified sample was greater compared to control (Fig. 2.16c). Apart from

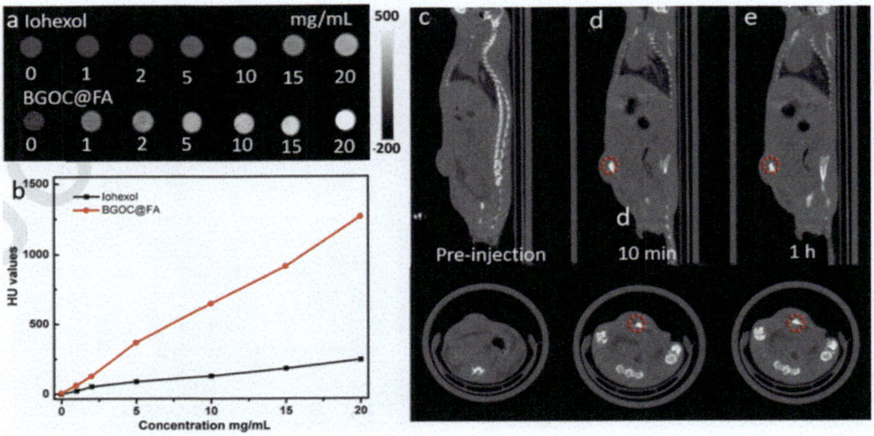

Fig. 2.15 CT images of **a** BGOC-FA nanoprobes and iohexol with different mass concentrations, CT values of **b** BGOC-FA nanoprobes and iohexol with different mass concentrations, In vivo CT imaging of H22 tumour-bearing mice with BGOC-FA via orthotopic injection: **c** pre-injection, **d** 10 min post-injection, **e** 1 h post-injection. Reprinted with permission from Ref. [59]. Copyright 2021 Elsevier Ltd.

tail vein injection, PEG-GAMG@Au was given to the healthy mouse by oral administration as well and the results prove the materials capability as an efficient probe in this case also (Fig. 2.16d). In vivo MRI and CT imaging were also performed after the injection of the prepared probe through the tail vein of the nude mouse (without and with MCF—7 tumour). The T_1—weighted MRI imaging resulted increased contrast in the liver of the mouse 40 min after injection (Fig. 2.17a, b). CT imaging also resulted in enhanced signal from the liver 2 h after injection (Fig. 2.17c–e). Thus, the material was proved to be an efficient triple modal imaging agent for combined MRI/CT/PersL imaging.

2.8 Conclusions

In summary, the chapter provides an overview of in vivo fluorescence imaging fundamentals and their diverse applications in multimodal imaging. It delves into an exploration of luminescent materials employed for in vivo imaging, encompassing various types of fluorescent probes, their limitations, and recent advancements in their application within medical imaging. Considering the growing research focus on utilizing inorganic fluorescent nanoparticles in clinical settings, particularly in in vivo imaging experiments, the significance of developing materials like upconversion nanomaterials, NIR fluorescent probes, and persistent luminescent materials cannot be overstated in the biomedical industry. Thus, the chapter examines the recent advancements

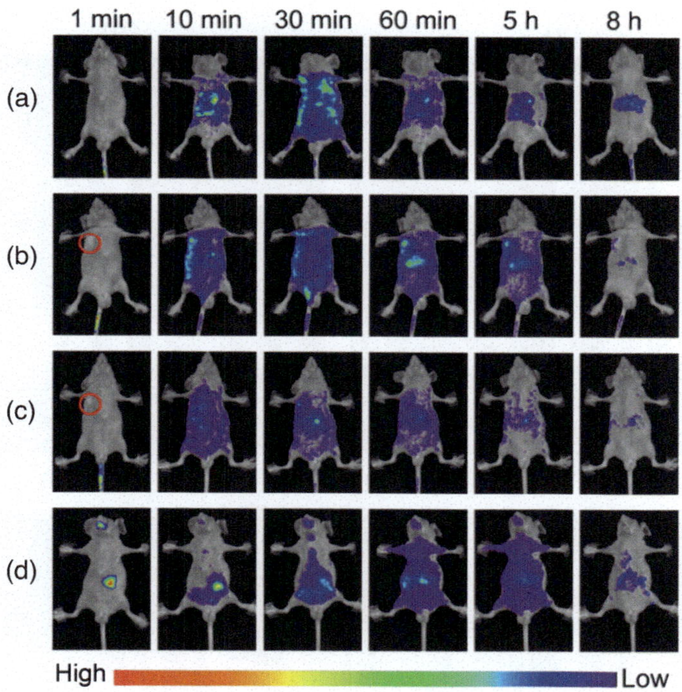

Fig. 2.16 **a–c** In vivo NIR luminescence assay via tail vein injection (0.4 mg, 5 min irradiation by the UV lamp (254 nm, 6 W) prior to injection): Pre-excited FA-PEG-GAMG@Au injected into the normal nude mouse (**a**); Pre-excited FA-PEG-GAMG@Au (**b**), and PEG-GAMG@Au (**c**) injected into the MCF-7 tumour-bearing nude mouse (red circles locate the tumour site); **d** Pre-excited GAMG@Au injected into the normal nude mouse via oral administration. Reprinted with permission from Ref. [61]. Copyright © 2016, American Chemical Society

in these materials concerning their utilization in in vivo imaging and their potential in multimodal imaging applications.

Various molecular imaging techniques such as MRI, CT, PET, SPECT, etc., are combined to provide different multimodal imaging platforms which takes the advantages of all the imaging modes combined. On the inclusion of fluorescence imaging with other molecular imaging techniques, the image resolution multiplies. The integration of persistent or upconversion luminescence, alongside MRI, CT, PET, SPECT, and other imaging modalities, has significantly revolutionized the field of medical bioimaging.

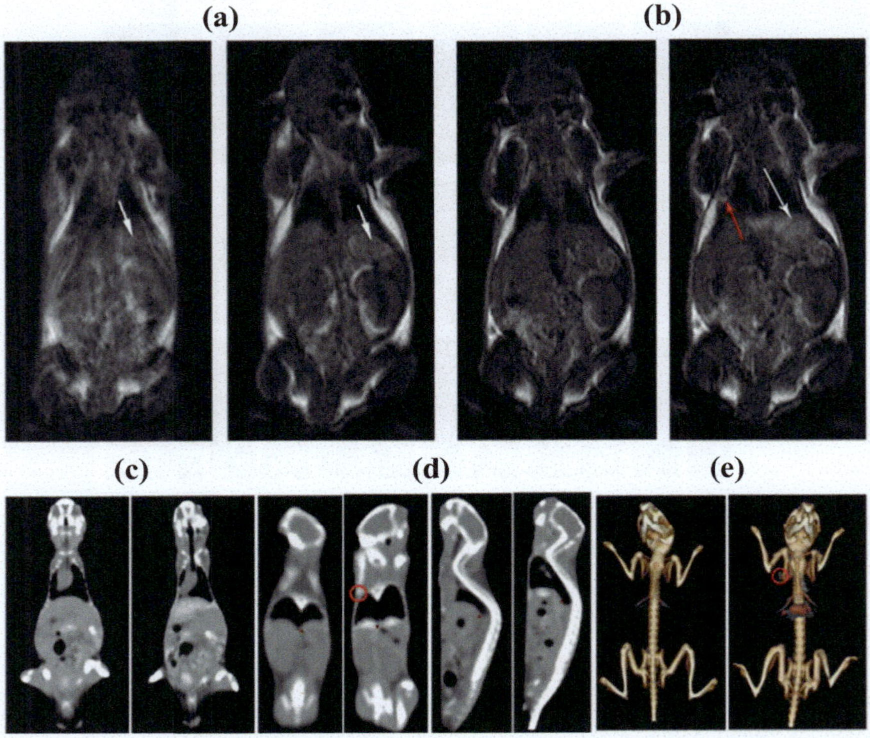

Fig. 2.17 **a** In vivo T1-weighted MR coronal images of a nude mouse before (left) and after (right) injection of the FA-PEG-GAMG@Au nanoprobe via tail vein; **b** In vivo T1-weighted MR coronal images of MCF-7 tumour-bearing mouse before (left) and after (right) intravenous injection of the FA-PEG-GAMG@Au nanoprobe; The white and red arrows indicate the liver and tumour site, respectively. **c** CT coronal view images of a nude mouse before injection of the FA-PEG-GAMG@Au (left) and at 2 h post-injection of the FA-PEG-GAMG@Au (right); **d** CT coronal (left) and sagittal (middle) view images of MCF-7 tumour-bearing mouse before injection of the FA-PEG-GAMG@Au (left) and at 1 h post-injection of the FA-PEG-GAMG@Au (right) along with the corresponding 3D renderings of in vivo CT image **e**; The red cycle indicated the tumour site. Reprinted with permission from Ref. [61]. Copyright © 2016, American Chemical Society

References

1. J. Gao et al., Ultrasmall near-infrared non-cadmium quantum dots for in vivo tumor imaging. Small **6** (2010)
2. H. Chen et al., Near-infrared persistent luminescence phosphors ZnGa2O4:Cr3 + as an accurately tracker to photothermal therapy in vivo for visual treatment. Mater. Sci. Eng. C **79** (2017)
3. Y. Zhong et al., In vivo molecular imaging for immunotherapy using ultra-bright near-infrared-IIb rare-earth nanoparticles. Nat. Biotechnol. **37** (2019)
4. P. Pei et al., X-ray-activated persistent luminescence nanomaterials for NIR-II imaging. Nat. Nanotechnol. **16** (2021)

5. J. Lifante, Y. Shen, E. Ximendes, E. Martín Rodríguez, D.H. Ortgies, The role of tissue fluorescence in in vivo optical bioimaging. J. Appl. Phys. **128** (2020)
6. B. Ballou, B.C. Lagerholm, L.A. Ernst, M.P. Bruchez, A.S. Waggoner, Noninvasive imaging of quantum dots in mice. Bioconjug. Chem. **15** (2004)
7. D.R. Larson et al., Water-soluble quantum dots for multiphoton fluorescence imaging in vivo. Science **300**(80) (2003)
8. X. Gao, Y. Cui, R.M. Levenson, L.W.K. Chung, S. Nie, In vivo cancer targeting and imaging with semiconductor quantum dots. Nat. Biotechnol. **22** (2004)
9. X. Sun et al., Self-illuminating 64Cu-Doped CdSe/ZnS nanocrystals for in vivo tumor imaging. J. Am. Chem. Soc. **136** (2014)
10. X. Michalet et al., Quantum dots for live cells, in vivo imaging, and diagnostics. Science **307** (2005)
11. E. Hemmer, A. Benayas, F. Légaré, F. Vetrone, Exploiting the biological windows: Current perspectives on fluorescent bioprobes emitting above 1000 nm. Nanoscale Horiz. **1** (2016)
12. J.V. Frangioni, In vivo near-infrared fluorescence imaging. Curr. Opin. Chem. Biol. **7** (2003)
13. G. Hong, A.L. Antaris, H. Dai, Near-infrared fluorophores for biomedical imaging. Nat. Biomed. Eng. **1** (2017)
14. T. Maldiney et al., In vivo optical imaging with rare earth doped Ca_2Si_5N_8 persistent luminescence nanoparticles. Opt. Mater. Express **2** (2012)
15. M. Palner, K. Pu, S. Shao, J. Rao, Semiconducting polymer nanoparticles with persistent near-infrared luminescence for in vivo optical imaging. Angew. Chemie Int. Ed. **54** (2015)
16. T. Maldiney et al., The in vivo activation of persistent nanophosphors for optical imaging of vascularization, tumours and grafted cells. Nat. Mater. **13** (2014)
17. Q.L.M. De Chermont et al., Nanoprobes with near-infrared persistent luminescence for in vivo imaging. Proc. Natl. Acad. Sci. U. S. A. **104** (2007)
18. X. Sun, J. Shi, X. Fu, Y. Yang, H. Zhang, Long-Term in vivo biodistribution and toxicity study of functionalized near-infrared persistent luminescence nanoparticles. Sci. Rep. **8** (2018)
19. Y. Chen, S. Wang, F. Zhang, Near-infrared luminescence high-contrast in vivo biomedical imaging. Nat. Rev. Bioeng. **1** (2023)
20. M. Kamimura, N. Kanayama, K. Tokuzen, K. Soga, Y. Nagasaki, Near-infrared (1550 nm) in vivo bioimaging based on rare-earth doped ceramic nanophosphors modified with PEG-b-poly(4-vinylbenzylphosphonate). Nanoscale **3** (2011)
21. L. Quan et al., Near-infrared emitting fluorescent BODIPY nanovesicles for in vivo molecular imaging and drug delivery. ACS Appl. Mater. Interfaces **6** (2014)
22. H. Watanabe et al., Molecular imaging of β-amyloid plaques with near-infrared boron dipyrromethane (BODIPY)-based fluorescent probes. Mol. Imaging **12** (2013)
23. I. Villa et al., 1.3 μm emitting SrF2:Nd3+ nanoparticles for high contrast in vivo imaging in the second biological window. Nano Res. **8** (2015)
24. U. Rocha et al., Neodymium-doped LaF3 nanoparticles for fluorescence bioimaging in the second biological window. Small **10** (2014)
25. N. Hadrup et al., The similar neurotoxic effects of nanoparticulate and ionic silver in vivo and in vitro. Neurotoxicology **33** (2012)
26. E. Phillips et al., Clinical translation of an ultrasmall inorganic optical-PET imaging nanoparticle probe. Sci. Transl. Med. **6** (2014)
27. R. Xie, K. Chen, X. Chen, X. Peng, InAs/InP/ZnSe core/shell/shell quantum dots as near-infrared emitters: bright, narrow-band, non-cadmium containing, and biocompatible. Nano Res. **1** (2008)
28. Y. Zhang et al., Biodistribution, pharmacokinetics and toxicology of Ag2S near-infrared quantum dots in mice. Biomaterials **34** (2013)
29. G. Hong et al., In vivo fluorescence imaging with Ag2S quantum dots in the second near-infrared region. Angew. Chemie Int. Ed. **51** (2012)
30. H.S. Choi et al., Synthesis and in vivo fate of zwitterionic near-infrared fluorophores. Angew. Chemie Int. Ed. **50** (2011)

31. J. Zhou, Q. Liu, W. Feng, Y. Sun, F. Li, Upconversion luminescent materials: advances and applications. Chem. Rev. **115** (2015)
32. D. Song, S. Zhao, Z. Xu, Upconversion luminescent materials: Properties and luminescence mechanisms, in *Principles and Applications of Up-converting Phosphor Technology* (2019). https://doi.org/10.1007/978-981-32-9279-6_1
33. K.K. Markose, R. Anjana, M.K. Jayaraj, Upconversion nanophosphors: an overview, in *Materials Horizons: From Nature to Nanomaterials* (2020). https://doi.org/10.1007/978-981-15-3314-3_2
34. M. Nyk, R. Kumar, T.Y. Ohulchanskyy, E.J. Bergey, P.N. Prasad, High contrast in vitro and in vivo photoluminescence bioimaging using near infrared to near infrared up-conversion in Tm 3+ and Yb 3+ doped fluoride nanophosphors. Nano Lett. **8** (2008)
35. L.Q. Xiong et al., Synthesis, characterization, and in vivo targeted imaging of amine-functionalized rare-earth up-converting nanophosphors. Biomaterials **30** (2009)
36. L. Xiong et al., High contrast upconversion luminescence targeted imaging in vivo using peptide-labeled nanophosphors. Anal. Chem. **81** (2009)
37. Q. Liu et al., Sub-10 nm hexagonal lanthanide-doped NaLuF 4 upconversion nanocrystals for sensitive bioimaging in vivo. J. Am. Chem. Soc. **133** (2011)
38. S.A. Hilderbrand, F. Shao, C. Salthouse, U. Mahmood, R. Weissleder, Upconverting luminescent nanomaterials: application to in vivo bioimaging. Chem. Commun. (2009). https://doi.org/10.1039/b905927j
39. J. Pichaandi, J.C. Boyer, K.R. Delaney, F.C.J.M. Van Veggel, Two-photon upconversion laser (Scanning and Wide-Field) microscopy using Ln3+-doped NaYF4 upconverting nanocrystals: a critical evaluation of their performance and potential in bioimaging. J. Phys. Chem. C **115** (2011)
40. W. Feng, X. Zhu, F. Li, Recent advances in the optimization and functionalization of upconversion nanomaterials for in vivo bioapplications. NPG Asia Mater. **5** (2013)
41. X. Qiu et al., Hybrid nanoclusters for near-infrared to near-infrared upconverted persistent luminescence bioimaging. ACS Appl. Mater. Interfaces **9** (2017)
42. P. Zhao et al., Upconverting and persistent luminescent nanocarriers for accurately imaging-guided photothermal therapy. Mater. Sci. Eng. C **79** (2017)
43. Z. Li et al., Near-infrared light activated persistent luminescence nanoparticles via upconversion. Nano Res. **10** (2017)
44. Z. Xue et al., A 980 nm laser-activated upconverted persistent probe for NIR-to-NIR rechargeable in vivo bioimaging. Nanoscale **9** (2017)
45. N. Basavaraju et al., Red persistent luminescence in MgGa2O4 : Cr 3+; a new phosphor for in vivo imaging. J. Phys. D. Appl. Phys. **46** (2013)
46. L. Li et al., Mechanism of the trivalent lanthanides' persistent luminescence in wide bandgap materials. Light Sci. Appl. **11** (2022)
47. Y.M. Yang et al., X-ray-activated long persistent phosphors featuring strong UVC afterglow emissions. Light Sci. Appl. **7** (2018)
48. Z. Xue et al., X-ray-activated near-infrared persistent luminescent probe for deep-tissue and renewable in vivo bioimaging. ACS Appl. Mater. Interfaces **9** (2017)
49. K. Saatchi et al., Hyperbranched polyglycerols as trimodal imaging agents: Design, biocompatibility, and tumor uptake. Bioconjug. Chem. **23** (2012)
50. R. Madru et al., Development of a hybrid nanoprobe for triple-modality MR/SPECT/optical fluorescence imaging. Diagnostics **4** (2014)
51. J. Zhou, X. Zhu, M. Chen, Y. Sun, F. Li, Water-stable NaLuF 4-based upconversion nanophosphors with long-term validity for multimodal lymphatic imaging. Biomaterials **33** (2012)
52. X. Zhu et al., Core-shell Fe 3O 4@NaLuF 4: Yb,Er/Tm nanostructure for MRI, CT and upconversion luminescence tri-modality imaging. Biomaterials **33** (2012)
53. H. Xing et al., Multifunctional nanoprobes for upconversion fluorescence, MR and CT trimodal imaging. Biomaterials **33** (2012)

54. F. Liu et al., Conjugation of NaGdF4 upconverting nanoparticles on silica nanospheres as contrast agents for multi-modality imaging. Biomaterials **34** (2013)
55. J.W. Shen et al., Incorporation of computed tomography and magnetic resonance imaging function into NaYF4:Yb/Tm upconversion nanoparticles for in vivo trimodal bioimaging. Anal. Chem. **85** (2013)
56. A. Abdukayum et al., Gadolinium complexes functionalized persistent luminescent nanoparticles as a multimodal probe for near-infrared luminescence and magnetic resonance imaging in vivo. Anal. Chem. **86** (2014)
57. Y. Wang, C.X. Yang, X.P. Yan, Hydrothermal and biomineralization synthesis of a dual-modal nanoprobe for targeted near-infrared persistent luminescence and magnetic resonance imaging. Nanoscale **9** (2017)
58. T. Maldiney et al., Gadolinium-doped persistent nanophosphors as versatile tool for multimodal in vivo imaging. Adv. Funct. Mater. **25** (2015)
59. L. Jinlei, G. Jiaqing, L. Hao, Q. Junle, S. Jun, Simultaneous realization of persistent luminescence and CT dual-mode imaging by x-ray recharged Bi2Ga4O9:Cr nanoprobes in depth-independent tumors. Chem. Eng. J. **406** (2021)
60. Y.C. Lu, C.X. Yang, X.P. Yan, Radiopaque tantalum oxide coated persistent luminescent nanoparticles as multimodal probes for in vivo near-infrared luminescence and computed tomography bioimaging. Nanoscale **7** (2015)
61. J.M. Liu, Y.Y. Liu, D.D. Zhang, G.Z. Fang, S. Wang, Synthesis of GdAlO3:Mn4+,Ge4+@Au core-shell nanoprobes with plasmon-enhanced near-infrared persistent luminescence for in vivo trimodality bioimaging. ACS Appl. Mater. Interfaces **8** (2016)

Chapter 3
Magnetic Resonance Imaging/Fluorescence Molecular Tomography Hybrid Imaging: Advantages, Limitations, and Applications

Shenbhagaraman Ramalingam and Jenefar Sudarson

Abstract Magnetic resonance imaging (MRI) and fluorescence molecular tomography (FMT) is a hybrid imaging technique that combines MRI and FMT to visualize and characterize molecular and cellular processes in vivo, using both targeted and untargeted fluorescent agents along with magnetic resonance signal. This chapter explains the key role of hybrid MRI-FMT imaging in field of molecular imaging and medicine. The key factors such as components, advantages, disadvantages, and application of hybrid MRI–FMT imaging systems were described. The major advantages are precise imaging of location and nature of disease such as tumour, atherosclerosis, myocardial infarction, neuro disorders, etc., and disadvantages include operating cost, MRI, and FMT signal interference. Advancements in imaging technology and molecular probes are enhancing MRI-coupled FMT's capabilities, aiming to translate it from preclinical research to clinical applications for disease diagnosis and monitoring.

3.1 Introduction

Medical imaging relies heavily on anatomical imaging for diagnosis, surgical guidance, and treatment monitoring. Meanwhile, the quickly developing field of molecular imaging promises improvements in quantitation and specificity for early diagnosis, screening, targeted therapy, and early treatment follow-up [1]. "The capacity

R. Shenbhagaraman (✉)
Centre for Material Engineering and Regenerative Medicine, Bharath Institute of Higher Education and Research, Chennai, India
e-mail: shenbhagaraman@gmail.com

S. Jenefar
Department of Biochemistry, Saveetha Medical College, Saveetha Institute of Medical and Technical Sciences, Chennai, Tamil Nadu 602105, India
e-mail: sjjene@gmail.com

N. Kalarikkal et al. (eds.), *Multimodal Biomedical Imaging Techniques*,
Biological and Medical Physics, Biomedical Engineering,
https://doi.org/10.1007/978-981-96-1124-9_3

to observe and quantify biological and cellular processes in vivo" is the definition of molecular imaging [2]. With the use of this imaging approach, biological processes in humans and other living systems at the molecular and cellular levels can be seen, described, and measured [3]. Various imaging techniques include magnetic resonance spectroscopy (MRS), magnetic resonance imaging (MRI), nuclear medicine, ultrasound and Positron emission tomography–computed tomography (PET-CT) optical imaging.

3.2 Magnetic Resonance Imaging

The medical field primarily makes use of magnetic resonance imaging, more often known as MRI, to get high-quality images of the human body's internal structures [4]. Atomic magnetic resonance (NMR), a spectroscopic method for obtaining tiny physical and chemical information about molecules, provides the theoretical foundation for magnetic resonance imaging (MRI) and its underlying process. An MRI scanner is characterized by a large circular magnet around a tube. A typical MRI scan involves positioning the patient on a movable bed inside the magnet [5]. Using a strong magnetic field, a magnet is used to align the protons of hydrogen atoms before they are exposed to a stream of radio waves. This spins different protons in the body, which the MRI scanner's receiver part picks up on as a faint signal. A computer generates a picture by processing the data received by the receiver [6].

3.3 Fluorescence Molecular Imaging

Optical imaging, particularly fluorescence imaging, is standard technique used for preclinical molecular imaging due to its sensitivity, ability to detect low concentrations of target molecules, and affordability, making it a valuable tool in the biomedical research community [7, 8]. Optical imaging in small animals initially focused on qualitative or semi-quantitative 2D imaging of light distribution on the specimen's surface. However, non-linear light transport in tissue prevents accurate quantification. To derive quantitative information, two criteria must be met: accurately accounting for light transport in tissue using a theoretical model and knowledge of optical parameters, and gathering sufficient independent experimental data to extract the unknown reporter probe distribution [9]. Progress in understanding light/tissue interaction has led to three-dimensional quantitative fluorescence tomography.

Fluorescence molecular tomography (FMT) is a method that measures individual source- detector pairs by substituting wide field lighting with a sequential scan of focus light sources, and a broad field of illumination, allowing for more independent data samples and reduced measurement time [10]. A sensitive detector array and a narrow excitation beam are the imaging components of FMT, and they are scanned point-by-point over the area of interest by means of a scanning device.

FMT offers molecular and functional information like PET but has poor spatial resolution compared to CT and MRI. Its limited light penetration depth makes it difficult to allocate data to specific structures. FMT is mainly used with probes in the red to near infrared spectral range for pre-clinical imaging [11].

A compelling option to explain the restricted spatial resolution is to integrate low resolution molecular imaging methods (FMT or PET) with MRI or CT, which allows molecular information to be annotated onto anatomical structures. In addition, structural information may be used to better confine the reconstruction problem for the low resolution technique [12]. When it comes to FMT, structural data and optical properties unique to certain tissues may be added as a prior knowledge to the forward problem. This information can be obtained either concurrently or sequentially, with the latter needing hybrid imaging equipment. In pre-clinical applications, hybrid FMT approaches will be more crucial as they offer supplementary readouts of higher quality [13].

FMT is non-invasive and virtually transferable into the clinic, as are related methods like diffuse optical tomography (DOT), which detect intrinsic tissue qualities associated to the primary absorbers deoxyhemoglobin and haemoglobin. Among the established uses of DOT are assessment of the finger joint in osteoarthritis, newborn brain imaging, and breast imaging [14]. Even commercial DOT systems for breast imaging in clinics are available (ART Advanced Research Technologies Inc., CA and Imaging Diagnostic Systems Inc., US). However, the limited number of fluorescent dyes that have been licensed for therapeutic usage now hinders the practical implementation of FMT [15]. For instance, only investigations using fluorescein in ophthalmology and indocyanine green (ICG) for research on the female breast have been described. However, these fluorescent probes are non-specific, providing information solely on the vasculature with extremely strong background fluorescence [16, 17].

3.4 Fluorescence Probes for in vivo Imaging

Molecular imaging relies on probes that specifically target biological targets in order to provide information on dynamic cellular and molecular processes. Genetically encoded reporter systems and exogenous synthetic methods are the two main approaches to probe design. A receptor binding ligand or other targeting moiety and a reporter moiety are the two main components of an exogenous probe, which are designed to produce a signal detectable by the specific imaging modality [18]. Other fluorescent proteins with genetic encoding, such as green fluorescent protein (GFP), tiny synthetic chemical dyes like cyanine and oxazine compounds, and semiconductor solid-state nanoparticles like quantum dots make up the three primary categories of optical reporters [18]. The absorption/excitation spectrum is determined by the dimension of the structure that dye molecules and fluorescent proteins have, which consists of delocalized π-electrons. The longer the π-system, the more the spectrum is red-shifted, as a general rule. This property has been utilized by genetically designed red-shifted fluorescent proteins, red and near-infrared (NIR) dyes, and others [19]. The core frequency of the emission spectrum of quantum dots, which is dependent on their physical size as long as it stays below the electron–hole pair's Bohr radius, is determined by the quantum dot's wide excitation (as long it does not exceed the Bohr radius of the electron–hole pair). This means that distinct quantum dots may be stimulated using a single excitation wavelength, a feature that makes it useful for multiplexing research—that is, the simultaneous investigation of numerous processes. The challenge with these kinds of probes is the difficult spectral immixing method used to assign the different fluorescent wavelengths to the many processes that have been studied extensively [20]. Phosphor nanoparticles, which are composed of a crystalline matrix doped with lanthanide ions, are an additional intriguing type of nanoparticle. Their toxicity is lower compared to quantum dots, and they have the potential for up conversion, which means that emission can be recorded at a shorter wavelength than the excitation wavelength (i.e., with a bigger energy differential). Eliminating auto fluorescence entirely and improving imaging performance is the key benefit of these probes. Sharp emission lines, long lifetimes, and enhanced photo stability are benefits of the up conversion mechanism, which is based on repeated photon absorption [21].

3.5 Hybrid MRI/FMT Imaging Systems

Fluorescence molecular tomography (FMT) and magnetic resonance imaging (MRI) can be combined to create hybrid imaging systems that provide high-definition anatomical references for molecular information. This combination can also use prior tissue optical property information to improve resolution. The high sensitivity of fluorescence imaging can detect molecular processes in living organisms [22].

The first study came from Ntziachristos et al. where they used FMT and MRI to evaluate activation of cathepsin B (gliosarcoma in naked mice). The reconstructed image of protease activity, achieved by placing the gadolinium-enhanced MR picture over the distribution of fluorescence intensity helps us to accurately pinpoint the visual signal in the relevant MR brain slice [23]. In a similar work, McCann et al. used a commercial FMT equipment to visualise activity of protease in a mouse brain glioma model. The anatomical reference for the study was provided by the MRI data that was recorded after the FMT procedure. They explain that a combination of FMT and MRI data accurately predicts chemotherapeutic intervention effects by figuring out the ProSense ratio fluorescence to tumour volume [24].

3.6 Magnetic Resonance Imaging-Coupled Paired Agent Fluorescence Molecular Tomography (MRI-PAFT)

Magnetic resonance imaging-coupled paired agent fluorescence molecular tomography (MRI-coupled PAFT) is an advanced imaging technique that combines magnetic resonance imaging (MRI) with fluorescence molecular tomography (FMT) to enhance the visualization and characterization of cellular and molecular processes in vivo. This is a hybrid imaging technique that uses both targeted and untargeted fluorescent agents to noninvasively measure drug-target engagement in vivo [25]. MRI-coupled PAFT combines the anatomical and structural imaging capabilities of MRI with the functional and molecular imaging capabilities of FMT. This integration allows for more precise localization and quantification of molecular targets within the anatomical context provided by MRI [26]. Paired agent approach involves using two fluorescent agents: a target-specific agent that binds to a specific molecular target and a non-specific control agent. By comparing the signals from these agents, researchers can improve the accuracy and specificity of molecular imaging [27].

3.6.1 Dynamic Experimental Protocol for MRI-PAFT

Anaesthesia was administered to the animal before it was placed in the MRI-FMT coil and the optical fibres were brought in connection with its organ of interest especially zone of tumour to begin the MRI-PAFT imaging procedure. The optical fibres plane intersected the tumour, as confirmed by a number of MRI scout images, some of which necessitated repositioning of the animal. Upon achieving an acceptable position, a complete collection of pre and post-fluorescent probe contrast T1-weighted pictures will be obtained across the organ of interest [28].

The schematic representation of MRI–FMT system is shown in Fig. 3.1. Static PAFT imaging conditions require that the PAFT imaging be finished when the MR data was being recorded. To make things easier, PAFT images were initiated without repositioning the animal once the MR scans were finished. After injecting a mixture of the two fluorescent tracers into the tail vein, PAFT imaging would be started (0.2 nmol of each). For a total of 27 frames, these measures were taken continuously for sixty minutes following administration, at a pace of about two minutes for every frame. There were 56 source–detector pairs in every data frame (8 sources and 7 detectors). To correct internal heterogeneities in optical characteristics and in order to get the results in line with the model, standard static PAFT imaging conditions are used to acquire a succession of excitation measurements that match the fluorescence emission observations. Each animal was only measured twice for excitation: once before the tracer was given and again between frames 15 and 16 of the dynamic fluorescence emission measurements; this was done to reduce the amount of time spent collecting data per frame [29].

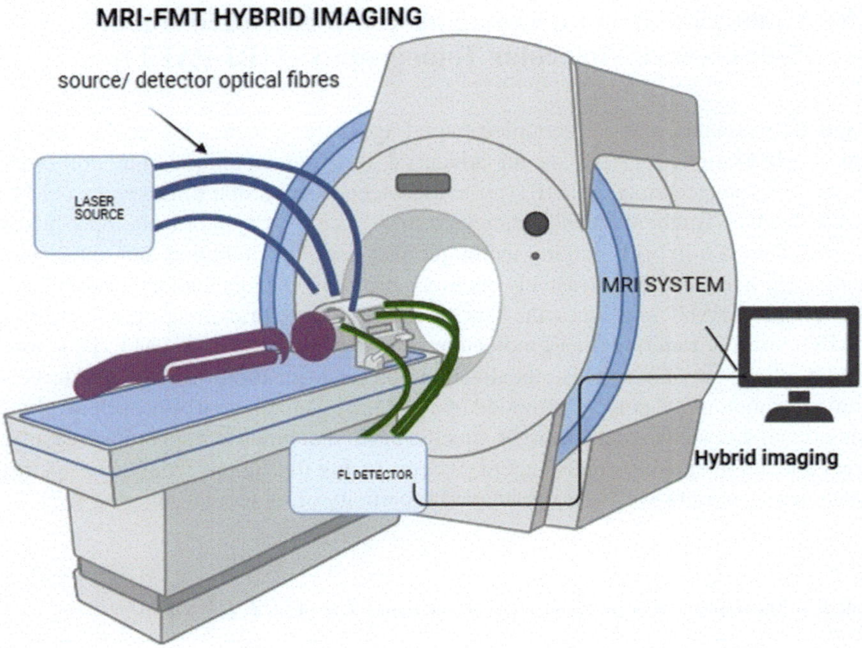

Fig. 3.1 Schematic MRI–FMT hybrid system

3.6.2 Image Recovery and Data Processing for MRI-PAFT

In order to retrieve the dual-tracer MRI-PAFT image, MRI and optical data must be processed and combined. Light is transmitted to and from the tissue via eight optical fibres that round the skull. With the help of MIMICS (Materialise Medical Software), the T1-weighted Gd-enhanced MRI images will be first segmented into three regions: a normal brain area, the tumour area as determined by Gd enhancement, and residual of surviving tissue [30]. The fluorescence emission spectra for a visual imaging component of the MRI-FMT sequence are made up of the tissue auto fluorescence and the coinciding emission spectra from the both fluorophores. Isolating the emission signals from the measured spectra allows one to visualise the internal localization of each tracer with contrast [31]. This spectrum discrimination was accomplished using a linear least-squares spectral fitting algorithm. This algorithm extracts individual fluorescent sources using premeasured basis spectra. The image recovery algorithm combines the optical and MRI data streams. In a nutshell, this procedure requires integrating optical data collected at the tissue border with a numerical model derived from the diffusion approximation of photon propagation in tissue. The MRI image stack is used to construct a predetermined element mesh, which then acts as a numerical pattern for recovering the optical pictures. Prior to reconstructing the image, the fluorescence data underwent model calibration through the utilization of excitation depths obtained through the imaging sequence—a procedure that has been previously documented [32]. The hard-priors' method, which works under the presumption that the fluorescence property is uniformly dispersed in a certain interior tissue region, was used to recover the images. This method drastically decreases the amount of unknowns involved in recovery of image. Every frame in the dynamic series had its image reconstructions for both tracers done again; however, as the animal remained still throughout the FMT and MRI scans, an MRI-derived finite element method (FEM) mesh was applied to every frame from an animal subject. The reference tracer model will extract binding potential values based on the values in the images [33].

3.6.3 Applications of MRI-Coupled PAFT

1. MRI-coupled PAFT can be used to visualize and quantify the expression of cancer biomarkers, monitor tumor progression, and assess the efficacy of targeted therapies.
2. This technique can help in studying brain function, neuroinflammation, and the distribution of molecular probes in neurological diseases.
3. MRI-coupled PAFT can be applied to study molecular processes in atherosclerosis, myocardial infarction, and other cardiovascular illnesses.

3.6.4 Advantages

- The combination of MRI's high spatial resolution with FMT's molecular sensitivity allows for detailed and accurate imaging.
- This technique is non-invasive, making it suitable for longitudinal studies in living subjects.
- One major benefit of in vivo MRI-FMT hybrid imaging is that it may detect diseases in sick tissues without the need for invasive biopsies or surgeries. This allows for a more tailored approach to treatment planning.
- The paired agent approach reduces background signal and enhances the specificity of molecular imaging.

3.6.5 Limitation

The integration of MRI and FMT requires sophisticated equipment and expertise, which can be expensive and complex to operate.

Managing the overlap and interaction between MRI and fluorescence signals can be technically challenging.

3.7 Conclusion and Future Directions

Progress in imaging technologies across time and molecular probes is anticipated to improve the capacities and application of MRI-coupled PAFT. Efforts are being made to translate this technology from preclinical research to clinical applications, where it can aid in diagnosis, treatment planning, and monitoring of diseases.

In summary, MRI-coupled paired agent fluorescence molecular tomography is a powerful imaging technique that leverages the strengths of both MRI and FMT to provide comprehensive anatomical and molecular information. It holds significant potential for advancing research and improving clinical outcomes in various fields of medicine.

References

1. S. Salih, A. Elliyanti, A. Alkatheeri, F. AlYafei, B. Almarri, H. Khan, The role of molecular imaging in personalized medicine. J. Personal. Med.
2. B. Altunay, A. Morgenroth, F.M. Mottaghy, Use of radionuclide-based imaging methods in breast cancer. Semin. Nucl. Med.
3. P.J. Cassidy, G.K. Radda, Molecular imaging perspectives. J. R Soc. Interface
4. A. Berger, Magnetic resonance imaging. BMJ **324**, 35 (2002)
5. Y. Lu, Y. Mazaheri, H. Hricak, A. Shukla-Dave, Clinical applications of magnetic resonance spectroscopic imaging. Compr. Biomed. Phys. **3**, 347 (2014)
6. S. Mastrogiacomo, W. Dou, J.A. Jansen, X.F. Walboomers, Magnetic resonance imaging of hard tissues and hard tissue engineered bio-substitutes. Mol. Imaging Biol. **21**(6), 1003 (2019)
7. D.L. Farkas, Biomedical applications of translational optical imaging: from molecules to humans. Molecules **26**, 6651 (2021)
8. T. Etrych, O. Janoušková, P. Chytil, Fluorescence imaging as a tool in preclinical evaluation of polymer-based nano-DDS systems intended for cancer treatment. Pharmaceutics **11**, 471 (2019)
9. H. Lee, J. Kim, H.H. Kim, C.S. Kim, J. Kim, Review on optical imaging techniques for multispectral analysis of nanomaterials. Nanotheranostics
10. Y. An, K. Wang, J. Tian, Recent methodology advances in fluorescence molecular tomography. Vis. Comput. Ind. Biomed. Art **1**, 1 (2018)
11. A. Walter et al., Correlated multimodal imaging in life sciences: expanding the biomedical horizon. Front. Phys.
12. T. Beyer, L. Bidaut, J. Dickson, M. Kachelriess, F. Kiessling, R. Leitgeb, J. Ma, L.K. Shiyam Sundar, B. Theek, O. Mawlawi, What scans we will read: imaging instrumentation trends in clinical oncology. Cancer Imaging
13. A.B. Konovalov, V.V. Vlasov, S.I. Samarin, I.D. Soloviev, A.P. Savitsky, V.V. Tuchin, Reconstruction of fluorophore absorption and fluorescence lifetime using early photon mesoscopic fluorescence molecular tomography: a phantom study. J. Biomed. Opt. **27** (2022)
14. X. Zhang, Instrumentation in diffuse optical imaging. Photonics **1**, 9 (2014)
15. N. MacRitchie, M. Frleta-Gilchrist, A. Sugiyama, T. Lawton, I.B. McInnes, P. Maffia, Molecular imaging of inflammation—current and emerging technologies for diagnosis and treatment. Pharmacol. Ther.. Ther. **211**, 107550 (2020)
16. C.L. Nguyen, N. Dayaratna, S. Graham, F. Azimi, C. Mak, C. Pulitano, S. Warrier, Evolution of indocyanine green fluorescence in breast and axilla surgery: an Australasian experience. Life **14** (2024)

17. Z.Y. Dai, C. Shen, X.Q. Mi, Q. Pu, The primary application of indocyanine green fluorescence imaging in surgical oncology. Front. Surg.
18. I.K. Cho, S. Wang, H. Mao, A.W. Chan, Genetic engineered molecular imaging probes for applications in cell therapy: emphasis on MRI approach. Am. J. Nucl. Med. Mol. Imaging **6**, 234 (2016)
19. C. Schirripa Spagnolo, S. Luin, Choosing the probe for single-molecule fluorescence microscopy. Int. J. Mol. Sci. **23** (2022)
20. D. Bera, L. Qian, T.K. Tseng, P.H. Holloway, Quantum dots and their multimodal applications: a review. Materials (Basel) **3**, 2260 (2010)
21. I. Rostami, H. Rezvani Alanagh, Z. Hu, S.H. Shahmoradian, Breakthroughs in medicine and bioimaging with up-conversion nanoparticles. Int. J. Nanomed. **14**, 7759 (2019)
22. W. Ren, J. Jiang, A.D.C. Mata, A. Kalyanov, J. Ripoll, S. Lindner, E. Charbon, C. Zhang, M. Rudin, M. Wolf, Multimodal imaging combining time-domain near-infrared optical tomography and continuous-wave fluorescence molecular tomography. Opt. Express **28**, 9860 (2020)
23. V. Ntziachristos, C.H. Tung, C. Bremer, R. Weissleder, Fluorescence molecular tomography resolves protease activity in vivo. Nat. Med. **8**, 757 (2002)
24. C.M. McCann, P. Waterman, J.-L. Figueiredo, E. Aikawa, R. Weissleder, J.W. Chen, Combined magnetic resonance and fluorescence imaging of the living mouse brain reveals glioma response to chemotherapy. Neuroimage **45**, 360 (2009)
25. B. Meng, M.R. Folaron, R.R. Strawbridge, K.S. Samkoe, K. Tichauer, S.C. Davis, Estimating drug delivery using hybrid system for simultaneous dynamic MRI and fluorescence tomography. Proc. SPIE Int. Soc. Opt. Eng. **11219** (2020)
26. B. Li, F. Maafi, R. Berti, P. Pouliot, E. Rhéaume, J.-C. Tardif, F. Lesage, Hybrid FMT-MRI applied to in vivo atherosclerosis imaging. Biomed. Opt. Express **5**, 1664 (2014)
27. C. Wang, X. Xu, S. Hodge, E.Y. Chen, P.J. Hoopes, K.M. Tichauer, K.S. Samkoe, Identification of a suitable untargeted agent for the clinical translation of ABY-029 paired-agent imaging in fluorescence-guided surgery. Mol. Imaging Biol. **25**, 97 (2023)
28. C. Hanusch, S. Hoeger, G.C. Beck, Anaesthesia of small rodents during magnetic resonance imaging. Methods **43**, 68 (2007)
29. B. Meng, N. Sadeghipour, M.R. Folaron, R.R. Strawbridge, K.S. Samkoe, K.M. Tichauer, S.C. Davis, Examining the feasibility of quantifying receptor availability using cross-modality paired-agent imaging. Mol. Imaging Biol. **24**, 23 (2022)
30. F. Wang, M. Nojima, Y. Inoue, K. Ohtomo, S. Kiryu, Assessment of MRI contrast agent kinetics via retro-orbital injection in mice: comparison with tail vein injection. PLoS One **10**, e0129326 (2015)
31. S.C. Davis, K.S. Samkoe, K.M. Tichauer, K.J. Sexton, J.R. Gunn, S.J. Deharvengt, T. Hasan, B.W. Pogue, Dynamic dual-tracer MRI-guided fluorescence tomography to quantify receptor density in vivo. Proc. Natl. Acad. Sci. U.S.A. **110**, 9025 (2013)
32. S. Cogliati et al., A spectral fitting algorithm to retrieve the fluorescence spectrum from canopy radiance. Remote Sens. **11** (2019)
33. S. Okawa, Y. Hoshi, A review of image reconstruction algorithms for diffuse optical tomography. Appl. Sci. **13** (2023)

Chapter 4
Rare Earth-Based Nanostructures for Medical Bio-imaging

K. R. Rakhimol, Sabu Thomas, and Nandakumar Kalarikkal

Abstract Rare earth-based nanomaterials are widely used in biomedical applications for bio-imaging. NIR emission, good biocompatibility, and photo stability are the main factors help them to act as a good bio-imaging agent. It can also be used as a multimodal imaging agent. It is used in magnetic resonance imaging, biosensing, etc. In this chapter, we try to explain the synthesis, functionalization strategies, and the mechanisms behind the luminescence emission of lanthanide-doped nanoparticles. This chapter also explains the use of up and downconversion nanophosphors in biological imaging with the energy transfer mechanism.

4.1 Introduction

Bio-imaging includes a wide range of imaging strategies, like magnetic resonance imaging (MRI), positron emission tomography (PET), ultrasound, and optical imaging. Optical imaging technology is of great significance for bioscience studies, alike imaging-guided surgery and post-surgery evaluations for complete removal of diseased section (Fig. 4.1) [1]. Similar approaches endured flourishing development in recent times due to their high temporal and spatial resolutions and the wide accessibility of luminescent materials, including small organic fluorophores, fluorescent doping nanoparticles (NPs), fluorescent proteins, quantum dots (QDs), and luminescent metal complexes. Despite great progress achieved with small organic fluorophores and fluorescent proteins in labeling and protein-interaction studies, some downsides of these fluorophores (e.g., poor photostability, high background noise (arising from autofluorescence)) have limited their usages in bio-imaging. To some

K. R. Rakhimol (✉) · S. Thomas
International and Inter University Centre for Nanoscience and Nanotechnology, Mahatma Gandhi University, Kottayam, India
e-mail: rakhimolkr@gmail.com

N. Kalarikkal
School of Pure and Applied Physics, Mahatma Gandhi University, Kottayam, India

N. Kalarikkal et al. (eds.), *Multimodal Biomedical Imaging Techniques*,
Biological and Medical Physics, Biomedical Engineering,
https://doi.org/10.1007/978-981-96-1124-9_4

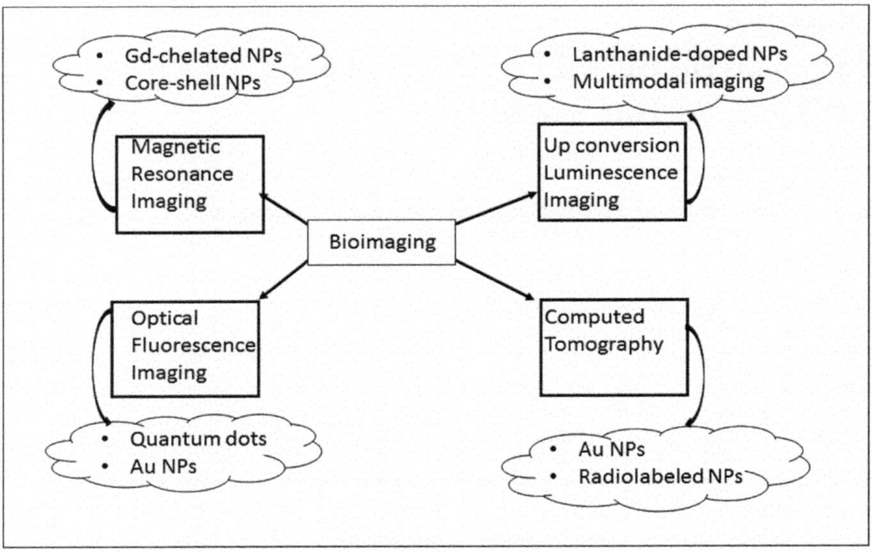

Fig. 4.1 Different types of imaging techniques used for biological imaging

degree, semiconducting QDs crushed these issues, but they also brought new problems, similar as cytotoxicity and photo blinking (intermittency of fluorescence); since QDs are made from toxic heavy metals, cytotoxicity test results indicated similar issues. Thus, there is still high provocation to search for further suitable phosphores with desirable characteristics, like high chemical stability and photostability, long excitation and emission wavelength, and good biocompatibility [2].

In recent times, rare-earth (RE)-metal-doped luminescent nanophosphores, specifically upconversion nanophosphores (UCNPs), were considered promising candidates for biomedical-imaging due to their unique optical properties, including photon upconversion, low cytotoxicity, high photostability, and inertness to environment conditions, such as pH, temperature, and coexisting chemicals [3]. The upconversion properties give lots of benefits in terms of imaging as with lower power excitation light, commonly at 980 nm, UCNPs generate UV/visible, and in some cases, even near-infrared (NIR) luminescence. The long wavelength light can penetrate samples more deeply without inducing autofluorescence. Moreover, the luminescence of UCNPs is very stable without photobleaching and photoblinking issues. When UCNPs are applied in bio-imaging, they therefore allow not only deeper penetration, no toxicity, and low autofluorescence, but also continuous monitoring. These advantages make UCNPs outstanding for bio-imaging [4].

Nevertheless, UCNPs normally have larger particle size than other conventional fluorophores, which leads to non-specific aggregation and reduces their applications in *in-vivo* biological imaging. Moreover, particles with size above 10 nm are hard to be eliminated from the body. Efforts have been devoted to decrease the size of UCNPs, but the surface defects present in decreased size particles in turn reduce the luminescence of the UCNPs. Ideally, UCNPs for bio-imaging should have sub-10 nm particle size while maintaining high luminescence [5].

UCNPs comprise a host matrix lattice and a mixture of RE metals as dopants. The luminescence of the UCNPs can be tuned by the host matrix used, the ratio of the doped metals, size and shape, and experimental conditions [6]. Careful control of synthesis conditions is therefore required in order to acquire the UCNPs with small particle size, uniform shape, and high luminescence. However, as most of the methods of synthesis applied use hydrophobic compounds as capping ligands, these UCNPs are always dispersible in non-polar organic solvents other than aqueous solutions or biological buffers. Surface modification is therefore required to endow hydrophilic and biocompatibility properties to these UCNPs being considered for use in biological systems. It is worth noting that the surface modification should not increase the size and or reduce the luminescence, giving more challenges [7].

4.2 Synthesis of Rare Earth-Based Nanostructures

Traditional solid-state synthesis is used for the synthesis of rare earth-based nanomaterials. However, the high-temperature annealing leads to the formation of irregular particles with broad size distribution, which restricts the bio-based applications [8]. To overcome this issue, wet chemical methods are mostly used for the synthesis of rare earth-based nanomaterials used for bio-imaging. Most common synthesis methods used for the effective synthesis are precipitation at atmospheric pressure (which can be done in aqueous or organic solvents), hydrothermal [9], or solvothermal methods [10].

To obtain phosphors with persistent luminescence, high-temperature annealing is required. Solid-state synthesis methods usually used for this result in the formation of heterogeneous microparticles [11]. Wet chemical synthesis methods followed by annealing sometimes result in the production of nanoparticles (for example sol–gel synthesis method). However, proper control on the morphology of the synthesized particles is very difficult in these methods. Most of the times, the heterogeneous size limits their bio-applications [12].

4.3 Functionalization of Rare Earth-Based Nanostructures

The change in the surface chemistry can be contributed by the functionalization of nanoparticles. The main objectives of the functionalization are (a) the preparation of water dispersible nanoparticles, (b) the introduction of electrostatic and steric repulsion to prevent the agglomeration inside the host, and (c) the attachment of different reactive groups such as amino groups and carboxylic groups to bind with biomolecules like antibodies, proteins, or drugs [13, 14]. If the nanoparticle is hydrophobic in nature, the functionalization is necessary. It can be done by the removal of hydrophobic layer through acid treatment or ultra-sonication using ethanol. However, these strategies make the further conjugation difficult or lead to the formation of aggregation. So, ligand exchange and polymer coating are generally used for the functionalization of hydrophobic nanoparticles. The ligand exchange comprises replacement of the original hydrophobic group with a hydrophilic group (PEG or citrate) [15, 16]. In polymer coating method, hydrophobic nanoparticles are coated using an amphiphilic polymer. Most commonly used one is PMA and its modifications. The interaction between the nanoparticle surface and polymer coating is hydrophobic interaction [15, 17–20]. Ligand exchange method gives nanoparticles with small size but the colloidal stability is very less. In polymer coating, the nanoparticle shows high colloidal stability but possess large size (Fig. 4.2) [21].

The synthesis of most hydrophilic nanoparticles takes place through one-pot synthesis. In this method, nanoparticle synthesis and functionalization can be done simultaneously. This is considered to be the most convenient method for the synthesis of highly stable nanoparticles in aqueous media. However, these techniques may not work for all systems. In such cases, other methods are used such as layer by layer (LbL) method. This involves the coating of the nanoparticle using polyelectrolyte with alternating charges. The stoichiometric concentration of polyelectrolytes, concentration of nanoparticle, and degree of polymerization of polyelectrolyte are the main variables to be adjusted [22, 23]. This method is commonly used for $REVO_4$: Bi^{3+}, Eu^{3+} [24]. This method is usually used for hydrophilic nanoparticles, but it can also be used for hydrophobic nanoparticle like $NaYF_4$:Yb3 + , Eu3 + nanoparticles after ligand exchange [15].

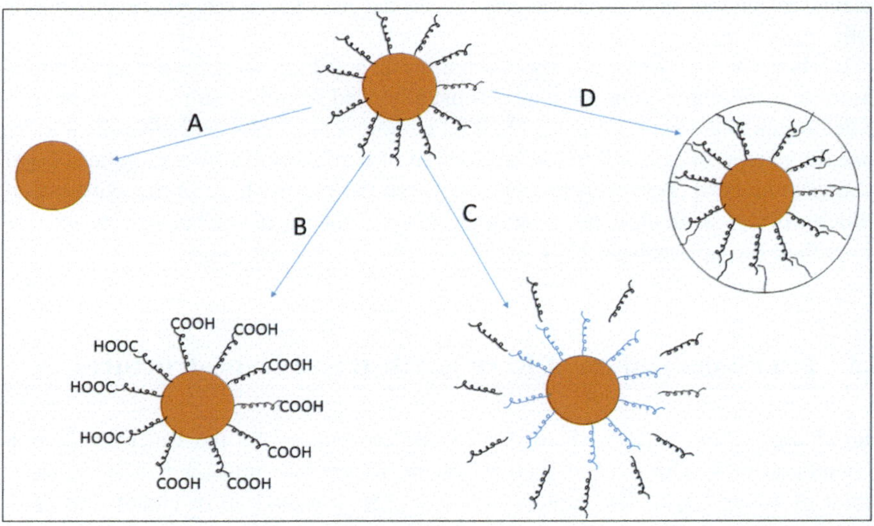

Fig. 4.2 Schematic representation of some functionalization methods for hydrophobic nanoparticles: **a** hydrophobic layer removal; **b** oxidation; **c** ligand exchange; and **d** polymer coating

Another commonly used method which is applicable to hydrophilic and hydrophobic nanoparticles is the encapsulation using amorphous silica [25, 26]. The encapsulation using functionalized silica shows many advantages especially related to biocompatibility and colloidal stability. Hydrophobic nanoparticles such as $NaYF_4$:Yb^{3+}, Tb^{3+} and YF_3. Mesoporous silica can also be used for hydrophobic β-$NaYF_4$ nanoparticle with CTAB as second surfactant [27–29].

4.4 Bio-imaging Through Rare Earth-Based Nanostructures

The electronic configuration, [Xe] $4f^n$, n = 0–14 is the basis of the luminescent properties of Ln ions. The 4f electronic levels are shielded by the 5s2 and 5p6 shells. The chemical environment has no effect on the emission of the material. So, the Ln cations show very sharp excitation and emission bands with similar wavelength in all the matrices where they are located. However, the intensity of the emission is dependent on the matrix [30].

Upconversion and downconversion phosphors can be obtained through the dopant cations. In the downconversion process, the high energy photons are converted into lower energy photons. Even though downconversion is expected, many Ln cations such as Tb^{3+} and Eu^{3+} produce the luminescence in red, yellow and green region [31–36]. The large energy difference between lower level of excited state and higher level of ground state helps to minimize the non-radiative processes in these Ln ions and it makes them a very good candidate for in vivo imaging [37]. Upconversion is a nonlinear optical process in which a higher energy photon is formed from the conversion of two or more low energy pump photons (Fig. 4.3). There are several types of upconversion mechanisms. They are excited-state absorption, photon avalanche, energy migration-mediated upconversion, energy transfer upconversion and co-operative upconversion. In theses, rare earth nanoparticles used for biomedical applications are developed mainly through energy transfer upconversion [38, 39].

A wide variety of inorganic materials are used as the matrices for Ln-based upconversion and downconversion nanoparticles. Fluorides are used because of their improved quantum efficiency. But, phosphates are considered due to their biocompatibility and biodegradability [40, 41]. Other matrices are also be used to enhance the emission of downconversion materials [42–44].

The intensity of luminescence produced by the Ln-doped nanophosphors is very low. The nature of the electronic transitions of the Ln^{3+} ions is the main reason for these low intensity transitions. Another reason for this is the surface quenching. This phenomenon is related with the surface Ln ions. Defects formed on the surface, impurities associated with the nanoparticle surface, and the solvent molecules in which the nanoparticle is dispersed tend to reduce the intensity of the emission of Ln cations present near to the surface of the nanoparticle [45, 46]. The size of the nanoparticle

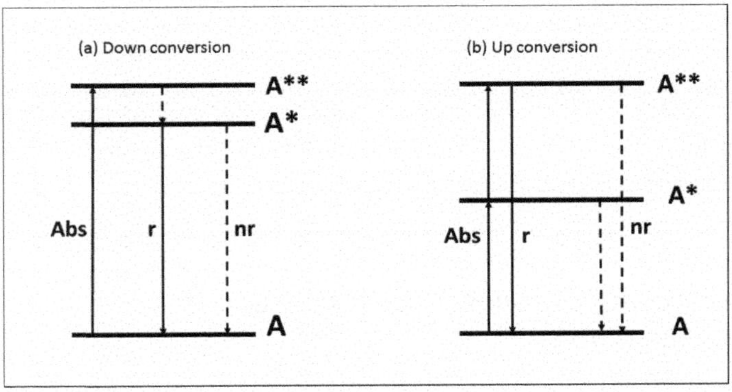

Fig. 4.3 The schematic representation of down and upconversion mechanisms

also affects the emission. Luminescence emission is found to be reduced with the nanoparticle size. Thus, the emission from the ions in the interior and surface together contribute to the global emission [47, 48]. Another factor which affects the fluorescence emission is the interaction of ions with its matrix. Matrices with high phonon energy raise the chances of non-radiative relaxation. This changes with the vibrational quanta needed to cross the energy gap between emitting and non-emitting level of lanthanide ions [49, 50]. Oxide-based and inorganic matrices are causing more non-radiative decay than fluoride-based matrices due to the low phonon energy of fluoride-based matrices, which leads to higher quantum yield for oxide and inorganic matrices. The concentration of active luminescent cations present in the system also affects the quenching mechanism. Normally, if more number of dopants are present, the more emitting centers will be available and luminescence enhancement will be happen. But, after certain level of doping, the energy transfer happening between Ln cations will lead to non-radiative relaxation which in turn reduces the emission intensity. This concentration quenching mainly depends on nanoparticle morphology, distribution, and chemical nature of the material and dopant. So, the optimization of dopant concentration is necessary to get maximum emission intensity [51, 52].

4.5 Luminescence Enhancement Strategies

To overcome the limitations in the application of Ln-doped nanoparticles, the improvement in the emission intensity is highly needed. There are mainly two mechanisms to enhance the luminescence emission of up- and downconversion nanoparticles. The first one is by enhancing absorption/charge transfer and the second one is by minimizing the quenching effects [53].

4.5.1 Energy Transfer in Downconversion Nanoparticles

In downconversion nanoparticles, the excitation of doped Ln cations through the energy transfer from neighboring ions give more efficiency in luminescence enhancement compared to the direct excitation. The donors for this energy transfer mechanism ranges from complex anion groups (vanadate, molybdate, etc.) to individual lanthanide ions (Bi^{3+}, Ce^{3+}, and Gd^{3+}) [54]. The most efficient way to obtain optimum luminescence is the energy transfer from anion groups. However, the energy transfer from anion groups has a serious limitation because of their absorption in UV region. This limits them from in vivo bio-imaging applications. This can be changed by the co-doping with materials which can extend the excitation wavelength to longer wavelength, which enable them to use in bio-imaging applications [55–57].

4.5.2 Energy Transfer in Upconversion Nanoparticles

Sensitizer ion for the energy transfer to the upconversion cations is needed for the development of upconversion nanophosphors. Yb^{3+} is the most commonly used sensitizer (Fig. 4.4). These sensitizer ions permit the excitation at longer wavelengths and improves the penetration depth. The absorption of Ln in near infra-red region minimizes the absorbance by biomolecules. Low concentration of these ions minimizes the quenching effects, and it is physically separated from the emitters by designing a core–shell structure [58–60].

Antenna effect is the common strategy used for the enhancement of Ln-doped nanoparticle's luminescence. In this, the light is absorbed by a fluorophore which is

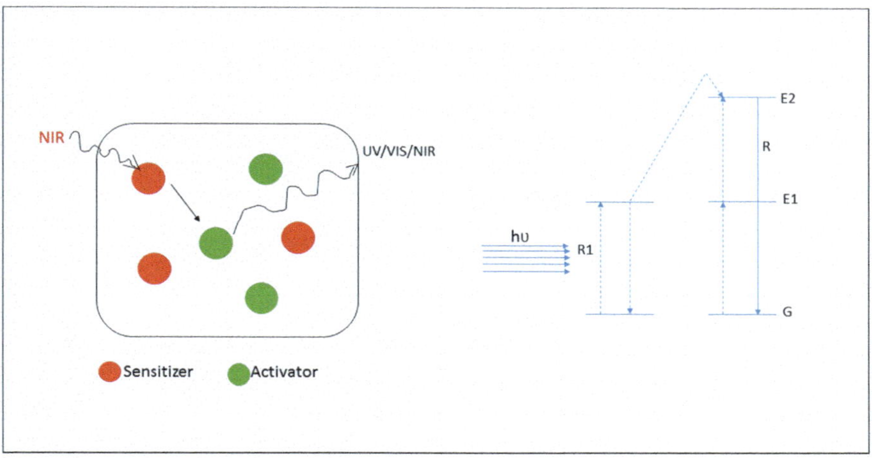

Fig. 4.4 Energy transfer in upconversion nanophosphors

attached to the nanoparticle and it is transferred to the Ln cation for the enhancement and spectral broadening of its absorption. This method can be used for both up and downconversion nanoparticles. Antenna compounds are very relevant in upconversion nanoparticles. It can be conjugated with NIR emitting organic dyes [61–63].

Core shell structure is also used to reduce the quenching by physically separating the surface defects and surface ligands from Ln emitters. This mechanism is also mainly adapted for upconversion nanoparticles [64]. Ln-doped nanoparticles can be coated with silica like amorphous material for functionalization. It also helps to reduce surface quenching. The core–shell structure can be created with same material as core and shell, in which the shell contains inert form of the core material. In this case, the shell can be fabricated with a different material to that of core, but the lattice parameters of the core and shell should be similar. A sensitizer can also be attached to the shell to form active shell (Fig. 4.5). This strategy provides better luminescence than the core–shell structure. The concentration and distribution of the dopant ions present in core and shell affect the luminescence quenching of upconversion nanoparticles. Plasmonic materials like gold or silver are also used as coating material for plasmonic properties and luminescence enhancement [65–69].

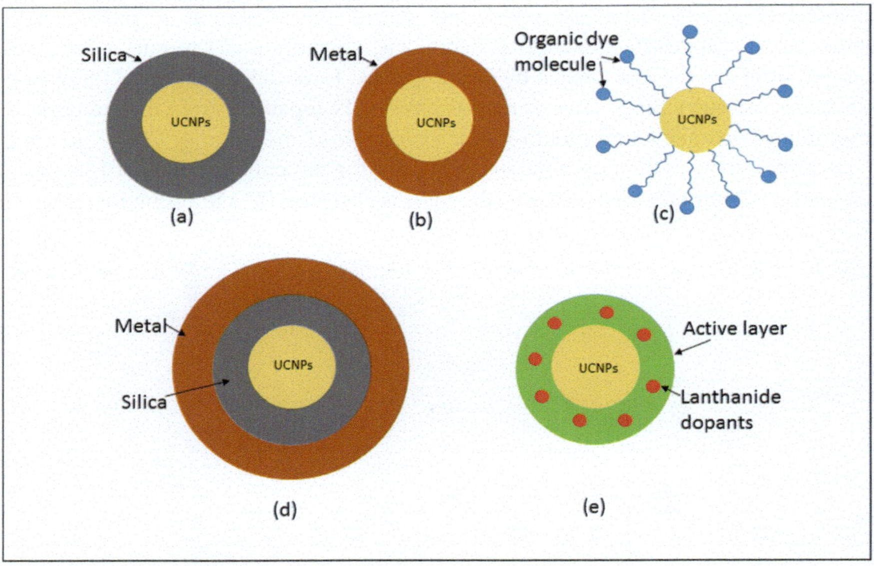

Fig. 4.5 Different types of core–shell structures. **a** UCNP-silica core–shell structure; **b** UCNPs: metal core–shell structure; **c** UCNPs coated with organic dye molecule; **d** UCNP: silica: metal core–shell-shell structure; and **e** Core–shell structure with active shell (layers containing active lanthanide layer

4.6 Luminescent Bio-imaging of Downconversion Nanophosphors

The downconversion nanophosphors based on Ln cations can be of two different types based on their excitation and emission. For the first one, excitation is by UV region and emission in near infra-red region (Eu^{3+} , Tb^{3+}). However, the second one is excited by and emitted in NIR region (doped with Nd^{3+}) [22]. The downconversion phosphors are less commonly used than upconversion nanophosphors because of its need for UV radiation for the excitation, which can harm biological systems. It also produces strong self-fluorescence [70]. Recently, the downconversion nanophosphors excited by near UV region or visible region is reported for bio-imaging application. But for this, codoping is needed to shift the original absorption band towards visible region. Fluorescence microscope or confocal laser scanning microscope can be used to excite these nanoparticles.

The downconversion nanophosphors with the excitation and emission in the near IR region can overcome these issues because they avoid the radiation in UV region. The deep penetration capability and better image contrast makes them a suitable candidate for in vivo biological imaging. In this region, the absorption and self-fluorescence of some biological molecules can also be avoided. Near IR excitation and emission can also reduce the photobleaching and phototoxicity ($NaGdF_4$:Nd^{3+} , $LiYF_4$:Nd^{3+} etc.) [71].

4.7 Luminescent Bio-imaging of Upconversion Nanophosphors

Upconversion nanophosphors are commonly used for in vitro imaging. After proper functionalization and attachment of targeting molecules (peptides, antibodies, growth factors, etc.) on the surface of the nanoparticles, it can also be used in in vivo imaging applications. The use of upconversion nanophosphors for the in vivo imaging in microorganisms (mainly in bacteria), plants and animals (mice models) are already reported [72–74]. It can also be used for single-particle imaging because of their brightness even after long-time continuous laser illumination. The quality of host material is one of the important factors which affect the efficiency of upconversion nanoparticles. So, the selection of host crystal that maintains excellent optical properties and chemical stability is necessary [75].

4.8 Persistent Luminescent Nanophsphors (PLNPs)

The materials which can emit light for long duration (minutes, hours or days) after stopping the excitation are known as persistent luminescent materials. These types of materials can be fabricated by the doping of inorganic matrix with activator metals containing specific traps or defects. When these materials are excited, the charged carrying electrons or holes are excited to high energy excited state; from there, they are taken to a metastable state. A very slow release is only possible for them to the low energy state and it can be controlled by various techniques (Fig. 4.6). Hence, a long lasting luminescence is possible with such materials [76, 77]. These metastable states are usually associated with the defects in the materials. These defects can be intentionally produced through the treatments with high temperature, electron beam treatment, laser treatment, etc. This property can also be produced by the addition of dopants or impurities to the system. The nature and concentration of the impurity or dopant can be optimized. In some cases, Ln acts as an activator to provide a persistent luminescence or emit light through slow energy transfer in a controlled manner. The persistent luminescent materials are commonly silicate, gallate, or aluminate-based materials. Recently in vivo imaging in small animals has been carried out using rare earth based PLNPs [78, 79]. The main parameters to be care about are the duration of luminescence after the removal of the source of excitation and the restoration possibility of the signal.

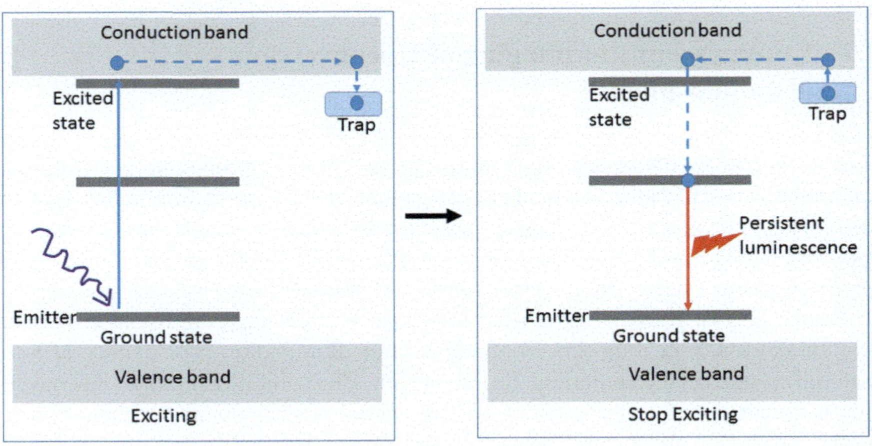

Fig. 4.6 Schematic diagram of persistent luminescence

4.9 Ln-based Nanoparticles for Magnetic Resonance Imaging

Magnetic resonance imaging is a type of non-invasive imaging technique to produce three-dimensional tomographic images of internal organs and tissues with high resolution and better contrast. In this, no ionizing radiation and radioactive agents are used for the imaging. The most commonly used MRI contrast agent is superparamagnetic iron oxide nanoparticles. It is because of their strong magnetic interaction and high stability. However, there is a chance to get false results from blood or any other metallic deposits in the body. To avoid these difficulties, Gd- or Mn-based materials can be used. However, they have a short life span in the vascular systems and which prevents them from long MRI scans [80, 81]. All these complications and limitations can be eliminated by the use of Ln-based nanophosphors. Ln-based nanophosphors are easy to functionalize and possess long circulation time in the biological systems [82]. It makes them an attractive candidate for MRI applications.

4.10 Ln-based Nanoparticles for X-ray Computed Tomography Imaging

Computed tomography is a tissue imaging technique, which is based on the attenuation of X-rays by the body tissues. It is measured in Hounsfield units (HU). Through the computed tomography, 3D details of the tissues with high resolution can be obtained. The temporal resolution is higher than that of MRI. It can complete and record a scan within several minutes. All these advantages make them a good diagnostic technique in hospitals. Classical computed tomography contrast agents approved for human use are mainly iodine-based chelates and barium sulfate. Inherent toxicity of Ba^{2+} ions prevents them from human use other than gastrointestinal imaging. The iodine-based compounds possess short circulation time and high osmolality; it also prevents them from broader applications. High amount of these contrast agents are needed for computed tomography [83]. So a contrast agent with reduced toxicity, long circulation time, capacity for maximum activity at less dosage, and reduced toxicity are needed as contrast agents for computed tomography. Ln-based nanophosphors are widely studied because they have all the above said properties. Most of the contrast agents based on Ln are fluorides because of their above-mentioned properties [84, 85]. Researches proved the efficiency of $LiLuF_4$, $LiGdF_4:Yb^{3+}$, etc. In vitro and in vivo imaging of the Ln-based contrast agents showed good results.

4.11 Ln-based Nanoparticles for Multimodal Imaging

Rare earth-based nanoparticles can be used for multimodal imaging techniques due to the possibility to combine different rare earth nanoparticles in the system. Their ability to show luminescence and magnetic/X-ray absorption makes them a suitable candidate. Dual up- and downconversion modal emission is possible by using luminescent nanoparticles doped with proper combination of rare earth nanoparticles [86, 87].

4.12 Ln-based Nanoparticles for Biosensing Applications

Ln-based sensing systems are a novel method for bioanalytical sensing. The main properties which make them a superior in biosensing application are their chemical stability, low toxicity, narrow emission peak, high quantum yield, etc. Many of these systems are reported as sensing probe; however, the research on this is in the infant stage itself. The structure and the surface properties determine the mechanism involved in the sensing. Depending on this, the sensing systems are divided into three groups. They are Ln-doped nanoparticle without any ligand on their surface in which only physical mechanism is taking place, Ln-doped nanoparticles with functionalized ligands on their surface to recognize the target analyte in which both physical and chemical processes are responsible for the response, and Ln ions embedded in metal organic framework or any polymer in which physical processes together with physico-chemical reactions result in the response [88–90].

Ln-doped nanoparticles are also used as nanothermometers because of their high sensitivity in wide temperature range [91, 92]. The difference in the fluorescence intensities of the emission from the lanthanides has been used for the temperature change monitoring in different types of cells such as HeLa cells. It helps to determine the thermal and physiological processes in cells and cellular death. The main limitation of these nanothermometers is based on its precision and sensitivity. So further researches are needed to improve the efficiency of the system. The possibility of wider functionalization makes them for a wide range of sensing application. Ln-doped nanoparticles do not have any intrinsic luminescence response against any analytes, and the functionalization with suitable materials helps them for sensing different analytes. Most commonly used functionalization materials are antibodies, polypeptides and DNA. Upconversion nanophosphors can be conjugated with energy acceptor molecules, which have the emission spectra which overlaps with the emission spectra of the NPs. Carbon-based nanoparticles, semiconductor quantum dots, organic fluorophores, metal nanoparticles, proteins, etc. can be used as energy acceptors [93–96].

4.13 Conclusion

In spite of the advanced developments in the biotechnological applications of rare earth based nanostructures with different structure, morphology and functionalization, some challenges and limitations are to be faced to improve the activity of these systems to use in clinical trials. The synthesis method commonly used is suitable for lab-scale synthesis only. Large-scale cost-effective synthesis methods having high efficiency need to be developed for commercial use. When the systems are for persistent luminescence, the nanoparticles with uniform size have to be developed. So the quicker synthesis method with better size control should be developed. Rare earth-based multifunctional systems need to be developed for various biotechnological applications. Most important study need to be addressed is the biocompatibility and long-term toxicity of the rare earth-based nanoparticles.

References

1. U. Mahmood, Near infrared optical applications in molecular imaging. IEEE Eng. Med. Biol. Mag. **23**(4), 58–66 (2004)
2. P. Sharma, S. Brown, G. Walter, S. Santra, B. Moudgil, Nanoparticles for bioimaging. Adv. Coll. Interface. Sci. **123**, 471–485 (2006)
3. V. Zepf, An overview of the usefulness and strategic value of rare earth metals. Rare Earths Ind., 3–17 (2016)
4. G. Wang, Q. Peng, Y. Li, Lanthanide-doped nanocrystals: synthesis, optical-magnetic properties, and applications. Acc. Chem. Res. **44**(5), 322–332 (2011)
5. A. Escudero, A.I. Becerro, C. Carrillo-Carrión, N.O. Nunez, M.V. Zyuzin, M. Laguna, W.J. Parak et al., Rare earth based nanostructured materials: synthesis, functionalization, properties and bioimaging and biosensing applications. Nanophotonics **6**(5), 881–921 (2017)
6. N. Feliu, J. Hühn, M.V. Zyuzin, S. Ashraf, D. Valdeperez, A. Masood, W.J. Parak et al., Quantitative uptake of colloidal particles by cell cultures. Sci. Total. Environ. **568**, 819–828 (2016)
7. Y. Chen, S. Ma, Microporous lanthanide metal-organic frameworks. Rev. Inorg. Chem. **32**(2–4), 81–100 (2012)
8. E. Matijevic, Preparation and properties of uniform size colloids. Chem. Mater. **5**(4), 412–426 (1993)
9. A. Rabenau, The role of hydrothermal synthesis in preparative chemistry. Angew. Chem. Int. Ed. Engl. **24**(12), 1026–1040 (1985)
10. R. Komban, R. Beckmann, S. Rode, S. Ichilmann, A. Kühnle, U. Beginn, M. Haase, Surface modification of luminescent lanthanide phosphate nanorods with cationic "quat-primer" polymers. Langmuir **27**(16), 10174–10183 (2011)
11. T. Maldiney, C. Richard, J. Seguin, N. Wattier, M. Bessodes, D. Scherman, Effect of core diameter, surface coating, and PEG chain length on the biodistribution of persistent luminescence nanoparticles in mice. ACS Nano **5**(2), 854–862 (2011)
12. Z. Huo, C. Chen, D. Chu, H. Li, Y. Li, Systematic synthesis of lanthanide phosphate nanocrystals. Chem. Eur. J. **13**(27), 7708–7714 (2007)
13. Y. Liu, K. Ai, J. Liu, Q. Yuan, Y. He, L. Lu, A high-performance ytterbium-based nanoparticulate contrast agent for in vivo X-ray computed tomography imaging. Angew. Chem. **124**(6), 1466–1471 (2012)

14. T. Pellegrino, S. Kudera, T. Liedl, A. Muñoz Javier, L. Manna, W.J. Parak, On the development of colloidal nanoparticles towards multifunctional structures and their possible use for biological applications. Small, **1**(1), 48–63 (2005)
15. S. Wilhelm, M. Kaiser, C. Würth, J. Heiland, C. Carrillo-Carrion, V. Muhr, T. Hirsch et al., Water dispersible upconverting nanoparticles: effects of surface modification on their luminescence and colloidal stability. Nanoscale **7**(4), 1403–1410 (2015)
16. A. Dong, X. Ye, J. Chen, Y. Kang, T. Gordon, J.M. Kikkawa, C.B. Murray, A generalized ligand-exchange strategy enabling sequential surface functionalization of colloidal nanocrystals. J. Am. Chem. Soc. **133**(4), 998–1006 (2011)
17. L.L. Li, R. Zhang, L. Yin, K. Zheng, W. Qin, P.R. Selvin, Y. Lu, Biomimetic surface engineering of lanthanide-doped upconversion nanoparticles as versatile bioprobes. Angew. Chem. (Intl. ed. in English) **51**(25), 6121 (2012)
18. G. Jiang, J. Pichaandi, N.J. Johnson, R.D. Burke, F.C. Van Veggel, An effective polymer cross-linking strategy to obtain stable dispersions of upconverting NaYF4 nanoparticles in buffers and biological growth media for biolabeling applications. Langmuir **28**(6), 3239–3247 (2012)
19. M. Deng, N. Tu, F. Bai, L. Wang, Surface functionalization of hydrophobic nanocrystals with one particle per micelle for bioapplications. Chem. Mater. **24**(13), 2592–2597 (2012)
20. C. Carrillo-Carrion, W.J. Parak, Design of pyridyl-modified amphiphilic polymeric ligands: Towards better passivation of water-soluble colloidal quantum dots for improved optical performance. J. Colloid Interface Sci. **478**, 88–96 (2016)
21. F. Zhang, E. Lees, F. Amin, P. Rivera_Gil, F. Yang, P. Mulvaney, W.J. Parak, Polymer-coated nanoparticles: a universal tool for biolabelling experiments. Small **7**(22), 3113–3127 (2011)
22. Y. Liu, D. Tu, H. Zhu, E. Ma, X. Chen, Lanthanide-doped luminescent nano-bioprobes: from fundamentals to biodetection. Nanoscale **5**(4), 1369–1384 (2013)
23. G. Schneider, G. Decher, Functional core/shell nanoparticles via layer-by-layer assembly. Investigation of the experimental parameters for controlling particle aggregation and for enhancing dispersion stability. Langmuir **24**(5), 1778–1789 (2008)
24. A. Escudero, C. Carrillo-Carrión, M.V. Zyuzin, S. Ashraf, R. Hartmann, N.O. Núñez, W.J. Parak et al., Synthesis and functionalization of monodisperse near-ultraviolet and visible excitable multifunctional Eu 3+, Bi 3+: REVO 4 nanophosphors for bioimaging and biosensing applications. Nanoscale **8**(24), 12221–12236 (2016)
25. A. Guerrero-Martínez, J. Pérez-Juste, L.M. Liz-Marzán, Recent progress on silica coating of nanoparticles and related nanomaterials. Adv. Mater. **22**(11), 1182–1195 (2010)
26. J.N. Liu, W.B. Bu, J.L. Shi, Silica coated upconversion nanoparticles: a versatile platform for the development of efficient theranostics. Acc. Chem. Res. **48**(7), 1797–1805 (2015)
27. H. Hu, L. Xiong, J. Zhou, F. Li, T. Cao, C. Huang, Multimodal-luminescence core–shell nanocomposites for targeted imaging of tumor cells. Chem. Eur. J. **15**(14), 3577–3584 (2009)
28. M. Darbandi, T. Nann, One-pot synthesis of YF3@ silica core/shell nanoparticles. Chem. Commun. **7**, 776–778 (2006)
29. O. Ehlert, R. Thomann, M. Darbandi, T. Nann, A four-color colloidal multiplexing nanoparticle system. ACS Nano **2**(1), 120–124 (2008)
30. G. Liu, B. Jacquier (eds.), *Spectroscopic Properties of Rare Earths in Optical Materials,* Vol. 83 (Springer Science & Business Media, 2005)
31. F.S. Richardson, Terbium (III) and europium (III) ions as luminescent probes and stains for biomolecular systems. Chem. Rev. **82**(5), 541–552 (1982)
32. L. Liping, Z. Minglei, T. Wenming, G. Xiangfeng, L. Guangshe, Y. Liusai, Preparation of cereal-like YVO_4: Ln^{3+}(Ln= Sm, Eu, Tb, Dy) for high quantum efficiency photoluminescence. Nanotechnology (Print), 21 (2010)
33. S.V. Eliseeva, J.C.G. Bünzli, Lanthanide luminescence for functional materials and biosciences. Chem. Soc. Rev. **39**(1), 189–227 (2010)
34. A. Riedinger, F. Zhang, F. Dommershausen, C. Röcker, S. Brandholt, G.U. Nienhaus, W.J. Parak et al., Ratiometric optical sensing of chloride ions with organic fluorophore–gold nanoparticle hybrids: a systematic study of design parameters and surface charge effects. Small **6**(22), 2590–2597 (2010)

35. A. Pandey, M.K. Roy, A. Pandey, M. Zanella, R.A. Sperling, W.J. Parak, H.C. Verma et al., Chloroform-and water-soluble sol-gel derived Eu/Y2O3 (Red) and Tb/Y2O3 (Green) nanophosphors: synthesis, characterization, and surface modification. IEEE Trans. Nanobiosci. **8**(1), 43 (2009)
36. H. Chen, B. Ding, J. Lin, Recent progress in upconversion nanomaterials for emerging optical biological applications. Adv. Drug Deliv. Rev., 114414 (2022)
37. J.C.G. Bünzli, C. Piguet, Taking advantage of luminescent lanthanide ions. Chem. Soc. Rev. **34**(12), 1048–1077 (2005)
38. F. Auzel, Upconversion and anti-stokes processes with f and d ions in solids. Chem. Rev. **104**(1), 139–174 (2004)
39. G. Chen, H. Qiu, P.N. Prasad, X. Chen, Upconversion nanoparticles: design, nanochemistry, and applications in theranostics. Chem. Rev. **114**(10), 5161–5214 (2014)
40. W. Zheng, S. Zhou, Z. Chen, P. Hu, Y. Liu, D. Tu, X. Chen et al., Sub-10 nm lanthanide-doped CaF2 nanoprobes for time-resolved luminescent biodetection. Angew. Chem. **125**(26), 6803–6808 (2013)
41. Z.L. Wang, H.L.W. Chan, H.L. Li, J.H. Hao, Highly efficient low-voltage cathodoluminescence of LaF3: Ln3+ (Ln= Eu3+, Ce3+, Tb3+) spherical particles. Appl. Phys. Lett. **93**(14) (2008)
42. M. Abdesselem, M. Schoeffel, I. Maurin, R. Ramodiharilafy, G. Autret, O. Clément, A. Alexandrou et al., Multifunctional rare-earth vanadate nanoparticles: luminescent labels, oxidant sensors, and MRI contrast agents. ACS Nano **8**(11), 11126–11137 (2014)
43. Y. Zeng, Z. Li, L. Wang, Y. Xiong, Controlled synthesis of Gd 2 (WO 4) 3 microstructures and their tunable photoluminescent properties after Eu 3+/Tb 3+ doping. CrystEngComm **14**(20), 7043–7048 (2012)
44. Y. Zhou, B. Yan, X.H. He, Controlled synthesis and up/down-conversion luminescence of self-assembled hierarchical architectures of monoclinic AgRE (WO 4) 2: Ln 3+(RE= Y, la, Gd, Lu; Ln= Eu, Tb, Sm, Dy, Yb/Er, Yb/tm). J. Mater. Chem. C **2**(5), 848–855 (2014)
45. R. Kumar, M. Nyk, T.Y. Ohulchanskyy, C.A. Flask, P.N. Prasad, Combined optical and MR bioimaging using rare earth ion doped NaYF4 nanocrystals. Adv. Func. Mater. **19**(6), 853–859 (2009)
46. J.C. Boyer, M.P. Manseau, J.I. Murray, F.C. Van Veggel, Surface modification of upconverting NaYF4 nanoparticles with PEG− phosphate ligands for NIR (800 nm) biolabeling within the biological window. Langmuir **26**(2), 1157–1164 (2010)
47. F. Wang, J. Wang, X. Liu, Direct evidence of a surface quenching effect on size-dependent luminescence of upconversion nanoparticles. Angew. Chem. Int. Ed. **49**(41), 7456–7460 (2010)
48. J.C. Boyer, F.C. Van Veggel, Absolute quantum yield measurements of colloidal NaYF 4: Er3+, Yb3+ upconverting nanoparticles. Nanoscale **2**(8), 1417–1419 (2010)
49. A.G. Griesbeck, *Essentials of Molecular Photochemistry* (VonA. Gilbert undJ. Baggott. Blackwell Scientific Publications, Oxford, 1991). XII, 538 S. kartoniert£ 17.95.-ISBN 0-632-02429-1. Angewandte Chemie **103**(11), 1554–1555 (1991)
50. C. Li, J. Lin, Rare earth fluoride nano-/microcrystals: synthesis, surface modification and application. J. Mater. Chem. **20**(33), 6831–6847 (2010)
51. N.S. Singh, R.S. Ningthoujam, M.N. Luwang, S.D. Singh, R.K. Vatsa, Luminescence, lifetime and quantum yield studies of YVO4: Ln3+ (Ln3+= Dy3+, Eu3+) nanoparticles: concentration and annealing effects. Chem. Phys. Lett. **480**(4–6), 237–242 (2009)
52. D. Wawrzynczyk, M. Nyk, A. Bednarkiewicz, W. Strek, M. Samoc, Morphology-and size-dependent spectroscopic properties of Eu 3+-doped Gd2O3 colloidal nanocrystals. J. Nanopart. Res. **16**, 1–13 (2014)
53. S. Han, R. Deng, X. Xie, X. Liu, Enhancing luminescence in lanthanide-doped upconversion nanoparticles. Angew. Chem. Int. Ed. **53**(44), 11702–11715 (2014)
54. K. Li, M. Shang, H. Lian, J. Lin, Recent development in phosphors with different emitting colors via energy transfer. J. Mater. Chem. C **4**(24), 5507–5530 (2016)
55. S.K. Gupta, M. Sahu, P.S. Ghosh, D. Tyagi, M.K. Saxena, R.M. Kadam, Energy transfer dynamics and luminescence properties of Eu3+ in CaMoO4 and SrMoO4. Dalton Trans. **44**(43), 18957–18969 (2015)

56. G. Blasse, B.C. Grabmaier, G. Blasse, B.C. Grabmaier, *A General Introduction to Luminescent Materials* (Springer, Berlin Heidelberg, 1994), pp.1–9
57. G. Blasse, A. Bril, Investigations on Bi3+-activated phosphors. J. Chem. Phys. **48**(1), 217–222 (1968)
58. Q. Liu, W. Feng, T. Yang, T. Yi, F. Li, Upconversion luminescence imaging of cells and small animals. Nat. Protoc. **8**(10), 2033–2044 (2013)
59. R. Weissleder, A clearer vision for in vivo imaging. Nat. Biotechnol. **19**(4), 316–317 (2001)
60. F. Jia, G. Li, B. Yang, B. Yu, Y. Shen, H. Cong, Investigation of rare earth upconversion fluorescent nanoparticles in biomedical field. Nanotechnol. Rev. **8**(1), 1–17 (2019)
61. J. Goetz, A. Nonat, A. Diallo, M. Sy, I. Sera, A. Lecointre, L.J. Charbonnière et al., Ultrabright lanthanide nanoparticles. ChemPlusChem **81**(6), 526–534 (2016)
62. W. Zou, C. Visser, J.A. Maduro, M.S. Pshenichnikov, J.C. Hummelen, Broadband dye-sensitized upconversion of near-infrared light. Nat. Photonics **6**(8), 560–564 (2012)
63. G. Chen, J. Damasco, H. Qiu, W. Shao, T.Y. Ohulchanskyy, R.R. Valiev, P.N. Prasad et al., Energy-cascaded upconversion in an organic dye-sensitized core/shell fluoride nanocrystal. Nano Lett. **15**(11), 7400–7407 (2015)
64. M. Wang, C.C. Mi, W.X. Wang, C.H. Liu, Y.F. Wu, Z.R. Xu, S.K. Xu et al., Immunolabeling and NIR-excited fluorescent imaging of HeLa cells by using NaYF4: Yb Er upconversion nanoparticles. ACS Nano **3**(6), 1580–1586 (2009)
65. D.M. Wu, A. García-Etxarri, A. Salleo, J.A. Dionne, Plasmon-enhanced upconversion. J. Phys. Chem. Lett. **5**(22), 4020–4031 (2014)
66. K. Aslan, C.D. Geddes, Metal-enhanced fluorescence: progress towards a unified plasmon-fluorophore description. Met. Enhanced Fluoresc., 1–23 (2010)
67. E.L. Ru, J. Grand, N. Félidj, J. Aubard, G. Lévi, A. Hohenau, P.G. Etchegoin et al., Spectral profile modifications in metal-enhanced fluorescence. Met.Enhanced Fluoresc., 25–65 (2010)
68. D.J. Ross, N.P. Pieczonka, R.F. Aroca, The role of plasmonic engineering in potential surface-enhanced fluorescence. Met. Enhanced Fluoresc. **23**, 67–90 (2010)
69. K. Munechika, Y. Chen, J.M. Smith, D.S. Ginger, Importance of spectral overlap: fluorescence enhancement by single metal nanoparticles. Met. Enhanced Fluoresc., 91–118 (2010)
70. J. Qiu, Y. Li, Y. Jia, *Persistent Phosphors: From Fundamentals to Applications* (Woodhead Publishing, 2020)
71. E. Hemmer, A. Benayas, F. Légaré, F. Vetrone, Exploiting the biological windows: current perspectives on fluorescent bioprobes emitting above 1000 nm. Nanoscale Horiz. **1**(3), 168–184 (2016)
72. C.F. Chan, M.K. Tsang, H. Li, R. Lan, F.L. Chadbourne, W.L. Chan, K.L. Wong et al., Bifunctional up-converting lanthanide nanoparticles for selective in vitro imaging and inhibition of cyclin D as anti-cancer agents. J. Mater. Chem. B **2**(1), 84–91 (2014)
73. N. Bogdan, E.M. Rodríguez, F. Sanz-Rodríguez, M.C.I. de la Cruz, Á. Juarranz, D. Jaque, J.A. Capobianco et al., Bio-functionalization of ligand-free upconverting lanthanide doped nanoparticles for bio-imaging and cell targeting. Nanoscale **4**(12), 3647–3650 (2012)
74. E.A. Grebenik, A. Nadort, A.N. Generalova, A.V. Nechaev, V.K. Sreenivasan, E.V. Khaydukov, A.V. Zvyagin et al., Feasibility study of the optical imaging of a breast cancer lesion labeled with upconversion nanoparticle biocomplexes. J. Biomed. Opt. **18**(7), 076004–076004 (2013)
75. M.H. Alkahtani, F.S. Alghannam, C. Sanchez, C.L. Gomes, H. Liang, P.R. Hemmer, High efficiency upconversion nanophosphors for high-contrast bioimaging. Nanotechnology **27**(48), 485501 (2016)
76. H.F. Brito, J. Hölsä, T. Laamanen, M. Lastusaari, M. Malkamäki, L.C. Rodrigues, Persistent luminescence mechanisms: human imagination at work. Opt. Mater. Express **2**(4), 371–381 (2012)
77. P.F. Smet, D. Poelman, M.P. Hehlen, Focus issue introduction: persistent phosphors. Opt. Mater. Express **2**(4), 452–454 (2012)
78. T. Maldiney, A. Lecointre, B. Viana, A. Bessière, M. Bessodes, D. Gourier, D. Scherman et al., Controlling electron trap depth to enhance optical properties of persistent luminescence nanoparticles for in vivo imaging. J. Am. Chem. Soc. **133**(30), 11810–11815 (2011)

79. A.S. Paterson, B. Raja, G. Garvey, A. Kolhatkar, A.E. Hagström, K. Kourentzi, R.C. Willson et al., Persistent luminescence strontium aluminate nanoparticles as reporters in lateral flow assays. Anal. Chem. **86**(19), 9481–9488 (2014)
80. M. Rogosnitzky, S. Branch, Gadolinium-based contrast agent toxicity: a review of known and proposed mechanisms. Biometals **29**(3), 365–376 (2016)
81. S.S. Laha, N.D. Thorat, G. Singh, C.I. Sathish, J. Yi, A. Dixit, A. Vinu, Rare-earth doped iron oxide nanostructures for cancer theranostics: magnetic hyperthermia and magnetic resonance imaging. Small **18**(11), 2104855, 91 (2022)
82. J. Li, J. You, Y. Dai, M. Shi, C. Han, K. Xu, Gadolinium oxide nanoparticles and aptamer-functionalized silver nanoclusters-based multimodal molecular imaging nanoprobe for optical/magnetic resonance cancer cell imaging. Anal. Chem. **86**(22), 11306–11311 (2014)
83. Y. Liu, K. Ai, L. Lu, Nanoparticulate X-ray computed tomography contrast agents: from design validation to in vivo applications. Acc. Chem. Res. **45**(10), 1817–1827 (2012)
84. X. Zhai, P. Lei, P. Zhang, Z. Wang, S. Song, X. Xu, H. Zhang et al., Growth of lanthanide-doped LiGdF4 nanoparticles induced by LiLuF4 core as tri-modal imaging bioprobes. Biomaterials **65**, 115–123 (2015)
85. X. Han, K. Xu, O. Taratula, K. Farsad, Applications of nanoparticles in biomedical imaging. Nanoscale **11**(3), 799–819 (2019)
86. J. Zhou, N. Shirahata, H.T. Sun, B. Ghosh, M. Ogawara, Y. Teng, J. Qiu et al., Efficient dual-modal NIR-to-NIR emission of rare earth ions co-doped nanocrystals for biological fluorescence imaging. J. Phys. Chem. Lett. **4**(3), 402–408 (2013)
87. Z. Wei, Y. Liu, B. Li, J. Li, S. Lu, X. Xing, H. Zhang et al., Rare-earth based materials: an effective toolbox for brain imaging, therapy, monitoring and neuromodulation. Light Sci. Appl. **11**(1), 175 (2022)
88. S. Guo, X. Xie, L. Huang, W. Huang, Sensitive water probing through nonlinear photon upconversion of lanthanide-doped nanoparticles. ACS Appl. Mater. Interfaces **8**(1), 847–853 (2016)
89. A. Skripka, D. Mendez-Gonzalez, R. Marin, E. Ximendes, B. Del Rosal, D. Jaque, P. Rodríguez-Sevilla, Near infrared bioimaging and biosensing with semiconductor and rare-earth nanoparticles: recent developments in multifunctional nanomaterials. Nanoscale Adv. **3**(22), 6310–6329 (2021)
90. M.K. Hossain, M.I. Khan, A. El-Denglawey, A review on biomedical applications, prospects, and challenges of rare earth oxides. Appl. Mater. Today **24**, 101104 (2021)
91. C.D. Brites, P.P. Lima, N.J. Silva, A. Millán, V.S. Amaral, F. Palacio, L.D. Carlos, Thermometry at the nanoscale. Nanoscale **4**(16), 4799–4829 (2012)
92. B. Zhang, X. Guo, Z. Zhang, Z. Fu, H. Zheng, Luminescence thermometry with rare earth doped nanoparticles: status and challenges. J. Lumin., 119110 (2022)
93. A. Sedlmeier, D.E. Achatz, L.H. Fischer, H.H. Gorris, O.S. Wolfbeis, Photon upconverting nanoparticles for luminescent sensing of temperature. Nanoscale **4**(22), 7090–7096 (2012)
94. F. Vetrone, R. Naccache, A. Zamarrón, A.J. de la Fuente, F. Sanz-Rodrıguez, L.M. Maestro, Temperature sensing using fluorescent nanothermometers. ACS Nano **4**, 3254–3258 (2010)
95. L.H. Fischer, G.S. Harms, O.S. Wolfbeis, Upconverting nanoparticles for nanoscale thermometry. Angew. Chem. Int. Ed. **50**(20), 4546–4551 (2011)
96. Q. Zhang, S. O'Brien, J. Grimm, Biomedical applications of lanthanide nanomaterials, for imaging, sensing and therapy. Nanotheranostics **6**(2), 184 (2022)

Chapter 5
Review of Various Synthesis Routes of Ultrasmall Renal Clearable Nanoparticles for Medical Imaging Applications

B. C. Bhadrapriya, Bosely Anne Bose, Nandakumar Kalarikkal, Murukeshan Vadakke Matham, Parasuraman Padmanabhan, and Sabu Thomas

Abstract Ultrasmall nanoparticles have a significant role in revolutionizing healthcare, particularly in the realms of diagnostics and personalized medicine. This chapter discusses the various synthesis routes of ultra-small renal clearable nanoparticles for such applications. The focus is on nanoparticles with sizes below 10 nm, which are crucial for efficient renal clearance and minimizing long-term toxicity. The chapter explores different synthesis methods, including chemical, biological, and physical approaches, highlighting their advantages and limitations. The unique properties of metal, alloy, metal oxide, rare earth-based, and carbon-based nanoparticles are discussed in the context of their suitability for medical imaging modalities such as MRI, CT, PET, and fluorescence imaging. Emphasis is placed on the

---

B. C. Bhadrapriya · B. A. Bose · N. Kalarikkal (✉)
School of Pure and Applied Physics, Mahatma Gandhi University, Kottayam 686560, Kerala, India
e-mail: nkkalarikkal@mgu.ac.in

B. C. Bhadrapriya · B. A. Bose · N. Kalarikkal · S. Thomas
International and Inter University Centre for Nanoscience and Nanotechnology, Mahatma Gandhi University, Kottayam 686560, Kerala, India

N. Kalarikkal
School of Nanoscience and Nanotechnology, Mahatma Gandhi University, Kottayam 686560, Kerala, India

M. V. Matham · P. Padmanabhan
Imaging Probe Development Platform, Nanyang Technological University, 59 Nanyang Drive, Singapore 636921, Singapore

M. V. Matham
School of Mechanical and Aerospace Engineering, Centre for Optical and Laser Engineering (COLE), Nanyang Technological University, 50 Nanyang Avenue, Singapore 639798, Singapore

P. Padmanabhan
Cognitive Neuroimaging Centre, Nanyang Technological University, 59 Nanyang Drive, Singapore 636921, Singapore

N. Kalarikkal et al. (eds.), *Multimodal Biomedical Imaging Techniques*, Biological and Medical Physics, Biomedical Engineering,
https://doi.org/10.1007/978-981-96-1124-9_5

importance of size control, surface modification, and biocompatibility to ensure the nanoparticles' effectiveness and safety in clinical applications. These ultra-small nanoparticles, characterized by their high surface area, enhanced reactivity, and prolonged circulation time, are particularly suited for medical imaging applications, offering improved contrast and resolution across modalities like MRI, CT, PET, and multimodal imaging.

5.1 Introduction

5.1.1 Nanoscience and Nanotechnology: Properties and Potential of Nanoparticles in Health Care

Nanotechnology is the science of manipulating material at the nanoscale, typically less than 100 nm. This discipline has created new and intriguing potential in healthcare. At the nanoscale, quantum mechanical effects become significant, leading to unique behaviours of materials. For example, nanoparticles can exhibit different optical, electrical, and magnetic properties than their bulk counterparts. As materials are scaled down to the nanoscale, properties such as melting point, conductivity, and reactivity can change significantly. Nanomaterials exhibit a high surface-to-volume ratio, rendering them particularly suitable for applications where surface properties play a critical role, such as catalysts and drug delivery systems. The rapid advancement of nanoscience and the use of nanotechnology in medicine are changing how we approach disease prevention, diagnosis, and treatment. Nanoparticles have shown great promise in healthcare due to their unique properties and potential applications. Their small size, large surface area-to-volume ratio, and tunable characteristics make them versatile tools for various healthcare-related purposes. Some potential applications of nanoparticles in healthcare include drug delivery, imaging and diagnosis, cancer therapy, gene therapy, vaccines, tissue regeneration, antimicrobial agents, personalized medicine, diagnostics, minimally invasive surgery, etc. Nanomaterials have gained significant attention in recent years as novel contrast agents due to their unique properties and advantages. Contrast agents play a crucial role in medical imaging by enhancing the visibility of anatomical and functional structures within the body. Nanomaterial-based contrast agents have several advantages over traditional contrast agents, including their ability to generate high-resolution, high-contrast images and integrate multiple properties within a single agent. These properties make nanomaterials highly promising for advancing medical imaging and personalized medicine, as they provide greater versatility and precision in diagnosis and therapy. However, their safety, toxicity, and regulatory considerations are essential factors that need to be carefully addressed in their development and clinical use [1].

5.1.2 Synthesis Techniques at the Nanoscale—An Overview

5.1.2.1 Top-Down Techniques

Top-down techniques, in the context of nanotechnology and material science, refer to methods of fabricating or manipulating nanoscale structures or materials by starting with larger structures and then reducing them in size. It involves the fabrication of nanoparticles by breaking down larger materials into smaller structures through physical methods such as milling, lithography, or etching. Top-down techniques are essential for manufacturing integrated circuits, microelectromechanical systems (MEMS), and many other nanoscale devices. They offer precise control over the size and shape of structures and are widely used in industries ranging from electronics and photonics to materials science and biotechnology.

Top-down techniques, such as electron beam evaporation, physical vapour deposition (PVD), laser ablation, mechanical milling, focused ion beam lithography, and electro- and thermal deposition, offer advantages in precisely controlling particle positions and accommodating various substrates. However, they may result in nanoparticles with higher defect densities and less uniformity than bottom-up approaches.

The physical vapour deposition (PVD) method involves the evaporation of a material to form a vapour cloud in the vacuum chamber. The vaporized material travels through the vacuum chamber, comes into contact with the substrate surface, condenses, and forms a thin film [2]. Thermal evaporation, electron beam evaporation, sputtering, pulsed laser deposition (PLD), etc. are different types of PVD techniques. In the thermal and electron beam evaporation technique, the material to be deposited, typically a metal or semiconductor, is loaded into a crucible or boat made of a refractory material such as tungsten or molybdenum. The crucible is heated using resistive heating coils, an electron beam, or a focused ion beam heating until it sublimates and forms a vapour cloud [3]. Sputtering involves bombarding a target material (metals, semiconductors, oxides, and nitrides) with energetic ions, causing atoms or molecules to be ejected from the target surface and deposited onto a substrate [4]. In the PLD method, a high-power pulsed laser beam (typically in the ultraviolet or near-infrared wavelength range) is focused onto a target material, causing rapid heating and vaporization of the target surface. This generates a plasma plume composed of ions, atoms, and molecules from the target material. The plasma plume expands and travels through the vacuum chamber, eventually reaching and depositing onto a substrate surface. In the pulsed laser ablation (PLA) technique, a laser beam is focused onto a target immersed in a solvent, resulting in the formation of a plasma plume. Later, the plasma plume cools down, and the formation of nanoparticles in the solvent occurs [5]. Figure 5.1 shows a schematic representation of the PLD and PLA process [5].

Mechanical milling or ball milling involves using a rotating container filled with grinding balls (made of stainless steel, tungsten carbide or ceramics like ZrO_2), which collide with the material to be milled, exerting mechanical impact and frictional

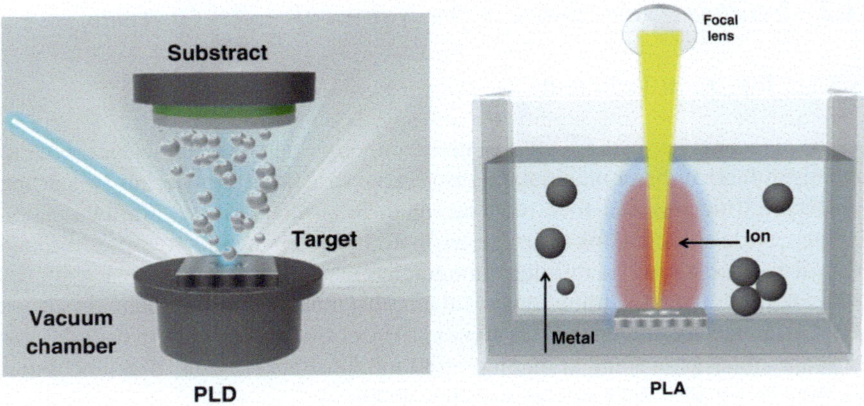

Fig. 5.1 PLD and PLA synthesis processes

forces on the particles. This mechanical action results in the reduction of particle size through processes such as crushing, fracturing, and abrasion [1]. Several factors such as size of the balls, type of the balls used, speed, milling time, solvent used, etc. will affect the properties of the nanoparticles formed [2]. The ball mills are mainly of two types, based on their speed of rotation, namely high-energy and low-energy ball mills. Attritor ball mills, shaker mills, and planetary ball mills are high-energy ball mills. The Tumbler ball mill is an example of a low-energy ball mill. Figures 5.2 a and b show the photograph of a planetary ball mill and the schematic representation of the process of ball milling, respectively [Reproduced from [2]].

5.1.2.2 Bottom-Up Techniques

Bottom-up techniques, in the context of nanotechnology and material science, refer to methods of fabricating or assembling nanoscale structures or materials by starting with individual atoms or molecules and building them up into larger, more complex structures, and involve building up nanoparticles from atomic or molecular precursors through chemical or biological processes.

Chemical methods include reduction of metal ions in solution, self-assembly of molecules, or precipitation reactions to form nanoparticles. Biological methods utilize living organisms, biomolecules, or biological extracts to mediate nanoparticle synthesis through enzymatic or metabolic pathways. Bottom-up approaches are often associated with chemical or green procedures that rely on environmentally friendly solvents, reducing agents, and stabilizers. These methods tend to produce nanoparticles with well-defined sizes, shapes, and monodispersity, as they allow precise control over reaction conditions and nanoparticle growth. They also tend to result in fewer defects compared to top-down methods. These techniques are at the

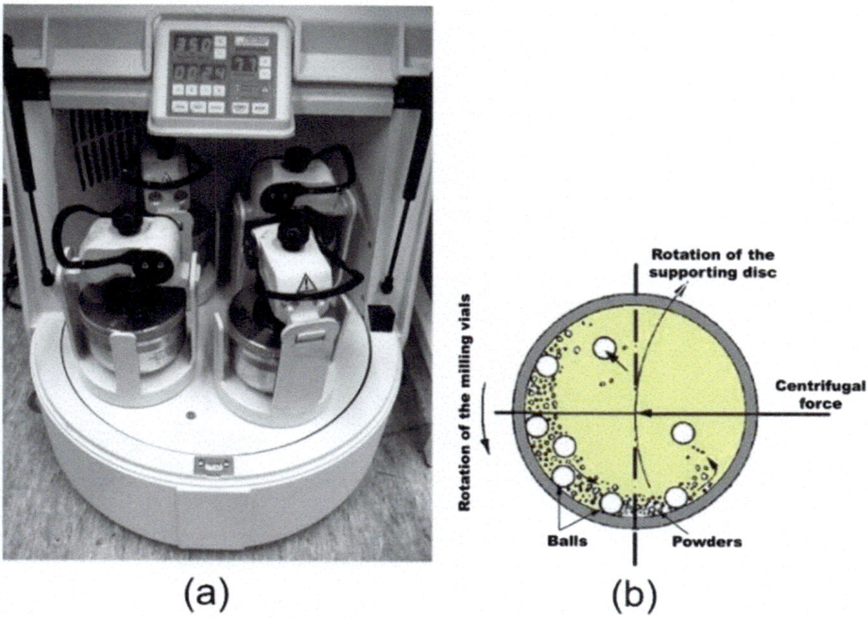

Fig. 5.2 **a** Planetary ball mill **b** schematic representation of ball milling process

forefront of nanotechnology research and have led to significant advancements in various fields, from electronics and materials science to medicine and energy.

Chemical methods include the chemical reduction method, chemical vapour deposition (CVD), chemical vapour condensation (CVC), sol–gel technique, microwave-assisted synthesis, polyol synthesis, hydrothermal method, etc. In the chemical reduction method, a precursor solution containing metal salt is prepared. A reducing agent is added to the solution, initiating the reduction reaction. The reducing agent transfers electrons to the metal ions, converting them into metal atoms with a lower oxidation state. Common reducing agents include sodium borohydride ($NaBH_4$), sodium citrate, hydrazine [3], and ascorbic acid (vitamin C) [4]. Once reduced, the metal atoms nucleate and aggregate to form nanoparticles. The kinetics of nucleation and growth are influenced by factors such as temperature, reaction time, concentration of reactants, and the presence of stabilizing agents. Control over these parameters allows for the tuning of nanoparticle size, shape, and dispersity. Figure 5.3 shows the chemical reduction method of synthesis of gold nanoparticles [5].

Both CVD and CVC involve chemical reactions in the vapour phase to deposit materials; CVD is primarily used for thin film deposition, whereas CVC is focused on the synthesis of nanoparticles. In CVD technique, gaseous precursor compounds containing the elements to be deposited are introduced into a reaction chamber. These precursor gases may be organometallic compounds, metal halides, metal

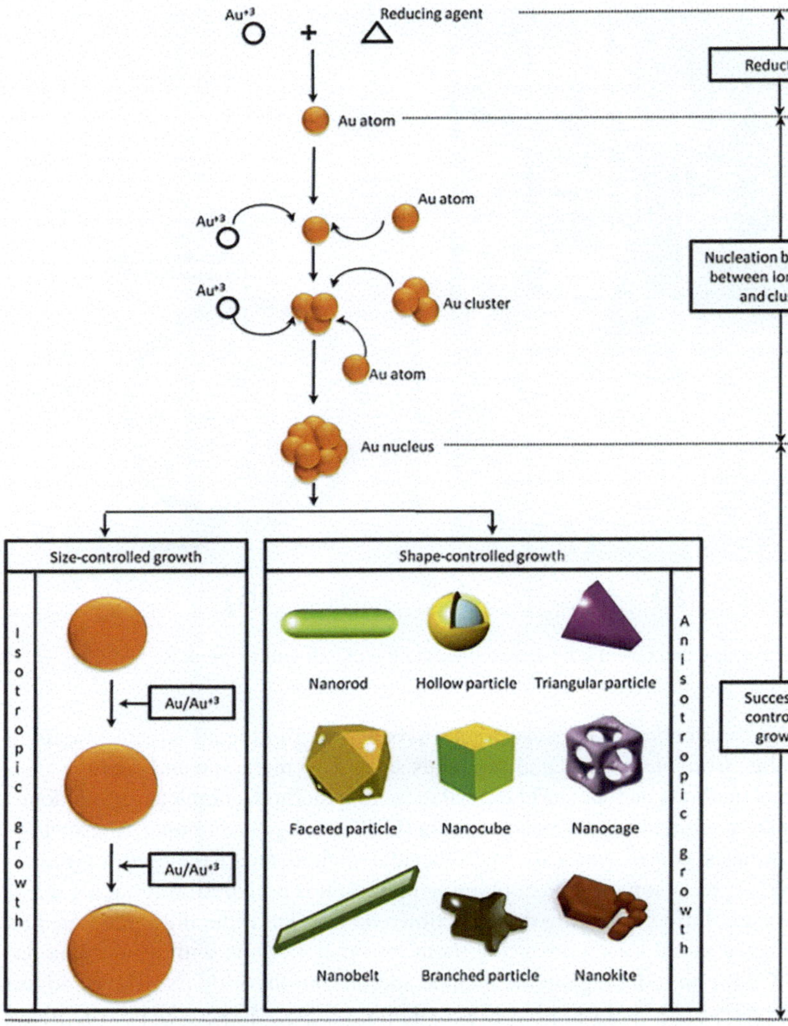

Fig. 5.3 Chemical reduction method of synthesis of gold nanoparticles

carbonyls, or other volatile compounds. The precursor gases undergo chemical reactions, typically initiated by heating or plasma excitation, to decompose and release the desired elements onto the substrate surface. These reactions may involve processes like thermal decomposition, chemical reduction, or chemical vapour transport. The released atoms or molecules from the precursor gases condense onto the substrate surface, forming a thin film layer. The substrate may be heated to promote adhesion and film growth.

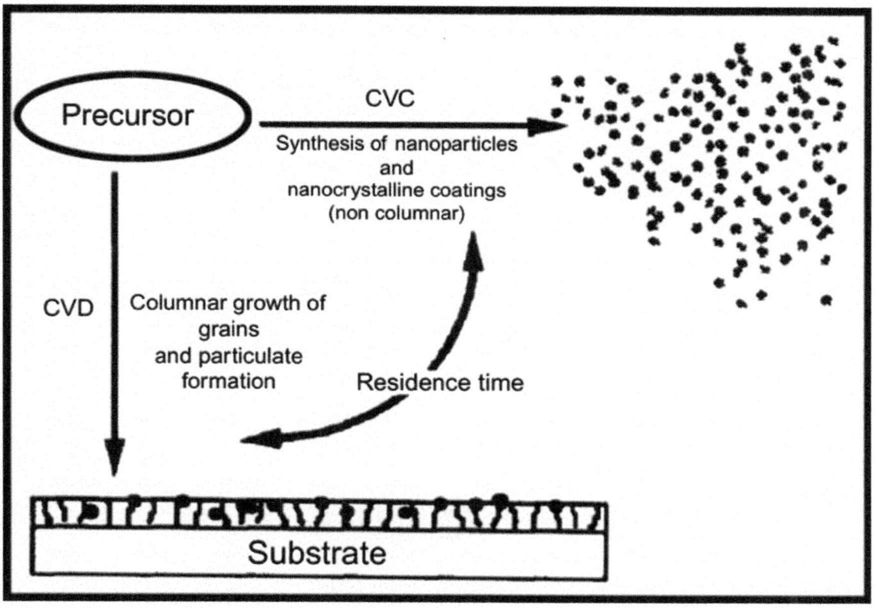

Fig. 5.4 Schematic representation of CVD and CVC processes

The CVC method is a modified CVD method considering conditions like the reactor temperature, the partial pressure of precursors, the use of substrate, the time of reaction, etc. [6]. Figure 5.4 shows the schematic representation of CVD and CVC processes.

The sol–gel method is a versatile wet chemical synthesis technique used to produce a wide range of materials, including ceramics, glasses, and thin films, from solutions of precursors. The nanoparticles are formed from the precursor solution by undergoing several steps such as sol formation, gel formation, ageing, and drying. Figure 5.5 shows the process of formation of thin films and nanoparticles via the sol–gel method. The sol–gel process begins with preparing a sol, which is a stable dispersion of solutes in a liquid solvent. The solutes are typically metal alkoxides, metal chlorides, or metal salts, which serve as precursors for the desired nanoparticles. The solvents used can be water, alcohol, or a combination of both, depending on the solute and desired properties of the final nanoparticles. The sol is typically formed by mixing the solute and solvent under controlled temperature, pH, and concentration conditions, followed by stirring or sonication to achieve homogeneity [7]. The next stage is the gelation process, by which the soul transforms into a gel. During gelation, the sol undergoes chemical reactions such as hydrolysis of metal alkoxide groups and condensation of hydroxyl groups, forming a network structure. After gelation, the gel is typically subjected to aging, a process during which it is allowed to undergo further chemical reactions and structural rearrangements. Aging helps improve the mechanical strength, homogeneity, and stability of the gel

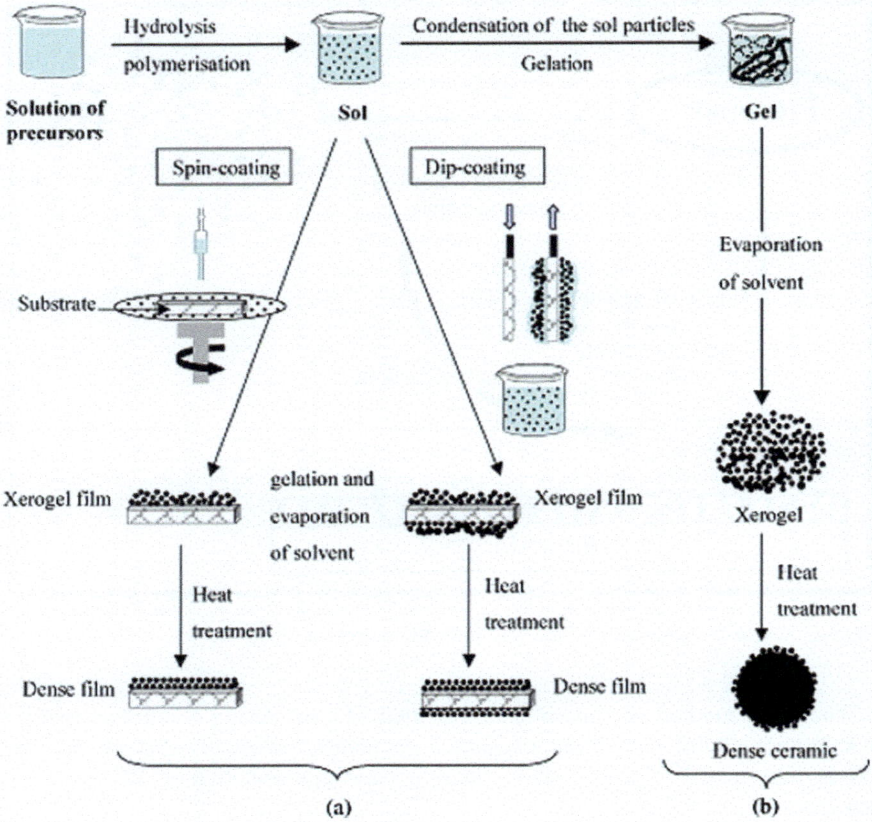

Fig. 5.5 Sol–gel process of synthesis of nanoparticles and thin films

network [8]. Following aging, the gel is dried to remove the liquid solvent and obtain a solid material. Drying can be carried out by various methods, including air drying, oven drying, freeze-drying (lyophilization), or supercritical drying, depending on the desired properties of the final material [9].

In microwave-assisted synthesis, the reaction mixture, containing the precursors and reagents necessary for the synthesis, is taken in a microwave reaction tube or flask and placed inside a microwave reactor. The reaction mixture is exposed to microwave irradiation for a predetermined duration, typically at a specified power and frequency. Microwave irradiation generates electromagnetic radiation with frequencies ranging from hundreds of MHz to tens of GHz. When materials are exposed to microwave radiation, the electromagnetic energy is converted into thermal energy through dielectric heating. The elevated temperatures achieved through microwave irradiation accelerate the rate of chemical reactions, enabling the synthesis of desired products in significantly shorter reaction times compared to conventional heating methods [10]. In conventional heating, the reaction vessel is heated up first and then

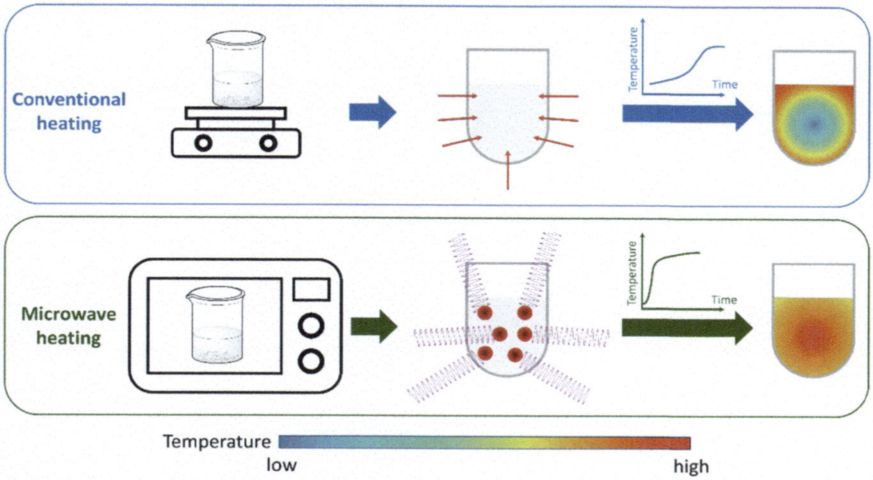

Fig. 5.6 A comparison of temperature of the reaction mixture during conventional and microwave-assisted synthesis

the heat is transferred into the reaction mixture. However, in microwave-assisted synthesis, the reaction mixture is directly heated up and uniformly. Figure 5.6 shows a comparison of the temperature of the reaction mixture during conventional and microwave-assisted synthesis [11].

Polyol synthesis is a versatile and widely used method for the synthesis of metal nanoparticles, particularly noble metal nanoparticles such as gold, silver, and platinum. This method involves the reduction of metal precursors in the presence of a polyol solvent and a stabilizing agent, resulting in the formation of nanoparticles with controlled size, shape, and surface properties. Polyols are a class of organic compounds which contain two or more hydroxyl groups (-OH) attached to aliphatic or aromatic carbon atoms. The number of hydroxyl groups and the carbon chain length can vary, leading to a wide range of polyol structures and properties. Common polyols include glycerol, ethylene glycol, propylene glycol, tetraethylene glycol, polyethylene glycol (PEG), butanediol, etc. The pioneering work by Fievet, Lagier, and Figlarz in 1989 introduced the concept of using polyols as both solvents and reducing agents for synthesizing small particles, particularly metallic nanoparticles. They termed this approach the "polyol process" or "polyol synthesis" [12, 13]. The choice of polyol solvent, reaction temperature, and reaction time could influence the nucleation, growth, and morphology of nanoparticles. Figure 5.7 shows the schematic representation of polyol synthesis of silver nanoparticles and the variation in morphology when polyol of different molecular weights is used [14].

Solvothermal is a versatile synthesis technique used to synthesise nanoparticles, under controlled temperature and pressure conditions in an organic solvent, typically in a sealed autoclave or reactor vessel. This method enables the synthesis of a wide

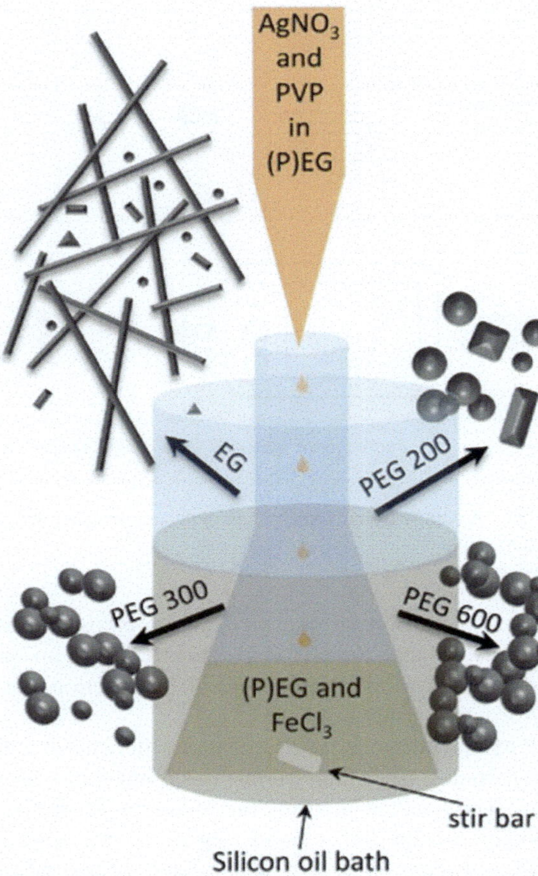

Fig. 5.7 Schematic representation of polyol synthesis of silver nanoparticles

range of materials, including nanoparticles, nanocrystals, thin films, and bulk crystals, with controlled size, shape, and properties. Depending on the desired material, the precursors used in solvothermal synthesis may include metal salts, metal oxides, metal complexes, or organic compounds. Organic solvents are chosen based on their ability to dissolve the precursors and facilitate the desired chemical reactions. The choice of solvent can influence reaction kinetics, solubility, and product morphology. The reaction vessel is heated to the desired temperature using an external heat source, such as an oven, oil bath, or microwave irradiation. Under solvothermal conditions, dissolved precursors react to form nuclei, which subsequently grow and aggregate to form nanoparticles, nanocrystals, or crystalline phases. The rate of precipitation, crystallization, and particle growth is influenced by factors such as temperature, pressure, solvent composition, precursor concentration, and reaction time. Controlling these parameters allows for the tuning of particle size, morphology, and phase purity. While similar to the solvothermal method, hydrothermal synthesis utilises deionized water

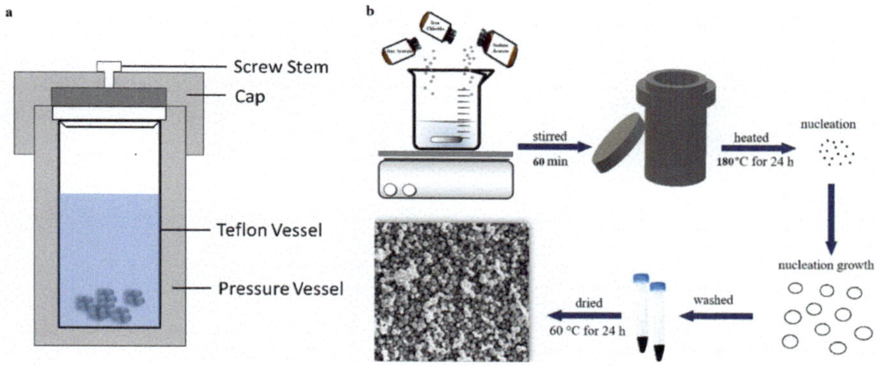

Fig. 5.8 Schematic of **a** solvothermal reactor vessel and **b** typical solvothermal synthesis

instead of organic solvents as the reaction medium. Figure 5.8 shows a schematic of the hydrothermal reactor vessel [14] and typical hydrothermal synthesis [15].

Biological methods include nanoparticle synthesis using plant extracts, bacteria, fungi, etc. In nanoparticle synthesis with plant extracts, the first step is to wash and prepare the plant parts. After that, boil the plant parts to extract phytochemicals and purify the extract by filtration and centrifugation. Common plant extracts include polyphenols and carotenoids [16]. Later, mix the extract with a desired metal salt solution and water. Monitor for colour changes during incubation, which is a direct indication of a reduction reaction. Wash and collect the nanoparticles by centrifuging at different speeds to remove unwanted components and large particles. Figure 5.9 [16] shows a schematic of the synthesis of nanoparticles from plant extracts.

In nanoparticle synthesis using microorganisms, there are two different methods such as extracellular synthesis and intracellular synthesis [17]. In the extracellular synthesis method, microorganisms are cultured under optimized temperature and pH conditions. Later, centrifuge the culture to remove biomass. Further, add a metal salt solution to the supernatant to synthesise the nanoparticles. The mixture will be incubated and monitored for colour changes indicating nanoparticle formation. Wash and collect the nanoparticles by centrifuging at different speeds to remove unwanted components and large particles. In the intracellular synthesis method, the first step is to culture microorganisms for an optimum period and collect biomass. Then, wash and dissolve the biomass in a metal salt solution. Monitor for colour changes during incubation. Later, ultrasonication and centrifugation were used to wash and remove the biomass. During these processes, cell walls break down and subsequently release the nanoparticles. Peptides, enzymes, etc. present in microorganisms have multiple benefits, including efficient reduction of metal ions, stabilization of nanoparticles, and enhancement of their stability and dispersibility [18, 19]. Figure 5.10 [20] shows a schematic of nanoparticle synthesis using microorganisms.

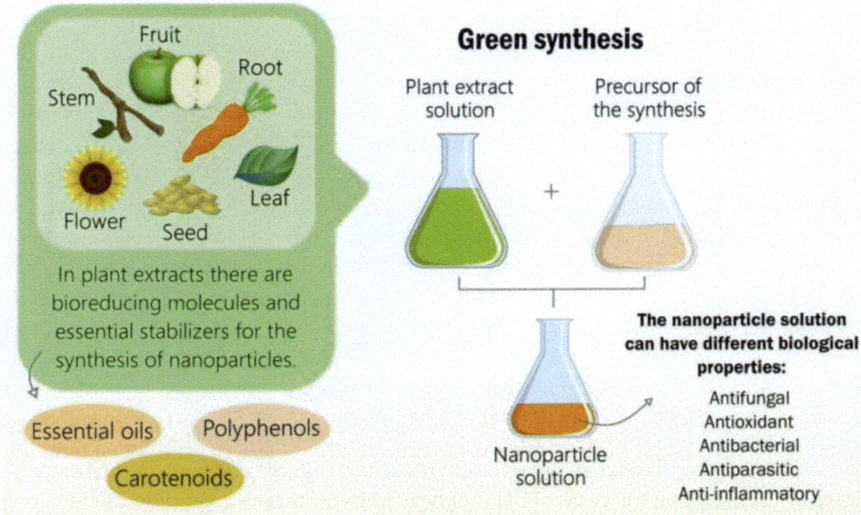

Fig. 5.9 Schematic representation of nanoparticle synthesis from plant extracts

5.2 Properties and Potential of Ultra-Small Nanoparticles in Medical Imaging

The origins of molecular imaging can be traced back to Coons [21] innovative use of antibodies as immune reagents, utilizing their exquisite specificity to direct a signal emitter, such as a fluorochrome, to targets in tissues for fluorescence microscopy. Today, the targeting group can be a variety of molecules, including antibodies, DNA, RNA, and specific ligands, while the signal emitter can encompass a broad range of materials suited to different imaging modalities, such as organic fluorochromes, metals, metal oxides, etc. Current imaging systems encompass direct optical imaging using various regions of the optical spectrum, as well as indirect imaging techniques that use computer software along with scanning systems to produce images. Techniques involving ultrasound, positrons, photons and magnetic fields are employed to create images that enable detection, diagnosis and therapy. In recent years, it has become evident that many of the challenges in molecular imaging can be addressed through the use of nanostructured materials. As a result, nanoparticles now play a crucial role in all areas of molecular imaging.

Ultra-small nanoparticles, typically measuring a few nanometers or less, offer unique properties and significant potential in the field of medical imaging. Their unique characteristics make them invaluable across various imaging modalities and diagnostic applications. With an exceptionally high surface area relative to volume that allows for a higher payload of imaging agents or contrast materials, thereby enhancing sensitivity and signal intensity in imaging. Moreover, their small size

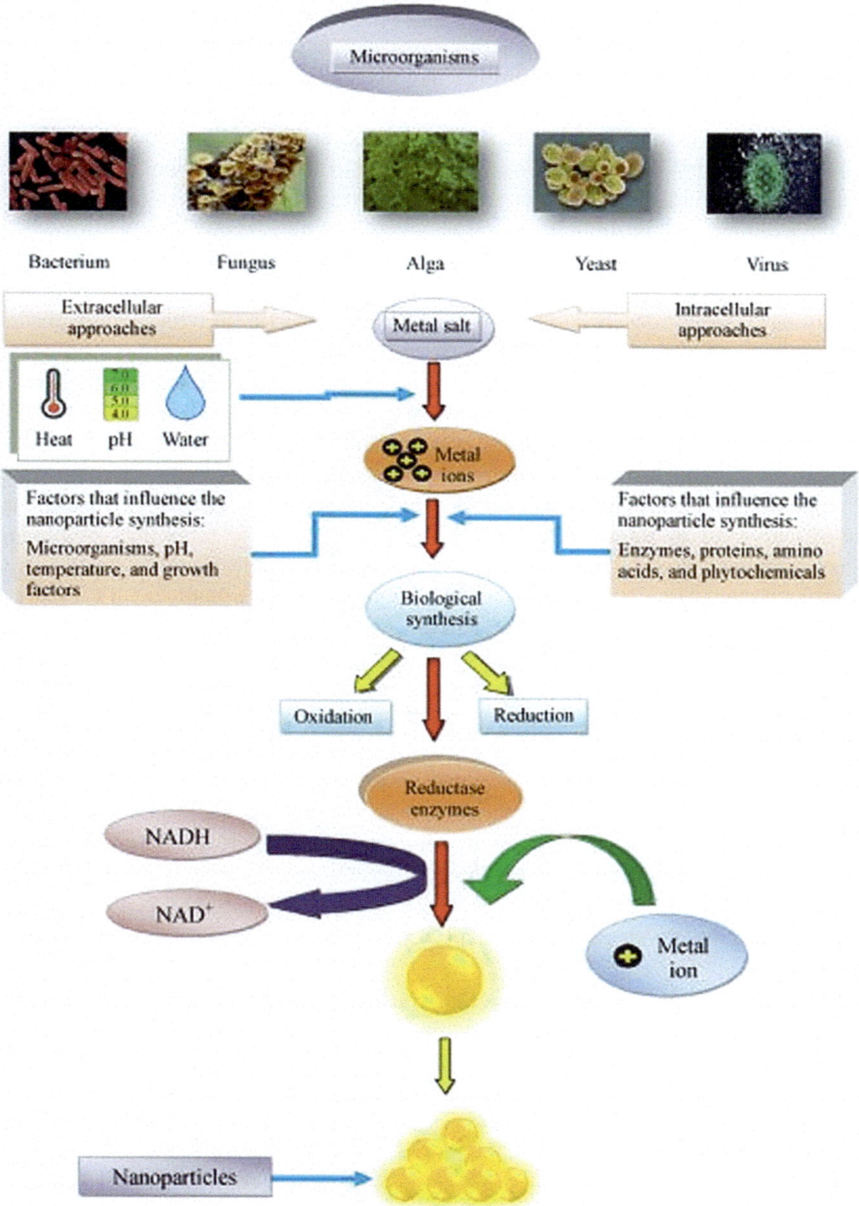

Fig. 5.10 Schematic representation of the process of synthesis of nanoparticles from microorganisms

and surface modifications enable prolonged circulation in the bloodstream, facilitating tracking and imaging of dynamic processes within the body. Surface modifications also enhance biocompatibility, making them suitable to use in biological systems without causing significant toxicity or immune responses [22]. Functionalization strategies involve altering the surface chemistry of nanoparticles (NPs) with three primary objectives. Firstly, they aim to make hydrophobic NPs water dispersible. Secondly, these strategies work to prevent agglomeration in biological environments by introducing electrostatic and/or steric repulsions. Thirdly, they provide specific and reactive groups, typically amino or carboxylic groups, which can serve as anchors for further conjugation of NPs with various biomedical molecules like antibodies, peptides, proteins, and drugs. Functionalization with specific groups or ligands enables precise targeting of disease markers or tissues, thus improving imaging accuracy and reducing off-target effects. Additionally, the small size of these functionalized NPs allows for minimally invasive delivery via intravenous injection or other routes, minimizing patient discomfort and risks of infection [23].

Ultra-small nanoparticles with magnetic properties can serve as contrast agents in magnetic resonance imaging (MRI). Their small size and surface functionalization can lead to improved relaxivity and tissue-specific targeting, enhancing image contrast. Studies have indicated that smaller particle sizes (<5 nm) correspond to higher relaxivity values, rendering them more suitable as T1 contrast agents. In the realm of superparamagnetism, when particle size falls below around 5 nm, thermal energy easily flips the magnetization. This phenomenon enhances T1 contrast, attributed to factors like magnetization suppression, increased exposure of surface iron centers, surface effects, and water diffusion. These unique characteristics, coupled with their diminutive size, have positioned these nanostructures as extensively investigated materials in biomedicine, notably as diagnostic, theranostic, or therapeutic tools [24]. Nanoparticles engineered with high X-ray attenuation serve as effective contrast agents for computed tomography (CT) scans. Their small size facilitates improved distribution within tissues. Compared to small molecule CT contrast agents, nanoparticles offer numerous advantages, including prolonged residence times in the bloodstream and potential applications in cell tracking and targeted imaging [24].

Nanoparticles possess unique optical, electrical, and magnetic properties, leading to extensive research in developing "nanoprobes" for molecular imaging. Superparamagnetic iron oxide (SPIO) and ultrasmall SPIO (USPIO) are utilized in MRI, while gold nanoparticles find applications in CT imaging. Quantum dots (QDs) and upconversion nanoparticles (UCNPs) are employed in optical imaging, emitting intense and tunable fluorescence signals for visualizing biomolecules, cells, or tissues with high sensitivity. Carbon nanotubes and gold nanorods serve as contrast agents in photoacoustic imaging, leveraging their excellent acoustic properties for enhanced imaging of blood vessels, tumours, and other structures. Combining ultra-small nanoparticles with different imaging agents or modalities enables multimodal imaging, providing complementary information and enhancing diagnostic accuracy. Moreover, ultra-small nanoparticles allow for real-time monitoring of cellular and molecular processes, facilitating early disease detection and treatment response assessment [25].

It is reported that more than 50% of ultrasmall nanoparticles injected were excreted within hours [27]. Despite their numerous advantages, it is essential to consider potential challenges associated with ultra-small nanoparticles, such as their clearance from the body and potential toxicity. Ongoing research focuses on optimizing their properties and developing safe and effective imaging agents for clinical use. Overall, these nanoparticles hold great promise for advancing the field of medical imaging and diagnostics, offering improved sensitivity, specificity, and therapeutic potential.

5.3 Synthesis of Ultra-Small Nanoparticles in Medical Imaging

The synthesis of ultra-small nanoparticles for medical imaging involves carefully controlled chemical and physical processes to create nanoparticles with precise size, shape, and surface properties. These nanoparticles serve as contrast agents in various imaging modalities such as MRI, CT, ultrasound, and fluorescence imaging.

5.3.1 Synthesis of Metal Nanoparticles

Metal nanoparticles have gained significant attention in medical imaging due to their unique optical, magnetic, and surface properties, which make them excellent contrast agents for various imaging modalities such as magnetic resonance imaging (MRI), computed tomography (CT), positron emission tomography (PET), Photoacoustic imaging (PAI), and optical imaging. Gold nanoparticles (AuNPs), silver nanoparticles (AgNPs), copper nanoparticles (CuNPs), palladium nanoparticles (PdNps), platinum nanoparticles (PtNPs), cobalt nanoparticles (CoNPs), nickel nanoparticles (NiNPs), etc. are some commonly used metal nanoparticles in medical imaging. The synthesis of metal nanoparticles for medical imaging involves various methods that depend on the desired properties and applications. The selection of the preparation method for metallic nanoparticles is crucial as it directly influences their morphology, stability, and physicochemical properties. Each synthesis method offers unique advantages and challenges, and understanding the kinetics and mechanisms involved in nanoparticle formation is essential for controlling these properties.

Gold nanoparticles (AuNPs) have become increasingly popular in biomedical applications due to their low toxicity, biocompatibility, and inertness. Their strong surface plasmon resonance (SPR) properties make them ideal for optical imaging techniques such as surface-enhanced Raman scattering (SERS) and photothermal therapy. They can also be functionalized with targeting ligands or imaging agents for specific molecular imaging applications. In addition, AuNPs are commonly used as CT contrast agents because of their high X-ray attenuation coefficient. These nanoparticles can be synthesized using various methods, including physical, chemical, and biosynthesis.

A typical physical method used for the synthesis of AuNPs is the laser ablation method. JP Sylvestre et al. [26] used femtosecond laser (Ti/sapphire laser with wavelength 800 nm) ablation technique for the synthesis of AuNPs. The gold target was immersed in different solvents like de-ionized water and aqueous solutions of NaCl, KCl, NaOH, and n-propylamine. The average particle size of AuNPs formed was 5.5, 7, 8, and 7 nm, respectively, for NaCl, KCl, NaOH, and n-propylamine solutions.

In chemical synthesis methods for the synthesis of gold nanoparticles, several techniques are used to restrict their size to a 1–10 nm size regime. Changing the ratio of reactants [27], introducing ligands, and other surface modifying agents [28] are a few of them. The Turkevich method, developed by J. Turkevich in the 1950s, is a well-known and widely used chemical reduction method for synthesizing metal nanoparticles, particularly gold nanoparticles [29]. Turkevich's work not only established a robust method for gold nanoparticle synthesis but also proposed a mechanism to explain the formation and growth of colloidal gold nanoparticles. Reduction of Au salt using a strong reducing agent can result in the formation of ultrasmall AuNPs. Capping agents such as alginate, CTAB, Citrate, chitosan, etc., can also be added to the reaction mixture. Ruks et al. [30] synthesized 2 nm gold nanoparticles with a fluorescent peptide surface modification. The cell membrane was impermeable to the peptide used for the modification. However, peptide-modified AuNPs could penetrate the cell membrane and be used to target intracellular proteins. A. Silvestri et al. modified the Turkevich method to synthesise AuNPs functionalized with heterobifunctional thiolated polyethylene glycol. The obtained AuNPs are stable and showed potential to act as CT contrast agents [31].

Silver nanoparticles (AgNPs) possess strong localized surface plasmon resonance (LSPR) properties, making them suitable for various optical imaging techniques, including SERS and surface plasmon resonance imaging (SPRI). They can be used as contrast agents for optical coherence tomography (OCT) and photoacoustic imaging (PAI) due to their high absorption and scattering properties.

Synthesizing ultrasmall silver nanoparticles (AgNPs) for MRI imaging applications requires careful control over their size, stability, and surface properties to ensure biocompatibility and efficient contrast enhancement. The physical method, chemical reduction method, photochemical method, and green synthesis method are commonly used to synthesise AgNPs.

YH Chen et al. synthesized colloidal solutions of ultrasmall AgNPs using laser (Nd:YAG laser with a wavelength of 532 nm) ablation technique [32]. When the laser intensity was 120 mJ/pulse, AgNPs of sizes 4.2 ± 1.9 nm and 7.8 ± 4.5 nm were obtained, respectively, using surfactants sodium dodecylsulfuric acid (SDS) and cetyltrimethylammonium bromide (CTAB). When the laser intensity was reduced to half the former value, the size of AgNPs using surfactants SDS and CTAB was, respectively, 6.8 ± 2.7 nm and 9.4 ± 5.9 nm. Kumar et al. [33] reported the synthesis of AgNPs of 4–8 nm sizes using cryo-milling (ball milling at cryogenic temperature $-160 \pm 10°C$) method with the aid of liquid nitrogen (LN2). This technique enables the production of ultra-high purity silver nanoparticles (Ag NPs) with impurity levels below 35 parts per billion (ppb) without the silver powder or milling media coming into contact with LN2. The process involves cooling the silver powder and milling media with LN2, which evaporates harmlessly as nitrogen gas. The method avoids the use of hazardous chemicals, reducing agents, and capping agents, ensuring that the Ag NPs remain uncontaminated by oxidation, nitridation, or debris from the milling tools.

In the chemical reduction method, silver precursor (e.g., silver nitrate, $AgNO_3$) is dissolved in a suitable solvent (water or organic solvent) to form a silver precursor solution. To this solution, a stabilizing agent (such as polyvinylpyrrolidone (PVP) to prevent nanoparticle aggregation), a reducing agent (such as sodium borohydride ($NaBH_4$) or sodium citrate ($Na_3C_6H_5O_7$) to initiate the reduction of silver ions) are added [34]. Stirring time, temperature, ratio of reactants, etc. affect the size and properties of the synthesized silver nanoparticles. Surface modification of the AgNPs with biocompatible coatings or targeting ligands can further enhance their stability, biocompatibility, and targeting specificity for MRI imaging applications. Agnihotri et al. [35] synthesized silver nanoparticles of size below 10 nm using chemical reduction method with sodium borohydride and trisodium citrate as primary and secondary reductants. S.V Kumar et al. used microwave-assisted synthesis method to synthesise ultrasmall AgNps with silver nitrate, glucose, and starch as silver precursor, reducing agent and stabilizing agent respectively [36]. The mean particle size of AgNPs was 3.0 ± 1.2 nm for the reaction mixture with 0.012 mol L^{-1} $AgNO_3$.

In green synthesis methods, the reducing and stabilizing agents will be replaced by plant extracts, microorganisms or other biological entities [37]. This method harnesses the natural reducing capabilities of microbial enzymes and metabolites, providing an eco-friendly alternative to traditional chemical and physical synthesis techniques. Vigneshwaran et al. [38] reported that when exposed to a silver nitrate solution, the fungus *Aspergillus flavus* accumulated silver nanoparticles on its cell wall within 72 h. These nanoparticles, dislodged through ultrasonication, exhibited an absorption peak at 420 nm in the UV–visible spectrum, indicative of the plasmon resonance characteristic of silver nanoparticles. Transmission electron microscopy of the dislodged nanoparticles in an aqueous solution revealed that the fungus produced reasonably monodisperse silver nanoparticles with an average size of 8.92 ± 1.61 nm. The synthesis of silver nanoparticles (Ag NPs) can be achieved using plant extracts from various types of plants, including xerophytes (plants that are adapted to dry environments), mesophytes (plants that grow in moderate conditions, neither too wet nor too dry), and hydrophytes (plants adapted to grow in water or very moist conditions). A K Jha et al. prepared AgNPs by using plant extracts of healthy plant/plant part from *Bryophyllum* sp.(xerophytes), *Cyperus* sp. (mesophytes), and *Hydrilla* sp.(hydrophytes) [39]. AgNPs synthesized from all these plant extracts were in the size range 2–5 nm. Xerophytes utilize CO_2 fixation and organic acid metabolism along with emodin (an anthraquinone) tautomerization for silver nanoparticle synthesis. Mesophytes employ quinone (cyperoquinone, dietchequinone, and remirin) tautomerization for nanoparticle transformation, activated by warming and incubation. Hydrophytes use ammonia as a reducing agent and antioxidant systems involving catechol and protocatechuic acid for nanoparticle synthesis.

Copper nanoparticles are another widely explored metal nanoparticle. Copper nanoparticles provide a cost-effective alternative to silver or gold nanoparticles without significantly compromising performance. Various synthesis routes are adopted for the synthesis of CuNPs, such as chemical reduction, electrochemical synthesis, laser ablation, etc. Fernández-Arias et al. [40] synthesized CuNPs of average particle size 3.9 nm using nanosecond laser ablation of copper foil in methyl alcohol solvent. In the chemical reduction method, usually, copper salts such as copper sulphate, copper nitrate, copper chloride, or copper (II) acetylacetonate are treated with a reducing agent like hydrazine, sodium borohydride, glucose, hexadecane-1,2-diol, mercaptopropionic acid, citric acid, or ascorbic acid [41]. CuNPs of average particle size 7.6 ± 3.3 nm were synthesized using a microwave-assisted synthesis method [42]. Feng et al. used chemical reduction method using hydrazine hydrate as a reducing agent, copper nitrate as the precursor and CTAB, PEG, and Tween as organic modifiers. CuNPs of average particle sizes 5.9, 7.6 and 9.8 nm were obtained, respectively, when CTAB, PEG, and Tween were used [43]. Jin et al. reported the synthesis of ultrasmall CuNPs by the reduction of copper sulphate using dendrimers of higher generation [44]. Dendrimers are highly branched, tree-like polymers with a well-defined, monodisperse structure. They comprise a central core, repetitive branching units, and numerous terminal functional groups. Dendrimers with Trimesyl Core (G3–G6) act as stabilizing and templating agents, where G3–G6 refer to the generations of dendrimers with increasing size and complexity. CuNPs of smaller sizes and better dispersion are obtained when higher-generation dendrimers are used. The metal to dendrimer concentration ratio also affects the size of metal nanoparticles. Figure 5.11 [44] shows the relationship between size of CuNPs and dendrimer generation and concentration.

The synthesis of platinum nanoparticles (PtNPs) involves various methods, including chemical reduction, polyol, and green synthesis. These nanoparticles can be functionalized to improve their biocompatibility and targeting ability for use in MRI imaging. PtNPs enhance MRI contrast, providing detailed and high-quality images for better disease diagnosis and monitoring, particularly in oncology. Their multifunctional capabilities also open avenues for theranostic applications, combining diagnosis and treatment in a single platform. N.Wolff et al. synthesized ultrasmall PtNPs of average particle diameter 2 ± 0.4 nm [45] using simple reduction method using $NaBH_4$. The tripeptide molecule glutathione was used as a capping agent. Ji et al. [46] adopted a hydrothermal method to synthesise PtNPs in the size range 1–3 nm using poly(vinyl pyrrolidone) as a capping agent. AKAA Saidi et al. [47] reported the synthesis of ultrasmall NPs using the polyol method with triethylene glycol as the solvent. They used three different biocompatible polymers as capping agents. OT Fanoro et al. [48] synthesized PtNPs using the leaf extract of *Combretum erythrophyllum* (CE) plant. The average particle diameter of synthesized PtNPs is 1.04 ± 0.26 nm.

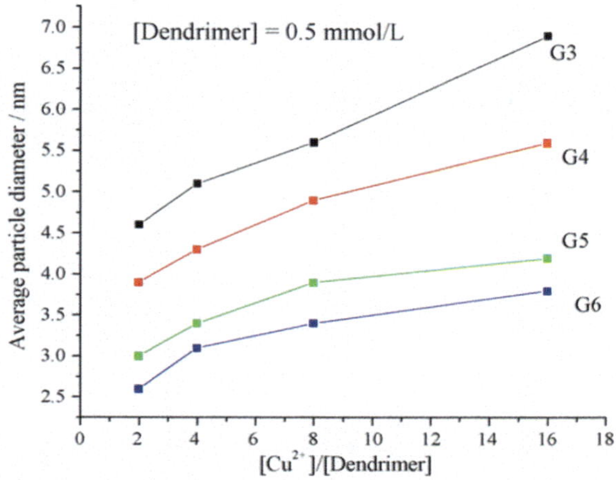

Fig. 5.11 Relationship between CuNP particle size and dendrimer generation and Cu^{2+}/dendrimer concentration ratio

Palladium nanoparticles (PdNps) have emerged as promising agents in the field of nanomedicine, offering significant advantages for both imaging and therapy. Their unique properties, including high photothermal conversion efficiency, excellent biocompatibility, and stability in physiological environments, make them suitable for various biomedical applications. As research continues to advance, Pd-based nanomaterials are expected to play an increasingly important role in the development of innovative diagnostic and therapeutic technologies. M. Chen et al. synthesized Pd nanosheets of 1.8 nm thickness and diameters 5 nm, 13 nm, 30 nm and 80 nm and capped with thiol-polyethylene glycol for biocompatibility [49]. They observed that the 5 nm sheets exhibited the highest blood circulation half-life, caused lesser alteration in the biological processes, showed better accumulation at tumour sites, and cleared more rapidly from the body through urine. Figure 5.12 shows the effect of different-sized Pd nanosheets on pharmacokinetics [49].

Tang et al. [50] synthesized glutathione functionalized Pd nanosheets of 4.4 nm size. Due to their small size and unique structure, palladium nanosheets can efficiently convert near-infrared (NIR) light into heat. This high photothermal conversion efficiency makes them ideal candidates for photothermal therapy (PTT), where localized heating can be used to destroy cancer cells with minimal impact on surrounding healthy tissues. Also, they can be efficiently cleared from the body through renal excretion. This optimal clearance characteristic reduces the risk of long-term toxicity and makes them safer for in vivo applications. Biological methods utilizing the extracellular electron transfer (EET) ability of electroactive bacteria (EAB) show promise for sustainable metal nanoparticle synthesis. R Jimenez-Sandoval et al. [51] presented a straightforward biological method for synthesizing size-controlled palladium nanoclusters using the EET capability of *Geobacter sulfurreducens*. By adjusting the parameters like concentration of metal precursors, time of incubation, etc., they tuned the size of Pd nanoclusters from 1.0 ± 0.6 to 4.8 ± 1.4 nm.

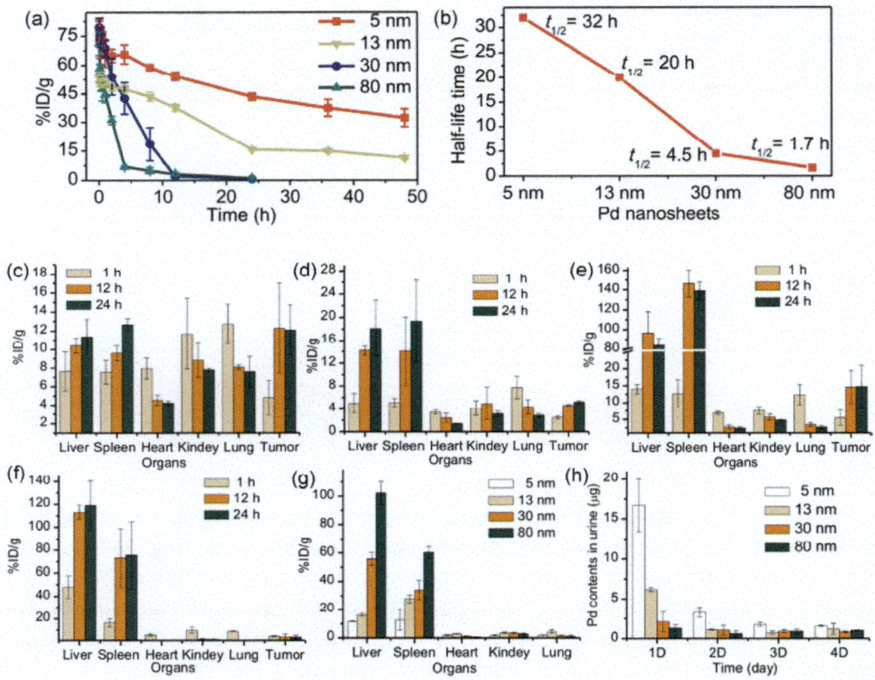

Fig. 5.12 The effect of different-sized Pd nanosheets on pharmacokinetics: **a** blood circulation curves, **b** comparison of blood circulation half-lives, **c–f** biodistribution of 5-nm, 13-nm, 30-nm and 80-nm Pd nanosheets, **g** Biodistribution after 30 days, and **h** Urine clearance after 4 days

Cobalt nanoparticles (CoNPs), with their strong magnetic properties and ability to enhance MRI contrast, hold significant promise for medical imaging applications. Various synthesis methods, including thermal decomposition, chemical reduction, and solvothermal synthesis, allow the synthesis of CoNPs with controlled size and properties. Functionalization of these nanoparticles further enhances their biocompatibility and targeting capabilities, making them valuable tools in imaging and diagnosis applications. Veregue et al. [52] adopted a chemical reduction method using $NaBH_4$ to synthesise CoNPs of average particle size 3.07 ± 0.59 nm. Tannic acid was used as a capping agent. Another chemical reduction method proposed by Lu et al. [53] utilizes $NaBH_4$ as a reducing agent and alkyl thioether end-functionalized poly(methacrylic acid) (PMAA-DDT) ligands to control the size and shape of the nanoparticles. By adjusting the polymer concentration, the size of the nanoparticles can be precisely tuned, making this method highly versatile. By altering the polymer concentration between 0.006 mM and 0.48 mM, the average particle size can be tuned between 7.5 ± 0.7 nm and 2.1 ± 0.2 nm. The resultant Co nanoparticles are not only readily water-soluble but also immediately applicable for biomedical uses such as MRI contrast enhancement, demonstrating significant potential in medical imaging technologies [54].

Ultrasmall metal alloy nanoparticles are gaining attention for their potential in biomedical imaging due to their unique properties, such as enhanced magnetic, optical, and catalytic behaviours compared to their bulk counterparts or single-metal nanoparticles. These nanoparticles typically have sizes less than 10 nm and often exhibit improved penetration and retention in biological tissues, making them ideal candidates for imaging applications. Commonly explored metal alloys include Iron-Platinum (FePt) alloys, Gold-Silver (Au–Ag) alloys, Gold-Copper (Au-Cu) alloys, etc. SW Chou et al. synthesized FePt nanoparticles of tunable sizes (3 nm, 6 nm, and 12 nm) using the polyol method and capped with cysteamine. Targeting Her2/neu oncogene expression in mice confirmed their dual modality imaging capability. Biodistribution studies showed clearance from the body in about a week, with 12 nm FePt having the highest serum concentration and imaging contrast, while 3 nm FePt was ideal for brain imaging [55]. Ultrasmall FePt nanoparticles were synthesized using solvothermal method and reduction method by Meng et al. and encapsulated the nanoparticles in metal–organic framework (MOF) [56]. Later, FePt-MOF NPs were functionalized with a tumour-targeting peptide tLyp-1. Functionalized with tLyp-1, these FePt-MOFs demonstrated high stability in physiological conditions and efficiently targeted tLyp-1-receptor positive tumour cells, enhancing drug accumulation at tumour sites. The MOF-based drug delivery system effectively targeted tumour cells via intravenous injection, releasing FePt nanoparticles upon uptake by the tumour cells. In vitro studies showed that these nanocomposites generated reactive oxygen species (ROS) by releasing active Fe ions and catalyzing intracellular hydrogen peroxide (H_2O_2), promoting ferroptosis. A schematic of ferroptosis and multimodal imaging is shown in Fig. 5.13 [56]. The WST-1 assay indicated a decrease in cancer cell survival with increased FePt concentration, significantly suppressing tumour growth in vitro and in vivo. Additionally, FePt-MOF nanocomposites served as MR and CT contrast agents for precise imaging-guided cancer treatment.

Ngyen et al. [57] proposed a simple laser ablation method to synthesise Au–Ag NPs of an average diameter of 4 nm. They mixed Ag and Au nanoparticle colloidal solutions in a 1:1 volume ratio in 0.02 M PVP solution, and the mixture was exposed to a 532 nm Nd:YAG laser at an average power density of 1.6 W/cm^2 for 20 min to obtain Au–Ag alloy. A new method for synthesizing Au/Cu@PDA nanocomposite with an Au:Cu atomic ratio of 84:16 was developed by Perlman et al. [58]. The synthesis occurred in mild aqueous conditions via co-reduction of $HAuCl_3$ and CuO nanoparticles within a polydopamine (PDA) matrix. This active PDA matrix could be useful for creating versatile multi-functional materials. Characterization and imaging experiments showed improved contrast in MRI and CT using phantom and ex vivo poultry heart models. Also, in vitro tests indicated low cytotoxicity.

5.3.2 *Synthesis of Metal Oxide Nanoparticles*

Metal oxide nanoparticles have gained significant attention due to their stability, relatively low production costs, and versatile properties that are beneficial for various

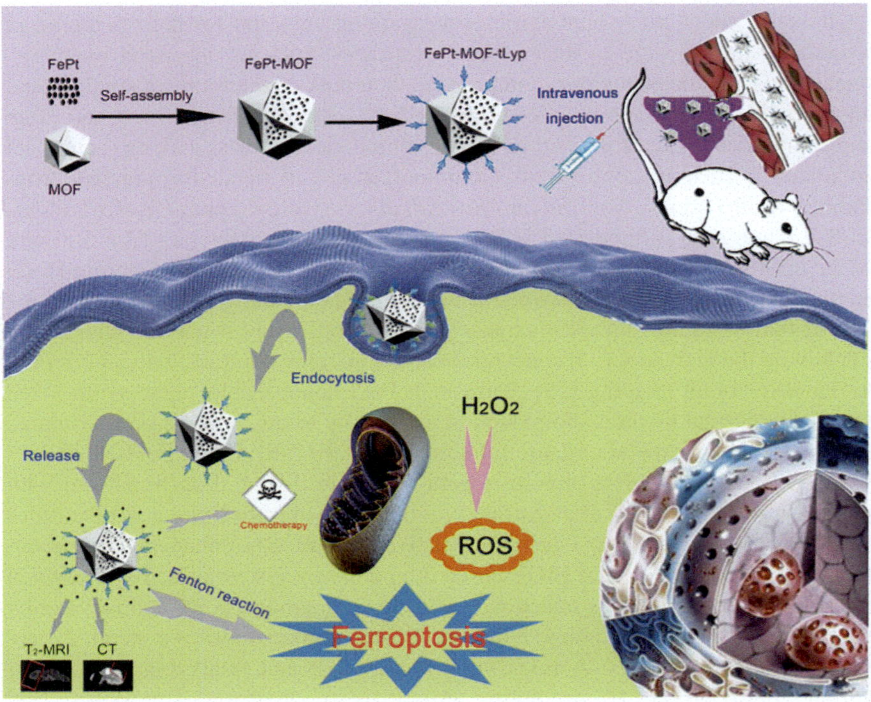

Fig. 5.13 A schematic of FePt-MOF synthesis, intravenous injection, release inside a tumour, ferroptosis, and application as multimodal imaging agent

medical applications. For example, magnetic metal oxide nanoparticles like iron oxide, manganese oxide, gadolinium oxide, and ferrite nanoparticles have been extensively researched for their use as MRI (magnetic resonance imaging) contrast agents. Tantalum oxide nanoparticles have also been studied as contrast agents for X-ray fluoroscopy and computed tomography (CT). Additionally, metal oxides such as ceria and titania nanoparticles are gaining interest for their catalytic effects in scavenging reactive oxygen species (ROS) and photodynamic therapy (PDT), respectively. Iron oxide nanoparticles have been employed as catalysts in Fenton reactions to address cancer hypoxia and enhance tumour therapy effectiveness. As many potential applications of metal oxide nanoparticles remain unexplored, there are opportunities for broader applications in large animals and humans when combined with large-scale synthesis methods [59].

The sol–gel method is indeed a versatile technique used for synthesizing materials like metal oxide nanoparticles (MONPs) and mixed oxide composites. The process begins with the hydrolysis of precursor molecules (such as metal alkoxides) in a solvent (usually an alcohol like ethanol or methanol) containing water. This hydrolysis reaction breaks down the precursor molecules into metal hydroxides or oxides, forming a sol (a suspension of colloidal particles in a liquid). After hydrolysis, the metal hydroxides or oxides start to undergo polycondensation. This involves the formation of chemical bonds between the hydrolyzed species, leading to the growth of a three-dimensional network or gel. This step is critical for building the desired structure of the final material. The gel obtained from polycondensation is then aged or allowed to mature. During aging, the gel undergoes further structural rearrangements and transformations, which enhance the properties of the resulting material (Fig. 5.14). Aging can occur at ambient or elevated temperatures over a period ranging from hours to days. Once adequately aged, the gel is dried to remove the solvent from the pores of the gel network. Drying can be achieved through various methods, such as air drying, freeze-drying (lyophilization), or supercritical fluid extraction. This step is crucial for stabilizing the gel structure and preventing undesired shrinkage or cracking. Finally, the dried gel is subjected to thermal decomposition or calcination at elevated temperatures. This helps in further densification of the material by removing any residual organics, decomposing precursors, and crystallizing the metal oxide phases. The specific temperature and duration of calcination depend on the desired properties and composition of the final product. Using this approach, uniform-sized nanoparticles of various materials such as titania, manganese ferrite, and zinc oxide have been successfully synthesized. The nonhydrolytic sol–gel method offers advantages for controlling the particle size and morphology, making it a valuable technique for preparing nanoparticles with specific properties and applications [60]. Md Atiqur Rahman et al. synthesized $Mg_xCa_{(0.90-x)}Zn_{0.10}Fe_2O_4$ nanoparticles using the sol–gel chemical method with citric acid as a chelating agent. Their comprehensive investigations confirm that the developed nanoparticle-based flexible microwave substrates exhibit tunable electromagnetic properties. These substrates are deemed suitable for the development of flexible antennas used in microwave head imaging systems. [61].

Laser ablation is cost-effective, simple, environmentally friendly, and conducted in a single step, where nanoparticles (NPs) are synthesized directly in a liquid medium without contaminations or byproducts. The productivity of NP generation is strongly influenced by the laser ablation mechanism. Therefore, factors such as the nature of the target material, the liquid medium, laser parameters, and surface

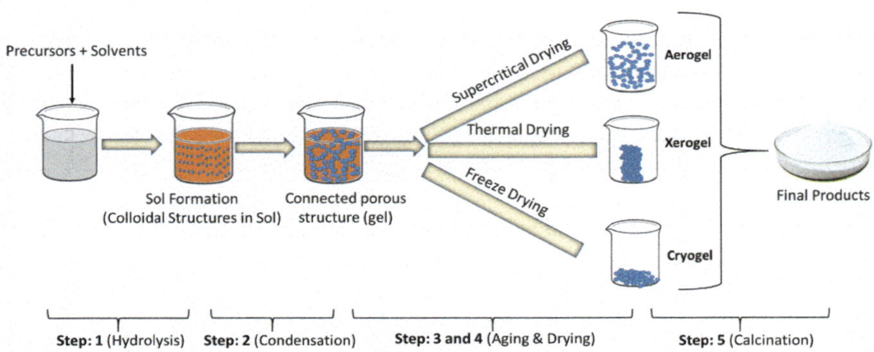

Fig. 5.14 Steps involved in sol–gel process to synthesize MONPs [3]

morphology play significant roles in determining the efficiency of laser ablation and its impact on productivity. The ablation rate shows an almost linear increase with pulse energy, laser beam area at a given fluence, and number of laser pulses. Additionally, the properties of the synthesized nanoparticles—including size, shape, composition, crystallinity, and morphology—can be controlled based on the quality of the ablation targets, the characteristics of the liquid medium, and the specific laser parameters employed. These factors collectively contribute to tailoring the desired properties of the synthesized nanoparticles to meet specific application requirements [62]. Khumaeniet et al. synthesized iron oxide nanoparticles using pulsed laser ablation (Nd: YAG laser, 1064 nm, 7 ns, 30 mJ) with the aim of developing a contrast agent for magnetic resonance imaging (MRI) applications. The laser was focused on a high-purity iron plate immersed in a liquid solution containing deionized water and polyvinylpyrrolidone (PVP) as a stabilizer. The resulting iron oxide nanoparticles, consisting of Fe_2O_3 and Fe_3O_4, exhibited excellent dispersibility and a spherical shape. These nanoparticles were subsequently evaluated as a contrast agent for MRI by varying their concentrations [63].

The co-precipitation technique involves the precipitation of oxo-hydroxide from a solution containing metal salt precursors (such as nitrates or chlorides) in a solvent like water using a precipitating agent. When the solution reaches a critical concentration of species, a rapid nucleation event occurs, followed by a growth phase. This method has been utilized for synthesizing metal oxides such as MnO_2 and CeO_2. Co-precipitation is commonly applied to prepare magnetic nanoparticles like magnetite by mixing ferrous and ferric salts dissolved in water with a base, typically NaOH or NH_4OH, in stoichiometric proportions. The choice of precipitating agent, including NaOH, KOH, and $(C_2H_5)_4NOH$, affects parameters such as pH, alkali nature, rate of alkaline solution addition, and powder drying method, which in turn influence the size, paramagnetic properties, and agglomeration degree of the synthesized magnetite nanoparticles. Furthermore, the use of surfactants has proven beneficial in optimizing surface characteristics. Key advantages of this method include its cost-effectiveness, mild reaction conditions (such as low synthesis temperature), the ability to perform direct synthesis in water, straightforward processing, ease of scalability, and flexibility in tuning both core and surface properties of the synthesized nanoparticles [64]. Yi-Hsuan Lin et al. conducted a one-pot facile synthesis of CeO_2 nanoparticles using a precipitation method with citric acid and EDTA as surfactants. The resulting highly monodispersed CeO_2 NPs are small in size (average size of 13 nm) and possess anti-oxidative/auto-reduction properties, along with stability in cell culture media. This promising approach demonstrates the potential of generating CeO_2 NPs for biomedical uses [65].

Polyols are highly suitable for synthesizing metal oxide nanoparticles due to their water-like properties and chelating capabilities. Simple metal salts such as halides, nitrates, acetates, and alkoxides can serve as starting materials for this process. Polyols function as solvents that resemble water but have high boiling points, making them ideal for dissolving various compounds, including inexpensive metal salts like halides, nitrates, and sulfates. One of the key advantages of using polyols is their ability to instantly coordinate with newly formed particle nuclei during synthesis, enabling precise control over particle size, size distribution, and agglomeration. Additionally, removing low-molecular-weight polyols (such as ethylene glycol, diethylene glycol, propanediol, butanediol, and glycerol) from the particle surface after synthesis is relatively simple. This can be achieved through methods such as washing (repeated centrifugation and re-dispersion) with water, exchanging with more strongly coordinating agents (like amines or carboxylates), or subjecting the particles to thermal treatment at temperatures ranging from 200 to 320 °C [66]. Jingjing Li et al. introduced a one-pot method to synthesize hydrophilic manganese oxide nanoparticles using a polyol-like approach with poly(ethylene glycol) (PEG) serving as both a solvent and surfactant. The resulting PEG-MnO NPs exhibited a high T1 relaxivity and a low r2/r1 ratio (12.942 s^{-1} mM^{-1} and 4.66) at 3.0 T magnetic field strength, which was three times higher than that of the clinically used contrast agent [67].

Zhang et al. [68] developed a dynamic simultaneous thermal decomposition (DSTD) strategy to synthesize ultrasmall metal ferrite nanoparticles (<4 nm) using ultrasmall manganese ferrite nanoparticles (UMFNPs) as a model. By pairing an iron-eruciate complex with a metal-oleate complex as precursors, DSTD facilitated nucleation-doping, ensuring uniform particle size and distribution. In contrast, dynamic nonsimultaneous thermal decomposition (DNSTD) under similar conditions resulted in nonuniform nanoparticles. The particle size can be varied from 2 nm to 3.9 nm by tuning the oleyl alcohol to oleic acid ratio and reaction temperature. The synthesized UMFNPs were surface functionalized with phosphorylated mPEG. UMFNPs demonstrated the highest r1 relaxivity (up to 8.43 mM^{-1} s^{-1}) among similar-sized ferrite nanoparticles, making them excellent multifunctional T1 MR probes for high-resolution blood pool and liver-specific MRI. In vivo studies showed UMFNPs are safe for potential clinical use. Figure 5.15 shows the comparison between T1 weighted MR images post-injection of UMFNPs and Omniscan (Gadodiamide, a commonly used Gd-based MRI contrast agent).

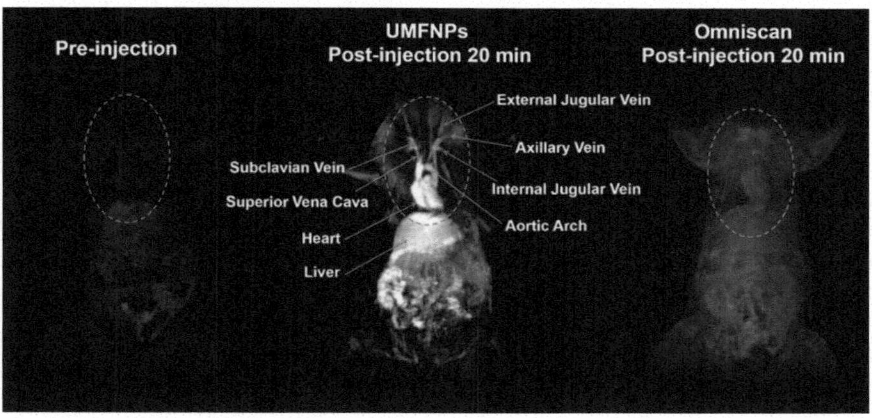

Fig. 5.15 Comparison between T1 weighted MR images post-injection of UMFNPs and Omniscan

5.3.3 *Synthesis of Rare Earth-Based Nanoparticles*

Rare earth-based nanoparticles hold great promise for advancing multimodal imaging techniques, providing comprehensive diagnostic capabilities and improving the accuracy and effectiveness of disease diagnosis and treatment monitoring. The synthesis of rare earth-based nanoparticles involves various methods tailored to achieve specific sizes, shapes, and surface functionalities. These nanoparticles are particularly valuable for applications in bioimaging, drug delivery, and other biomedical fields due to their unique optical, magnetic, and electronic properties. Rare earth elements, typically found in trivalent cation form, consist of lanthanides from lanthanum to lutetium, along with scandium and yttrium. Due to their f-orbital arrangements, lanthanide ions have the capacity to display distinct fluorescent emissions through intra-4f or 4f–5d transitions, making them extensively utilized as emitting components in numerous phosphors [69].

Solvothermal techniques involve reactions within sealed containers at pressures generated by the solvent's vapor above its boiling point. Often, rare earth chlorides and nitrates are the primary precursors in such processes [70]. The hydrothermal process involves utilizing either water or a mixture of alcohol and water as a solvent within a tightly sealed autoclave, which is subsequently heated to generate high temperature and pressure conditions. Typically, the reactants are insoluble or sparingly soluble under ambient conditions. Once heated, they dissolve and undergo reactions within the autoclave. The resulting intense convection, induced by temperature variations within the solution, forms a supersaturated solution, which subsequently crystallizes. This process enables the production of nanopowders characterized by uniform dispersion, a narrow range of particle sizes, well-defined crystal shapes, and adjustable appearances [71]. Xiao-yan Kuang et al. developed a two-step hydrothermal synthesis technique to synthesize $TbPO_4:Ce^{3+}@TbPO_4:Gd^{3+}$, a novel nanocomposite featuring a core–shell structure and dual fluorescence and magnetic resonance capabilities. The synthesized nanocomposites exhibit excellent crystallinity, high quantum yield rates, significant longitudinal relaxivity, and low toxicity. These characteristics suggest promising prospects for their use as dual-functional contrast agents in biomedical diagnosis and analysis [72].

Carbon-coated dysprosium oxide nanoparticles (DYO@C) of average particle size 3 nm were synthesized by Yue et al. [73] via polyol method. In vitro cellular cytotoxicity assays indicated that these nanoparticles were nearly non-toxic. The nanoparticles were stable in aqueous solutions due to the abundant hydroxyl groups on the carbon coating. The high r2/r1 ratio of DYO@C makes them highly effective as T2 MRI contrast agents, producing negative contrast enhancements in in vivo T2 MR images of mouse kidneys following intravenous administration. DYO@C has the potential for use as a multimodal imaging agent owing to the fluorescence of carbon in the visible region.

Zou et al. [74] synthesized PVP-coated $Gd_{0.8}Nd_{1.2}O_2S$ nanoparticles (rare earth oxy-sulphides) via ligand exchange, showing low and variable toxicity depending on the cell line. They exhibited an r1 of 2.42 $(mM)^{-1}\ s^{-1}$ and r2 of 2.83 $(mM)^{-1}\ s^{-1}$, which increased under a 7 T magnetic field to r1 of 9 $(mM)^{-1}$ s^{-1} and r2 of 57 $(mM)^{-1}$ s^{-1}, supporting bimodal MRI. PAI, USI, and CT had lower sensitivity than MRI but were more cost-effective and accessible.

Zhang et al. [75] explored the use of ultrasmall europium nanoparticles (uEuNP) as a biomedical probe. These nanoparticles, synthesized via thermal hydrolysis of europium(III) acetate in oleylamine and oleic acid under argon, were confirmed to be $\sim 7 \pm 2$ nm in size by TEM. To enhance fluorescence, Eu(II) was oxidized to Eu(III) using trimethylamine N-oxide (TMNO), increasing fluorescent intensity sixfold. The TMNO-modified uEuNPs showed promise as a biomedical probe. Europium, typically not used directly for in vivo optical imaging, can be paired with Cerenkov light to leverage its efficient fluorescent properties. TMNO-uEuNP@PEG is suitable for in vivo nuclear imaging (Cerenkov and PET) in mouse models for lymph node identification and tumour imaging. The novel nanoparticle design overcomes Cerenkov light deficiencies at longer wavelengths, creating a new nuclear-fluorescence multimodal imaging platform. Due to smaller sizes, they are rapidly excreted via the kidneys, reducing potential Eu toxicity and paving the way for potential human applications.

Ho et al. [76] synthesized polyacrylic acid functionalized ultrasmall Gd_2O_3 nanoparticles using the polyol method and coated with single (folic acid (FA)) and dual (cyclic arginylglycylaspartic acid (cRGD) and FA) tumour-targeting ligands. Both FA-PAA-Gd_2O_3 and cRGD/FA-PAA-Gd_2O_3 nanoparticles were nearly monodispersed with an average particle size of 1.7 nm. The hydrodynamic diameters were 11.4 nm for FA-PAA-Gd_2O_3 and 13.8 nm for cRGD/FA-PAA-Gd_2O_3, allowing renal excretion of nanoparticles. The nanoparticles exhibited r1 values approximately four times higher than commercially available Gd-chelates, with 12.0 and 11.2 s^{-1} mM^{-1} for FA-PAA-Gd_2O_3 and cRGD/FA-PAA-Gd_2O_3, respectively. cRGD/FA-PAA-Gd_2O_3 showed higher tumour site contrast due to their dual targeting effect. Both nanoparticle types displayed the highest contrast in the aorta, attributed to prolonged blood circulation facilitated by their optimal hydrodynamic sizes, which minimized opsonization and RES uptake while delaying renal excretion. Additionally, cRGD/FA-PAA-Gd_2O_3 nanoparticles exhibited higher contrast across all organs, especially in the aorta, due to the blood cell hitchhiking phenomenon, further extending their circulation time. Figure 5.16 [76] shows T1 MR images captured at a 3.0 T MR field at various time points before ("Pre") and after intravenous injection of aqueous FA-PAA-Gd_2O_3 and cRGD/FA-PAA-Gd_2O_3 nanoparticle suspensions into the tails of mice. The images highlight specific regions of interest with small dots.

5.3.4 Synthesis of Carbon-Based Nanoparticles

Carbon-based materials like carbon nanotubes, graphene, graphene quantum dots, carbon dots, nanodiamonds, and graphene oxide have gained significant attention

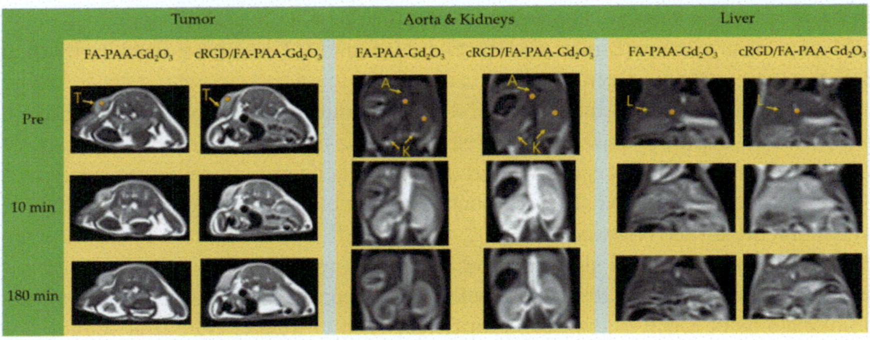

Fig. 5.16 T1 MR images captured at a 3.0 T MR field at various time points before ("Pre") and after intravenous injection of aqueous FA-PAA-Gd_2O_3 and cRGD/FA-PAA-Gd_2O_3 nanoparticle suspensions into the tails of mice

in various medical imaging applications due to their unique properties. These materials offer advantages such as excellent biocompatibility, low toxicity, and tuneable properties, making them promising candidates for improving imaging techniques [77]. Some popular methods used to synthesize carbon-based nanomaterials are laser ablation, arc discharge, and CVD (chemical vapor deposition).

Chemical vapor deposition (CVD) is a widely used technique for synthesizing carbon-based nanomaterials such as carbon nanotubes (CNTs) and graphene. In CVD, carbon-containing precursors are decomposed at high temperatures in the presence of a catalyst to form nanoparticles. As the carbon source breaks down, carbon atoms adhere to the catalyst surface. Because carbon has low solubility in these metallic particles at high temperatures, this causes carbon to precipitate, leading to the formation of carbon nanomaterials. By controlling the parameters like temperature, pressure, and catalyst composition allows for tuning the properties of the synthesized nanoparticles [78]. Ding and Wenzheng et al. studied ferromagnetic material-filled multi-walled carbon nanotubes (CNTs) synthesized via chemical vapor deposition (CVD) using a two-stage furnace system within a horizontal quartz tube reactor (20 mm I.D.; 600 mm length). These nanoparticles were explored as contrast agents for both photoacoustic imaging (PAI) and magnetic resonance imaging (MRI) in an in vivo tumour model. The promising results suggest that using a dual-modality probe could offer more precise imaging capabilities in clinical settings [79].

Hydrothermal or solvothermal methods involve the synthesis of carbon-based nanoparticles in a high-pressure, high-temperature aqueous or organic solvent environment. These conditions facilitate the formation of nanoparticles with controlled size, morphology, and surface properties. Carbon dots and graphene quantum dots are often synthesized using hydrothermal or solvothermal routes. Hydrothermal carbonization is recognized as a highly effective method for creating well-defined nanomorphologies. This sophisticated process involves wet thermochemical treatment using water, which possesses strong ionization capabilities and aids in the rapid disintegration of carbonaceous materials through complex chemical reactions. The reaction is typically carried out at elevated temperatures (150–350 °C) and processed for 2–24 h under pressure in an aqueous environment. Hydrothermal carbonization effectively breaks down and transforms carbonaceous materials into uniform carbon nanomaterials with specific nanostructures [80]. Vahid Ali et al. developed a one-pot hydrothermal method to synthesize magneto-fluorescent carbon quantum dots. These carbon quantum dots display a high r1 relaxivity value of 7.4 mM^{-1} S^{-1} and exhibit low cytotoxicity. Consequently, they are promising nano-contrast agents suitable for MRI imaging applications [81].

The arc discharge method is highly effective for producing defect-free carbon nanomaterials. Various types of carbon nanostructures, including graphene nanoplatelets (GNPs), multi-walled carbon nanotubes (MWCNTs), single-walled carbon nanotubes (SWCNTs), spherical carbon nanoparticles (SCNs), and carbon nanohorns (CNHs) can be collected from different regions within a direct current arc discharge reactor. In this process, a high-current electric arc is created between graphite electrodes within an inert atmosphere. The intense heat generated by the arc vaporizes the graphite electrodes, producing carbon vapor. As this vapor cools and condenses, it forms a soot-like material that contains the desired carbon nanostructures that are then collected [82]. H. Wang et al. synthesized single-walled carbon nanohorns (SWNHs) through arc discharge between two graphite electrodes submerged in liquid nitrogen [83].

Laser ablation is a versatile and efficient method used to synthesize carbon-based nanomaterials. In this process, a high-power laser beam is directed at a target material, usually graphite, within a controlled environment. The laser energy vaporizes the target material, creating a plasma plume. As the vapor condenses, it forms carbon nanoparticles, such as carbon nanotubes and graphene, depending on the specific synthesis conditions. Precise control of the laser energy enables tuning of the size, morphology, and surface characteristics of these nanoparticles, which are key factors influencing their properties [84]. L. Cui et al. synthesized carbon quantum dots (CQDs) by dual-beam laser cutting of low-cost carbon cloth using a Q-switched Nd:YAG laser operating at 10 Hz with a wavelength of 1064 nm, a pulse width of 6 ns, and a pulse energy of 20 mJ (Fig. 5.17). The ablation process lasted for 10 min. The resulting CQDs exhibited good stability and excellent anti-jamming performance, making them suitable for HeLa cell bioimaging applications [85]. Gaurav Kumar Yogesh's work demonstrates successful synthesis of non-toxic fluorescent graphene oxide nanoparticles through laser ablation of graphene solution, specifically for bioimaging applications [86].

Microwave irradiation can be used to accelerate the synthesis of carbon-based nanoparticles. Microwave-assisted synthesis involves using microwave irradiation to heat the reaction mixture. Microwave radiation is a type of electromagnetic energy consisting of electric and magnetic components. The electric component primarily induces heating through two mechanisms: dipolar polarization and ionic conduction. In dipolar polarization, the dipoles within the sample align with the applied electric field during microwave irradiation and attempt to follow its oscillations. This movement generates heat through molecular friction and dielectric loss. In ionic conduction, charged particles or ions within the sample oscillate back and forth in response to

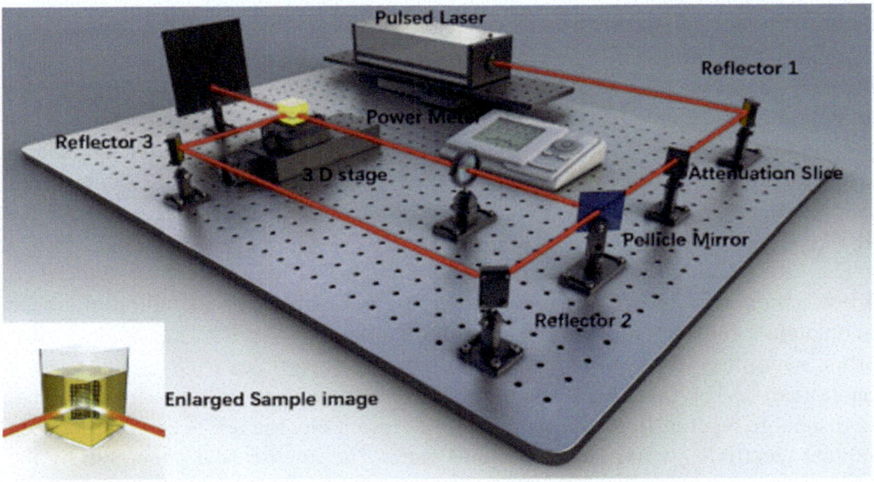

Fig. 5.17 Schematic illustration of the preparation process to synthesize CQDs by dual-beam pulsed laser ablation

the microwave field. As these particles collide with neighboring molecules or atoms, the temperature rises due to the energy generated from these collisions and frictions. Microwave-assisted methods are often employed to rapidly synthesize carbon dots and graphene quantum dots, offering advantages such as shorter reaction times and improved yield [87]. Hasan Eskalen et al. synthesized water-dispersed fluorescent carbon quantum dots (CDs) with an average particle diameter of 10 nm using a microwave-assisted process suitable for cancer cell imaging applications [88].

Template-assisted synthesis involves using templates or scaffolds to control the size, shape, and structure of carbon-based nanoparticles. Various templates, including porous materials, polymers, and biological molecules, can be employed to guide the nucleation and growth of nanoparticles with tailored properties. Template-based synthesis offers a clear advantage in producing high-quality products. However, the quality heavily relies on the structure of the template, which can be challenging to design according to desired structures and properties. Additionally, removing the template without damaging the product remains a significant challenge in template-based synthesis processes [89]. Clara Deeney et al. utilized template synthesis to prepare luminescent carbon fibers derived from beverage-related precursors, demonstrating potential applications in imaging [90].

Each synthesis method has its advantages and limitations, and the choice of technique depends on factors such as desired nanoparticle properties, scalability, and cost-effectiveness. Additionally, post-synthesis functionalization and surface modification techniques may be employed to further tailor the properties of carbon-based nanoparticles for specific medical imaging applications.

5.4 Conclusions

In conclusion, nanoparticles offer a wide range of applications in healthcare, from improving drug delivery to enhancing diagnostics and personalized medicine. Continued research and development in this field hold the promise of revolutionizing healthcare by providing more effective treatments and diagnostic tools.

This chapter discusses the synthesis of ultrasmall nanoparticles, involving the use of many materials, including metals, alloys, metal oxides, rare earth-based nanoparticles and carbon-based ones. Through various innovative methods of synthesis, or by combining some methods, it is possible to control nanoparticle size and shape at an atomic level, thus producing particles usually less than 10 nm in diameter.

Their unique characteristics, such as high surface area, improved reactivity and long blood circulation time make these ultra-small nanoparticles suitable for medical imaging purposes. These discussed particles possess great potentiality in improving contrast in various imaging modalities including MRI, CT, PET, and multimodal imaging techniques. These ultrasmall nanoparticles specifically provide a higher resolution image with better specificity and sensitivity required for early detection as well as monitoring of diseases.

Different functional ligands and coatings have also been incorporated into these nanoparticles, resulting in targeted imaging and therapeutic applications. For instance, conjugation of targeting ligands such as peptides or antibodies allows for precise localization of nanoparticles to tumour sites, improving the effectiveness of imaging and treatment.

Overall, the advancements in the synthesis of ultrasmall nanoparticles and their application in medical imaging represent a significant step forward in the field of nanomedicine. These developments hold great promise for the future of diagnostic imaging and targeted therapy, paving the way for more accurate, efficient, and personalized medical treatments.

References

1. M.S. El-Eskandarany, A. Al-Hazza, L.A. Al-Hajji, N. Ali, A.A. Al-Duweesh, M. Banyan, F. Al-Ajmi, Mechanical milling: a superior nanotechnological tool for fabrication of nanocrystalline and nanocomposite materials. Nanomaterials **11**, 2484 (2021)
2. D. Memarvar, S. Yaqoubi, H. Hamishehkar, M. Lam, A. Nokhodchi, Impact of grinding balls on the size reduction of aprepitant in wet ball milling procedure. Pharm. Dev. Technol. **29**, 353 (2024)
3. G. Vilardi, N. Verdone, R. Bubbico, Combined production of metallic-iron nanoparticles: exergy and energy analysis of two alternative processes using hydrazine and NaBH4 as reducing agents. J. Taiwan Inst. Chem. Eng. **118**, 97 (2021)
4. M.H. Hussain, N.F. Abu Bakar, A.N. Mustapa, K.F. Low, N.H. Othman, F. Adam, Synthesis of various size gold nanoparticles by chemical reduction method with different solvent polarity. Nanoscale Res. Lett. **15**, 140 (2020)
5. D.T. Nguyen, D.-J. Kim, K.-S. Kim, Controlled synthesis and biomolecular probe application of gold nanoparticles. Micron **42**, 207 (2011)

6. C. Dhand, N. Dwivedi, X.J. Loh, A.N. Jie Ying, N.K. Verma, R.W. Beuerman, R. Lakshminarayanan, S. Ramakrishna, Methods and strategies for the synthesis of diverse nanoparticles and their applications: a comprehensive overview. RSC Adv. **5**, 105003 (2015)
7. T.K. Tseng, Y.S. Lin, Y.J. Chen, H. Chu, A review of photocatalysts prepared by sol-gel method for VOCs removal. Int. J. Mol. Sci. **11**, 2336 (2010)
8. Q. Lei, J. Guo, A. Noureddine, A. Wang, S. Wuttke, C.J. Brinker, W. Zhu, Sol–gel-based advanced porous silica materials for biomedical applications. Adv. Funct. Mater. **30**, (2020)
9. D. Bokov, A. Turki Jalil, S. Chupradit, W. Suksatan, M. Javed Ansari, I.H. Shewael, G.H. Valiev, E. Kianfar, Nanomaterial by sol-gel method: synthesis and application. Adv. Mater. Sci. Eng. **1** (2021)
10. A. Kumar, Y. Kuang, Z. Liang, X. Sun, Microwave chemistry, recent advancements, and eco-friendly microwave-assisted synthesis of nanoarchitectures and their applications: a review. Mater. Today Nano **11**, 100076 (2020)
11. E. Gabano, M. Ravera, Microwave-assisted synthesis: can transition metal complexes take advantage of this "green" method? Molecules **27**, 4249 (2022)
12. F. Fievet, J.P. Lagier, M. Figlarz, Preparing monodisperse metal powders in micrometer and submicrometer sizes by the polyol process. MRS Bull. **14**, 29 (1989)
13. F. Fievet, J. Lagier, B. Blin, B. Beaudoin, M. Figlarz, Homogeneous and heterogeneous nucleations in the polyol process for the preparation of micron and submicron size metal particles. Solid State Ionics **32–33**, 198 (1989)
14. J.I. Tracey, S. Aziz, D.M. O'Carroll, Investigation of the role of polyol molecular weight in the polyol synthesis of silver nanoparticles. Mater. Res. Express **6**, 115067 (2019)
15. Z.M. Abdulhamid, A.A. Dabbawala, T. Delclos, R. Straubinger, M. Rueping, K. Polychronopoulou, D.H. Anjum, Synthesis, characterization, and preliminary insights of ZnFe2O4 nanoparticles into potential applications, with a focus on gas sensing. Sci. Rep. **13**, 19705 (2023)
16. S. Antunes Filho, M.S. dos Santos, O.A.L. dos Santos, B.P. Backx, M.L. Soran, O. Opriş, I. Lung, A. Stegarescu, M. Bououdina, Biosynthesis of nanoparticles using plant extracts and essential oils. Molecules **28**, 3060 (2023)
17. P. Singh, Y.-J. Kim, D. Zhang, D.-C. Yang, Biological synthesis of nanoparticles from plants and microorganisms. Trends Biotechnol. **34**, 588 (2016)
18. X. Zhang, S. Yan, R.D. Tyagi, R.Y. Surampalli, Synthesis of nanoparticles by microorganisms and their application in enhancing microbiological reaction rates. Chemosphere **82**, 489 (2011)
19. M.A. Alghuthaymi, H. Almoammar, M. Rai, E. Said-Galiev, K.A. Abd-Elsalam, Myconanoparticles: synthesis and their role in phytopathogens management. Biotechnol. Biotechnol. Equip. **29**, 221 (2015)
20. B.S. Das, A. Das, A. Mishra, M. Arakha, Microbial cells as biological factory for nanoparticle synthesis. Front. Mater. Sci. **15**, 177 (2021)
21. A.H. Coons, The beginnings of immunofluorescence. J. Immunol. **87**, 499 (1961)
22. K. Zarschler, L. Rocks, N. Licciardello, L. Boselli, E. Polo, K.P. Garcia, L. De Cola, H. Stephan, K.A. Dawson, Ultrasmall inorganic nanoparticles: state-of-the-art and perspectives for biomedical applications. Nanomed. Nanotechnol. Biol. Med. **12**, 1663 (2016)
23. A. Escudero, A.I. Becerro, C. Carrillo-Carrión, N.O. Núñez, M.V. Zyuzin, M. Laguna, D. González-Mancebo, M. Ocaña, W.J. Parak, Rare earth based nanostructured materials: synthesis, functionalization, properties and bioimaging and biosensing applications. Nanophotonics **6**, 881 (2017)
24. S. Caspani, R. Magalhães, J.P. Araújo, C.T. Sousa, Magnetic nanomaterials as contrast agents for MRI. Materials (Basel) **13**, 2586 (2020)
25. X. Li, X.-N. Zhang, X.-D. Li, J. Chang, X. Li, X.-N. Zhang, X.-D. Li, J. Chang, Multimodality imaging in nanomedicine and nanotheranostics. Cancer Biol. Med. **13**, 339 (2016)
26. J.-P. Sylvestre, S. Poulin, A.V. Kabashin, E. Sacher, M. Meunier, J.H.T. Luong, Surface chemistry of gold nanoparticles produced by laser ablation in aqueous media. J. Phys. Chem. B **108**, 16864 (2004)

27. X.-D. Zhang, D. Wu, X. Shen, J. Chen, Y.-M. Sun, P.-X. Liu, X.-J. Liang, Size-dependent radiosensitization of PEG-coated gold nanoparticles for cancer radiation therapy. Biomaterials **33**, 6408 (2012)
28. R. Yan, Y. Zhao, H. Yang, X. Kang, C. Wang, L. Wen, Z. Lu, Ultrasmall Au nanoparticles embedded in 2D mixed-ligand metal–organic framework nanosheets exhibiting highly efficient and size-selective catalysis. Adv. Funct. Mater. **28** (2018)
29. J. Turkevich, Colloidal Gold. Part I Gold Bull. **18**, 86 (1985)
30. T. Ruks, K. Loza, M. Heggen, O. Prymak, A.L. Sehnem, C.L.P. Oliveira, P. Bayer, C. Beuck, M. Epple, Peptide-conjugated ultrasmall gold nanoparticles (2 nm) for selective protein targeting. ACS Appl. Bio Mater. **4**, 945 (2021)
31. A. Silvestri, V. Zambelli, A.M. Ferretti, D. Salerno, G. Bellani, L. Polito, Design of functionalized gold nanoparticle probes for computed tomography imaging. Contrast Media Mol. Imaging **11**, 405 (2016)
32. Y.-H. Chen, C.-S. Yeh, Laser ablation method: use of surfactants to form the dispersed Ag nanoparticles. Colloids Surf. A Physicochem. Eng. Asp. **197**, 133 (2002)
33. N. Kumar, K. Biswas, R.K. Gupta, Green synthesis of Ag nanoparticles in large quantity by cryomilling. RSC Adv. **6**, 111380 (2016)
34. Q.H. Tran, V.Q. Nguyen, A.-T. Le, Silver nanoparticles: synthesis, properties, toxicology, applications and perspectives. Adv. Nat. Sci. Nanosci. Nanotechnol. **4**, 033001 (2013)
35. S. Agnihotri, S. Mukherji, S. Mukherji, Size-controlled silver nanoparticles synthesized over the range 5–100 nm using the same protocol and their antibacterial efficacy. RSC Adv. **4**, 3974 (2014)
36. S.V. Kumar, A.P. Bafana, P. Pawar, A. Rahman, S.A. Dahoumane, C.S. Jeffryes, High conversion synthesis of <10 Nm starch-stabilized silver nanoparticles using microwave technology. Sci. Rep. **8**, 5106 (2018)
37. S. Iravani, Green synthesis of metal nanoparticles using plants. Green Chem. **13**, 2638 (2011)
38. N. Vigneshwaran, N.M. Ashtaputre, P.V. Varadarajan, R.P. Nachane, K.M. Paralikar, R.H. Balasubramanya, Biological synthesis of silver nanoparticles using the fungus aspergillus flavus. Mater. Lett. **61**, 1413 (2007)
39. A.K. Jha, K. Prasad, K. Prasad, A.R. Kulkarni, Plant system: nature's nanofactory. Colloids Surf. B Biointerfaces **73**, 219 (2009)
40. M. Fernández-Arias, M. Boutinguiza, J. Del Val, C. Covarrubias, F. Bastias, L. Gómez, M. Maureira, F. Arias-González, A. Riveiro, J. Pou, Copper nanoparticles obtained by laser ablation in liquids as bactericidal agent for dental applications. Appl. Surf. Sci. **507**, 145032 (2020)
41. M.B. Gawande, A. Goswami, F.-X. Felpin, T. Asefa, X. Huang, R. Silva, X. Zou, R. Zboril, R.S. Varma, Cu and Cu-based nanoparticles: synthesis and applications in catalysis. Chem. Rev. **116**, 3722 (2016)
42. A.M. Raspolli Galletti, C. Antonetti, M. Marracci, F. Piccinelli, B. Tellini, Novel microwave-synthesis of Cu nanoparticles in the absence of any stabilizing agent and their antibacterial and antistatic applications. Appl. Surf. Sci. **280**, 610 (2013)
43. Y. Feng, A. Wang, H. Yin, X. Yan, L. Shen, Reduction of 3-nitro-4-methoxy-acetylaniline to 3-amino-4-methoxy-acetylaniline catalyzed by metallic Cu nanoparticles at low reaction temperature. Chem. Eng. J. **262**, 427 (2015)
44. L. Jin, S.-P. Yang, Q.-W. Tian, H.-X. Wu, Y.-J. Cai, Preparation and characterization of copper metal nanoparticles using dendrimers as protectively colloids. Mater. Chem. Phys. **112**, 977 (2008)
45. N. Wolff et al., Ultrastructure and surface composition of glutathione-terminated ultrasmall silver, gold, platinum, and alloyed silver-platinum nanoparticles (2 Nm). Inorg. Chem. **62**, 17470 (2023)
46. W. Ji, W. Qi, S. Tang, H. Peng, S. Li, Hydrothermal synthesis of ultrasmall Pt nanoparticles as highly active electrocatalysts for methanol oxidation. Nanomaterials **5**, 2203 (2015)
47. A.K.A. Al Saidi et al., Facile synthesis and x-ray attenuation properties of ultrasmall platinum nanoparticles grafted with three types of hydrophilic polymers. Nanomaterials **13**, 806 (2023)

48. O.T. Fanoro, S. Parani, R. Maluleke, T.C. Lebepe, R.J. Varghese, N. Mgedle, V. Mavumengwana, O.S. Oluwafemi, Biosynthesis of smaller-sized platinum nanoparticles using the leaf extract of combretum erythrophyllum and its antibacterial activities. Antibiotics **10**, 1275 (2021)
49. M. Chen, S. Chen, C. He, S. Mo, X. Wang, G. Liu, N. Zheng, Safety profile of two-dimensional Pd nanosheets for photothermal therapy and photoacoustic imaging. Nano Res. **10**, 1234 (2017)
50. S. Tang, M. Chen, N. Zheng, Multifunctional ultrasmall Pd nanosheets for enhanced near-infrared photothermal therapy and chemotherapy of cancer. Nano Res. **8**, 165 (2015)
51. R. Jimenez-Sandoval, S. Pedireddy, K.P. Katuri, P.E. Saikaly, Facile biological-based synthesis of size-controlled palladium nanoclusters anchored on the surface of geobacter sulfurreducens and their application in electrocatalysis. ACS Sustain. Chem. Eng. **11**, 1100 (2023)
52. F.R. Veregue, C.T. Pereira da Silva, M.P. Moisés, J.G. Meneguin, M.R. Guilherme, P.A. Arroyo, S.L. Favaro, E. Radovanovic, E.M. Girotto, A.W. Rinaldi, Ultrasmall cobalt nanoparticles as a catalyst for PET glycolysis: a green protocol for pure hydroxyethyl terephthalate precipitation without water. ACS Sustain. Chem. Eng. **6**, 12017 (2018)
53. L.T. Lu, L.D. Tung, I. Robinson, D. Ung, B. Tan, J. Long, A.I. Cooper, D.G. Fernig, N.T.K. Thanh, Size and shape control for water-soluble magnetic cobalt nanoparticles using polymer ligands. J. Mater. Chem. **18**, 2453 (2008)
54. L.M. Parkes, R. Hodgson, L.T. Lu, L.D. Tung, I. Robinson, D.G. Fernig, N.T.K. Thanh, Cobalt nanoparticles as a novel magnetic resonance contrast agent—relaxivities at 1.5 and 3 Tesla. Contrast Media Mol. Imaging **3**, 150 (2008)
55. S.-W. Chou, Y.-H. Shau, P.-C. Wu, Y.-S. Yang, D.-B. Shieh, C.-C. Chen, In vitro and in vivo studies of FePt nanoparticles for dual modal CT/MRI molecular imaging. J. Am. Chem. Soc. **132**, 13270 (2010)
56. Y. Meng, D. Zhang, X. Chen, Z. Dai, X. Yao, P. Cui, D. Yu, G. Zhang, X. Zheng, FePt nanoparticles embedded in metal-organic framework nanoparticles for tumor imaging and eradication. ACS Appl. Nano Mater. **3**, 4494 (2020)
57. T.B. Nguyen, T.D. Nguyen, T.D. Tran, T.H.N. Thi, Laser-Induced synthesis of Au–Ag alloy nanoparticles in polyvinylpyrrolidone (C6H9NO)n solution. J. Clust. Sci. **26**, 1787 (2015)
58. O. Perlman, A. Borodetsky, Y. Kauffmann, Y. Shamay, H. Azhari, I.S. Weitz, Gold/copper@polydopamine nanocomposite for contrast-enhanced dual modal computed tomography-magnetic resonance imaging. ACS Appl. Nano Mater. **2**, 6124 (2019)
59. H.J. Kwon, K. Shin, M. Soh, H. Chang, J. Kim, J. Lee, G. Ko, B.H. Kim, D. Kim, T. Hyeon, Large-scale synthesis and medical applications of uniform-sized metal oxide nanoparticles. Adv. Mater. **30** (2018)
60. M. Parashar, V.K. Shukla, R. Singh, Metal oxides nanoparticles via sol-gel method: a review on synthesis, characterization and applications. J. Mater. Sci. Mater. Electron. **31**, 3729 (2020)
61. M.A. Rahman, M.T. Islam, A. Hossain, M.J. Singh, S.M. Albadran, M.S. Soliman, M. Samsuzzaman, Preparation of new flexible antenna based on sol–gel synthesized MgxCa(0.9-x)Zn0.10Fe2O4 nanoparticle for microwave imaging applications. J. Mater. Res. Technol. **20**, 3579 (2022)
62. A. Naghilou, O. Bomati-Miguel, A. Subotic, R. Lahoz, M. Kitzler-Zeiler, C. Radtke, M.A. Rodríguez, W. Kautek, Femtosecond laser generation of bimetallic oxide nanoparticles with potential x-ray absorbing and magnetic functionalities for medical imaging applications. Ceram. Int. **47**, 29363 (2021)
63. A. Khumaeni, G. Kurniawan, E. Hidayanto, Iron oxide nanoparticles produced by a low-energy Nd: YAG laser ablation technique and their application as contrast agent for magnetic resonance imaging. J. Nano Res. **80**, 65 (2023)
64. S. Stankic, S. Suman, F. Haque, J. Vidic, Pure and multi metal oxide nanoparticles: synthesis, antibacterial and cytotoxic properties. J. Nanobiotechnol. **14**, 73 (2016)
65. Y.-H. Lin, L.-J. Shen, T.-H. Chou, Y. Shih, Synthesis, stability, and cytotoxicity of novel cerium oxide nanoparticles for biomedical applications. J. Clust. Sci. **32**, 405 (2021)
66. H. Dong, Y.-C. Chen, C. Feldmann, Polyol synthesis of nanoparticles: status and options regarding metals, oxides, chalcogenides, and non-metal elements. Green Chem. **17**, 4107 (2015)

67. J. Li, C. Wu, P. Hou, M. Zhang, K. Xu, One-pot preparation of hydrophilic manganese oxide nanoparticles as T1 nano-contrast agent for molecular magnetic resonance imaging of renal carcinoma in vitro and in vivo. Biosens. Bioelectron. **102**, 1 (2018)
68. H. Zhang et al., Ultrasmall ferrite nanoparticles synthesized via dynamic simultaneous thermal decomposition for high-performance and multifunctional T1 magnetic resonance imaging contrast agent. ACS Nano **11**, 3614 (2017)
69. S. Gai, C. Li, P. Yang, J. Lin, Recent progress in rare earth micro/nanocrystals: soft chemical synthesis, luminescent properties, and biomedical applications. Chem. Rev. **114**, 2343 (2014)
70. A. Rabenau, The role of hydrothermal synthesis in preparative chemistry. Angew. Chemie Int. Ed. Engl. **24**, 1026 (1985)
71. A.S. Patil, A.V. Patil, C.G. Dighavkar, V.A. Adole, U.J. Tupe, Synthesis techniques and applications of rare earth metal oxides semiconductors: a review. Chem. Phys. Lett. **796**, 139555 (2022)
72. X. Kuang, H. Liu, W. Hu, Y. Shao, Hydrothermal synthesis of core-shell structured TbPO 4: Ce 3+ @TbPO 4: Gd 3+ nanocomposites for magnetic resonance and optical imaging. Dalt. Trans. **43**, 12321 (2014)
73. H. Yue et al., New class of efficient T2 magnetic resonance imaging contrast agent: carbon-coated paramagnetic dysprosium oxide nanoparticles. Pharmaceuticals **13**, 312 (2020)
74. Q. Zou et al., PVP-coated ultrasmall Nd-doped Gd 2 O 2 S nanoparticles for multimodal imaging. Mater. Chem. Front. **7**, 4109 (2023)
75. Q. Zhang, E.C. Pratt, R. Tamura, A. Ogirala, H.-T. Hsu, N. Farahmand, S. O'Brien, J. Grimm, Ultrasmall downconverting nanoparticle for enhanced Cerenkov imaging. Nano Lett. **21**, 4217 (2021)
76. S.L. Ho et al., Mono and multiple tumor-targeting ligand-coated ultrasmall gadolinium oxide nanoparticles: enhanced tumor imaging and blood circulation. Pharmaceutics **14**, 1458 (2022)
77. Y. Zhang, M. Wu, M. Wu, J. Zhu, X. Zhang, Multifunctional carbon-based nanomaterials: applications in biomolecular imaging and therapy. ACS Omega **3**, 9126 (2018)
78. S. Deng, Q. Zhang, Y. Nie, H. Wei, B. Wang, J. Huang et al., Sorption mechanisms of perfluorinated compounds on carbon nanotubes. Environ. Pollut. **168,** 138 (2012)
79. W. Ding, C. Lou, J. Qiu, Z. Zhao, Q. Zhou, M. Liang, Z. Ji, S. Yang, D. Xing, Targeted Fe-filled carbon nanotube as a multifunctional contrast agent for thermoacoustic and magnetic resonance imaging of tumor in living mice, nanomedicine nanotechnology. Biol. Med. **12**, 235 (2016)
80. M.T.H. Siddiqui, S. Nizamuddin, H.A. Baloch, N.M. Mubarak, D.K. Dumbre, Inamuddin, A.M. Asiri, A.W. Bhutto, M. Srinivasan, G.J. Griffin, Synthesis of magnetic carbon nanocomposites by hydrothermal carbonization and pyrolysis. Environ. Chem. Lett. **16**, 821 (2018)
81. V. Ali, H. Kefayati, M. Shafiee Ardestani, A. Pourahmad, Synthesis and evaluation of new magneto-fluorescent carbon dot based on manganese citrate for MRI imaging. Magn. Reson. Mater. Phys. Biol. Med. **37**, 139 (2023)
82. G. Borand, N. Akçamlı, D. Uzunsoy, Structural characterization of graphene nanostructures produced via arc discharge method. Ceram. Int. **47**, 8044 (2021)
83. H. Wang, M. Chhowalla, N. Sano, S. Jia, G.A.J. Amaratunga, Large-scale synthesis of single-walled carbon nanohorns by submerged arc. Nanotechnology **15**, 546 (2004)
84. S. Krishna Saraswat et al., Carbon quantum dots: a comprehensive review of green synthesis, characterization and investigation their applications in bioimaging, Inorg. Chem. Commun. **162**, 112279 (2024)
85. L. Cui, X. Ren, J. Wang, M. Sun, Synthesis of homogeneous carbon quantum dots by ultrafast dual-beam pulsed laser ablation for bioimaging. Mater. Today Nano **12**, 100091 (2020)
86. G.K. Yogesh, E.P. Shuaib, P. Roopmani, M.B. Gumpu, U.M. Krishnan, D. Sastikumar, Synthesis, characterization and bioimaging application of laser-ablated graphene-oxide nanoparticles (NGOs). Diam. Relat. Mater. **104**, 107733 (2020)
87. M. Llenas, S. Sandoval, P.M. Costa, J. Oró-Solé, S. Lope-Piedrafita, B. Ballesteros, K.T. Al-Jamal, G. Tobias, Microwave-assisted synthesis of SPION-reduced graphene oxide hybrids for magnetic resonance imaging (MRI). Nanomaterials **9**, 1364 (2019)

88. H. Eskalen, S. Uruş, S. Cömertpay, A.H. Kurt, Ş Özgan, Microwave-assisted ultra-fast synthesis of carbon quantum dots from linter: fluorescence cancer imaging and human cell growth inhibition properties. Ind. Crops Prod. **147**, 112209 (2020)
89. M. Zhu, G. Diao, Review on the progress in synthesis and application of magnetic carbon nanocomposites. Nanoscale **3**, 2748 (2011)
90. M. Deeney, S.A. Belhout, B.J. Rodriguez, G. Redmond, S.J. Quinn, Template-assisted synthesis of luminescent carbon nanofibers from beverage-related precursors by microwave heating. Molecules **24**, 1455 (2019)

Chapter 6
MRI Contrast Agents Based on Iron Oxide Nanoparticles: Design, Biocompatibility and Biodegradability

Mridula Sreedharan, Nikhitha Elizabeth Thomas, Yves Grohens, Nandakumar Kalarikkal, and Sabu Thomas

Abstract The most commonly used imaging modality and a non-invasive tool for cancer diagnosis is magnetic resonance imaging (MRI). Contrast agents (CAs) create tissue contrast by magnifying the longitudinal (T1) and transverse (T2) relaxation rates. The use of iron oxide nanoparticles (IONP) as contrast agents for MRI is a better alternative than using gadolinium-based contrast agents due to the problem of toxicity. Iron oxide nanoparticles are less toxic and more cytocompatible and posses' unique characteristics such as magnetic properties. The unique size and modifications made on IONP makes it an excellent contrast agent for MRI. The designing of IONP plays the most crucial role in determining the efficiency and biocompatibility. Various modifications such as surface coating and attaching biomolecules modify the size and biodistribution, which eventually aids in target tissue accumulation and efficient imaging. This chapter deals with efficient and smart designing methods which make the IONP highly biocompatible. Various biocompatibility testing methods and how IONP behaves in physiological environment are also discussed. The biodegradation or clearance from the body is also very important when considering the designing

M. Sreedharan (✉) · N. E. Thomas · N. Kalarikkal (✉) · S. Thomas
International and Inter University Centre for Nanoscience and Nanotechnology, Mahatma Gandhi University, Kottayam, Kerala, India
e-mail: mridulanikhil@gmail.com; mridula.sreedharan@univ-ubs.fr

N. Kalarikkal
e-mail: nkkalarikkal@mgu.ac.in

N. E. Thomas
e-mail: nikkyelza99@gmail.com

S. Thomas
e-mail: sabuthomas@mgu.ac.in

M. Sreedharan · Y. Grohens
Université Bretagne Sud, South Brittany, Lorient, France
e-mail: yves.grohens@univ-ubs.fr

N. Kalarikkal
School of Pure and Applied Physics, Mahatma Gandhi University, Kottayam, Kerala, India

N. Kalarikkal et al. (eds.), *Multimodal Biomedical Imaging Techniques*,
Biological and Medical Physics, Biomedical Engineering,
https://doi.org/10.1007/978-981-96-1124-9_6

of IONP. Depending on the size and coating applied to IONP, the fate of the particle is determined. This chapter also explains the biodegradation route of the iron oxide nanoparticles when administered in vivo.

6.1 Introduction

MRI (magnetic resonance imaging) is a well-established imaging technique in various areas of medicine. With the great advantage of not using ionizing radiation, which avoids the biological damage associated with other three-dimensional imaging techniques like CT and CBCT, it has become essential for the non-invasive diagnosis of soft tissue diseases [1]. Contrast agents are essential for highlighting particular pathological conditions or improving the contrast between various tissues. Specific features such as chemical composition, which includes the presence or absence of metal atoms, administration route, magnetic properties, effect on the magnetic resonance image, biodistribution and imaging applications are used to categorize contrast agents for magnetic resonance imaging [2]. Because of their special qualities, including their biocompatibility, surface functionalization flexibility and ability, and effective contrast features, magnetic iron oxide nanoparticle (IONP) contrast agents may be among the most successful examples of inorganic nanoparticles used in medicine [3]. IONPs in particular, since they effectively reduce transverse relaxation durations, have been extensively studied as a T2 contrast agent for MRI. Notably, IONPs have proven to have flexible surface chemistry, extended blood half-lives and low toxicity [4]. Functionalized with epithelial growth factor receptor antibodies, aptamers, or short peptides such as RGD, IONPs have been suggested for the detection of several cancers, such as those of the breast, stomach, colon, kidney, liver, or brain [5]. The design of IONP-based contrast agents for MRI has undergone significant advancements in recent years, driven by the quest for improved sensitivity, specificity and biocompatibility. In this chapter, we explore the design principles of IONPs.

Magnetic Resonance Imaging diagnostic method can be used with and without contrast agents. But only by using a contrast agent, the specificity and sensitivity and thus tissue discrimination can be improved. So, there is a high demand for contrast agents for MRI and the materials used as contrast agents would adhere to a set of regulations as it is used in clinics. While designing a contrast agent, the efficiency and toxicity issues are to be considered. The contrast agents can be classified based on their chemical configuration, magnetic properties, occurrence of metal atoms, method for administration, consequence on the image, tissue distribution, etc. [2]. The mechanism by which contrast agents (CAs) enhance tissue visibility is that the relaxation rates of nearby water molecules will be enhanced (T1 is the longitudinal and T2 is the transverse relaxation). Positive CAs refers to T1 CA which aids in brighter contrast enhancement while negative CAs refers to T2 CA which helps in decreasing the signal intensity to create darker contrast enhancement. Gadolinium-based contrast agents which were commonly used in clinics for MRI imaging have shown to have toxic effects leading to nephrogenic systemic fibrosis, fractures or even death. Thus, iron oxide nanoparticles have become a potential candidate as contrast agent owing to its biocompatibility, biodegradability and low toxicity [6].

6.2 Designing of IONP as MRI Contrast Agent

Super paramagnetic IONP can reduce the T2 relaxation time and thus used as T2 contrast agents. Ultra small SPION can be used as T1 contrast agents. The relaxivity of the IONP determines whether it can be used as T1 or T2 contrast agent in MRI. Thus, the surface design significantly determines the fate of IONP as contrast agent. Size, magnetic properties, crystal structure and shape of IONPs account for their contrast capacity while surface modifications accounts for their performance in vivo. Usage of appropriate surface ligands aids in accumulation of these nanoparticles in tumour for efficient imaging of tumour tissues through MRI. The designing determines the imaging sensitivity of the contrast agent in MRI. To enhance the contrast of tumour images and aid in efficient imaging, certain features should be taken into consideration while designing. They are (i) to achieve increased concentration of the NPs in the tumour tissue and (ii) ability to adjust the imaging signal or imaging mode. Ligands which help in accumulation of IONP in tumour tissues are (Refer Fig. 6.1).

A. Proteins
B. Peptides
C. Antibodies
D. Aptamers

Responsive surface ligands are:

E. Glutathione
F. Light
G. pH
H. Enzymes [6]

6.3 Synthesis of IONPs

6.3.1 Co-precipitation

The most popular method for creating magnetic iron oxide nanoparticles (IONPs) is co-precipitation because it can be done easily in many labs using inexpensive, eco-friendly precursors and simple experimental procedures. The process is initiated by adding a base to an aqueous solution containing ferrous and ferric ions. However, this process frequently produces particles that are highly polydisperse and

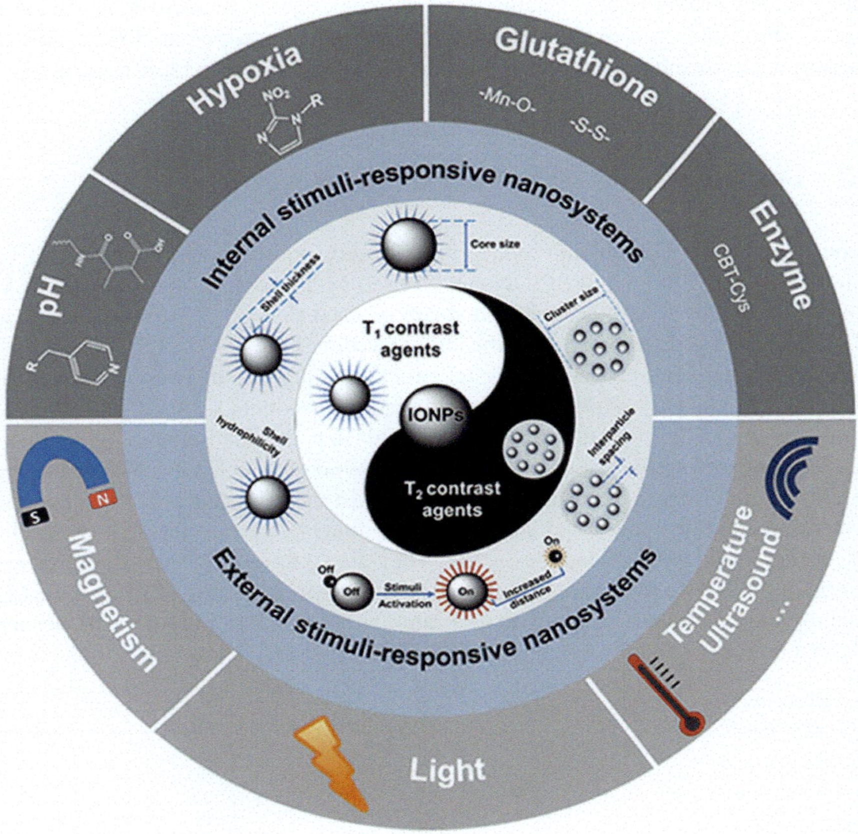

Fig. 6.1 Schematic diagram for designing criteria for IONP for MRI contrast agents Reproduced from [6] licensed under CC-BY-NC-ND 4.0

weakly magnetized, which makes stability and application difficult. Akaganeite and goethite are examples of intermediate phases that have been found during the production process; these phases are regulated by mixing efficiency and pH [7]. Research has demonstrated that changes in reaction circumstances might result in distinct paths for particle creation, which makes the mechanism more difficult to comprehend. Controlled mixing conditions are crucial because rapid phase shifts during synthesis make characterization and repeatability challenging. Recent research has concentrated on creating stabilizing agents for IONPs and using flow reactors to learn more about the first phases of co-precipitation synthesis. This has allowed for the continuous synthesis of extremely stable, monodisperse nanoparticles with repeatable characteristics [8].

6.3.2 Thermal Decomposition

Thermal decomposition of organometallic precursors in organic solvents combined with surfactant capping agents in an inert environment is a versatile synthesis platform that may be used to create a variety of oxide-based nanoparticles, including magnetic ones. These techniques have been used to control the nanoparticle size, polydispersity and shape. However, these reactions are usually conducted in an oxygen-free environment because of safety concerns over the high temperatures of the organic liquid and vapour phases used to perform them [9]. These pathways have been found to create a non-magnetic Wüstite phase at first, and then later on, as a result of oxidation upon exposure to oxygen, the magnetic phases of magnetite and maghemite [10]. Studies using high-resolution transmission electron microscopy have shown the existence of polycrystalline particles and defects, which may affect the magnetic and functional characteristics of the nanoparticles [11]. As a result, thermal-oxidative post-synthesis treatments have been devised to improve the magnetic characteristics of nanoparticles; they work particularly well for smaller particles, but not as well for larger ones or longer processing durations. These results imply that a major obstacle to the consistent production of oxide-based nanoparticles is the lack of oxygen in the thermal decomposition reaction media [9].

6.3.3 Microemulsions

Due to the effective control over the size, shape and content of the resulting nanoparticles, microemulsion procedures may be advantageous. Water-in-oil (W/O) microemulsions are made up of surfactant-stabilized aqueous microdomains dispersed across a continuous oil phase. These function as nanoreactors where a co-precipitation chemical reaction produces the necessary nanoparticles [12]. Chin and Yaacob reported synthesizing magnetic iron oxide nanoparticles (NPs) smaller than 10 nm using w/o microemulsion. Moreover, the particles generated by the microemulsion technique had higher saturation magnetization and smaller sizes than those produced by Massart's procedure. However, many washing procedures and further stabilizing treatments are often required for the generated NPs to aggregate, even in the presence of surfactants [13].

6.3.4 Surface Modification of IONP's

Surfactants and polymers are coated on the magnetic particles to improve their biocompatibility, avoid agglomeration and provide functionality. Nevertheless, the coatings may interact with the magnetic core's surface atoms to create a magnetically disordered layer, which would lower the overall magnetic phase concentration

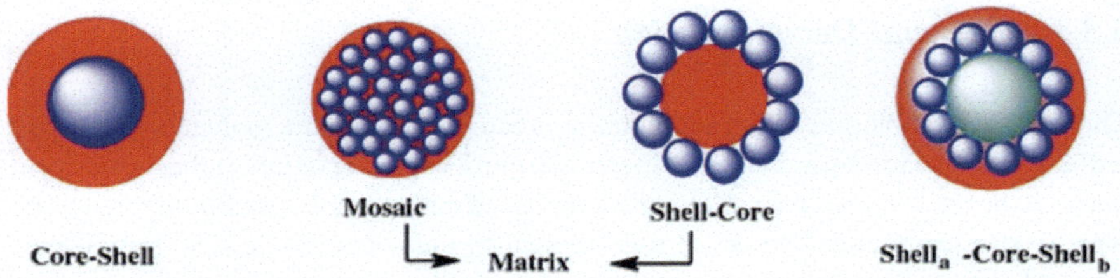

Fig. 6.2 Magnetic iron oxide NPs functionalized by the typical structure of organic materials (assuming iron oxide NPs were always considered as the core). Reproduced from [13]

which is a crucial characteristic in many applications [14]. The potential uses of organic compounds coated on iron oxide nanoparticles (NPs) are limitless since they combine all of the properties of organic molecules with the magnetic characteristics of iron oxides. The three primary structural types that these functionalized NPs usually display are matrix, shell-core–shell and core–shell (Fig. 6.2). Iron oxide cores encased in organic shells characterize core–shell formations, which permit varying compositions of cores (such as magnetite or maghemite) and shell materials. Matrix structures, including shell-core and mosaic, are composed of iron oxide nanoparticles (NPs) with an iron oxide shell and an organic core joined by chemical bonds. The flexible features of polymers make them especially popular for matrix embedding. Layer-by-layer method creates shell-core–shell structures, which are composed of organic shells and functionalized iron oxide nanoparticles with tunable characteristics [13].

6.4 Designing of Smart Contrast Agents

Depending on the modifications applied to IONP, these particles can be used for various applications of MRI imaging. Dual mode contrast agents can be used as T_1Wl and T_2Wl contrast agents. T_2WI contrast agents have the ability to shorten transverse relaxation time and enhance signal intensity. Thickness of the surface coating of the magnetic nanoparticles, their radius and saturation magnetization affect their transverse relaxivity r_2 as contrast agents. The size of the nanoparticles determines their magnetic properties and fate inside the body. Hydrodynamic diameters are usually considered for designing (Table 6.1).

IONP can also serve as T_1W contrast agents when they accelerate longitudinal relaxation rate. The ratio of longitudinal relaxivity (r_1) to transverse relaxivity (r_2) determines whether the CA is suitable to be used as T1 CA or T2 CA. The IONP can be used as T1 CA if r_2/r_1 is less than 5 and the value of r_1 is high. It is challenging to design IONP as T1 WI but when certain conditions are met, the value of r_2/r_1 can be reduced. Decreasing the size, crystallinity and clustering can end up in using IONP as T1 CA [6, 15].

Table 6.1 Effect of size of IONP as T_2 CA

Hydrodynamic diameter of the IONP	Distribution inside the body
Less than 5 nm	Pass through kidney and discharged through urine
Greater than 100 nm	Quickly cleared from circulation by phagocytosis
Greater than 80 nm	Stay in reticuloendothelial system and can be imaged easily

6.5 IONP Contrast Agents Based on Stimuli Response

Activation of contrast signal in response to various physiological or pathological signals can be achieved for better sensitivity and precision. The stimuli can be classified as internal and external stimuli where light and magnetic field are external and hypoxia, pH, enzymes, etc. are internal stimuli.

Tumour tissues have an acidic microenvironment when compared to normal tissues which can be utilized as stimuli for activating contrast signal. By changing the morphology or size or surface of IONP, the relaxation properties can be deployed. Thus, the IONP can be used as dual modal imaging for cancer diagnosis. When pH changes, there can occur changes in chemical bonds or degradation of metal oxides which can be considered as stimuli. When the size of the IONP is affected by change in pH, this affects the contrast imaging signal. As an example, IONP clusters when entering tumour cells will get disassembled giving rise to strong T1 signal whereas the disassembled larger IONP cluster will be taken up by normal cells giving strong T2 signals. This can improve the contrast between normal and tumour tissue. The disassembling of IONP in response to pH change can be brought by various designs like:

1. Small hepatocellular carcinoma could be imaged with high quality when IONP clusters were attached with a pH-sensitive DNA modifier linker with I motif.
2. IONP cross linked by small molecule ligands used to image tumour cells showed a disassembly of the nanoparticle system at lower tumour pH giving rise to hydrophilic ultra small iron oxide nanoparticles with strong T1 weighted contrast signal.
3. Enclosing IONP in hydrogels or polymeric micelles which degrades at lower pH releasing the IONP [6].

Long-term metabolic effects of these CAs in the liver should also be analysed well before applied to clinics.

Hypoxia-induced imaging signal has also been promising in research. Since tumour environment has a hypoxic condition, the IONP behave differently and thus contrast between tissues can be visualized in MRI. Major examples of designing IONP contrast agents responding to hypoxia are with addition of nitroimidazole derivatives which has a bioreductive property. IONP conjugated with oxygen-sensitive metronidazole ligand showed selective accumulation of nanoparticles in tumour tissue.

Glutathione (GSH) is a reducing agent which is produced in cancer cells in high quantities as a result of oxidative stress formed by ROS species. Decomposition of IONP clusters can occur due to breaking of GSH-sensitive chemical bonds. This breaking can induce significant change in imaging signal. Clustering of iron oxide nanoparticles in the tissue also can surge T2 enhancement. Core–shell structure of the contrast agent will be the potential design for stimuli responsive imaging. For example, a GSH-sensitive polymer as the shell and an omega-3 fatty acid modified IONP as the core, and this contrast agent when reaching glioma cells induced aggregation of IONP thereby increasing r2 value. Here the disulphide linkages are cleaved by GSH when exposed to tumour environment. IONPs can used to detect liver metastases. GSH-responsive hyaluronic acid-coated IONPs were able to detect small liver metastases by inducing aggregation of IONP in the target site [6, 16, 17].

Enzyme-sensitive IONPs are also a smart design to detect tumours. In the tumour microenvironment, certain enzymes will be elevated which can be made use as a stimulus for IONP induced imaging. Enzymes like matrix metalloproteinase, trypsin and caspase-3 will be enhanced in tumour tissues, and the enzyme detecting ligands can be attached to IONP through lick chemistry reactions. Tumour apoptosis can be detected by using caspase-3 responsive IONP. While designing enzyme responsive IONP, certain conditions have to be considered for efficient imaging.

a. Each tumour type and stage will have different levels of enzymes and also other diseases or physical state of the patient can also cause elevation in the level of certain enzymes. So, it is necessary to customize nanoparticles based on the above factors.
b. The liver contains substantial number of hydrolases, which can make the enzyme-responsive nanoprobe lose its potentiality (Fig. 6.3).

Light-responsive IONPs work when light alters the surface modifications changing relaxation properties. Magnetic-responsive CAs function with the super-paramagnetic property of the IONP when applied to a magnetic field. External static

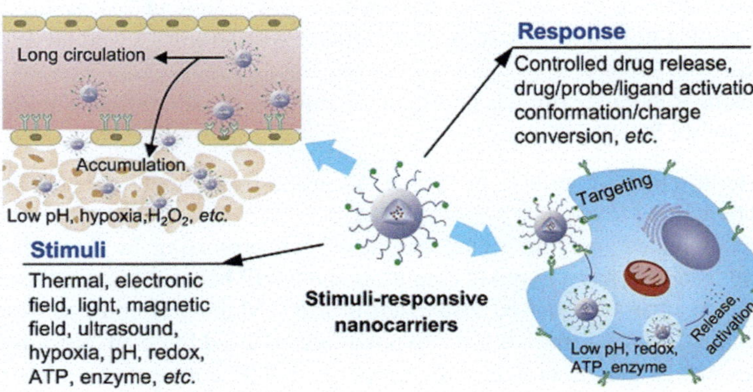

Fig. 6.3 Schematic representation of stimuli-responsive nanocarriers. Reproduced from [43]

magnetic field can be used to increase uptake of IONP in tumour sites. The limitation of this CA is its inefficiency in imaging deeper tissues [6, 16].

6.6 Coating of IONP to Improve Biocompatibility

The coating can be done to improve the biocompatibility of IONP. Two types of coating can be applied: inorganic and organic coating. Inorganic coating controls the release of iron ions and reactivity of the IONP. The chemically inert coating interacts with the surrounding and thus the behaviour of the inorganic shell to the physiological environment determines the reactivity of the IONPs making it more biocompatible. Silica and gold which are bio-inert materials can be used to coat IONP. Gold which is considered as a highly biocompatible material can make the IONP non-toxic to the body. Gold-coated IONP showed cytotoxicity 40% less than bare IONP. Silica which is regarded as safe material by FDA when coated to IONP makes it less toxic. It does not trigger the secretion of pro-inflammatory cytokines. Small sized silica-coated nanoparticles (size around 5 nm) show very high biocompatibility [18, 19].

Organic shell can be created by coating with proteins and polysaccharides like dextran, hyaluronic acid and bovine serum albumin. Dextran-coated IONP showed no cytotoxicity nor any superoxide production. Synthetic polymers like poly vinyl alcohol and poly vinyl pyrrolidone can be coated on IONP to reduce toxicity [18, 20].

6.7 Biocompatibility

The behaviour of IONP in physiological system should be well understood before designing IONP-based contrast agents. As soon as IONP enters the body, protein corona formation takes place where serum proteins adsorb to IONP thus decreasing its circulation time in blood and accumulation time in lesions. This can be overcome by attaching IONP with polymer like poly ethylene glycol (PEG) or zwitter ionic small molecules. Targeted MRI with enhanced contrast efficiency can be achieved if biomolecules like antibodies, aptamers, proteins or peptides are attached to IONP [21, 22].

6.7.1 Mechanism of Internalization of IONP

IONP when administered into physiological system enters into cells via endocytosis pathways depending on cell type and physicochemical features of the IONP. Pinocytosis and phagocytosis are the main mechanism by which these nanoparticles are engulfed by the cells. Opsonization is the first process that happens to any foreign material that enters bloodstream, where in opsonin proteins like IgG tag the

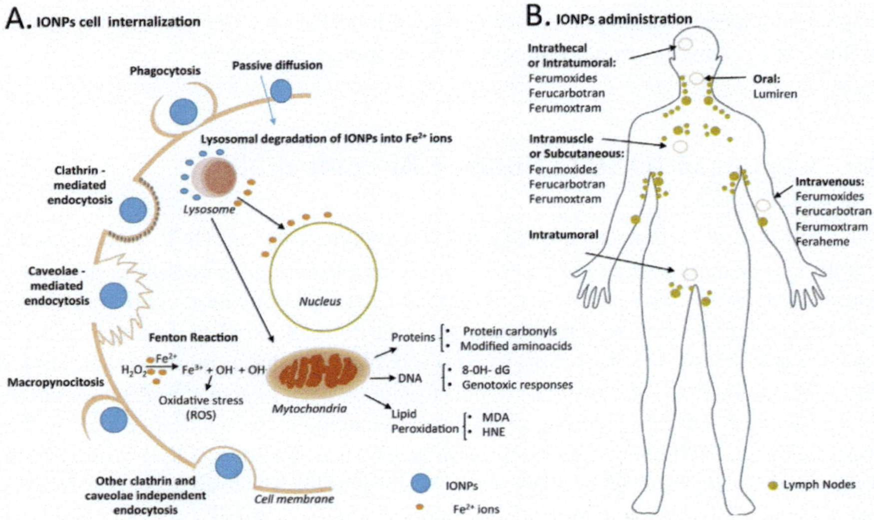

Fig. 6.4 **a** Different internalization pathways of IONPs in mammalian cells. **b** IONP administration in the human body, such as intrathecal, intratumorally, intravenous and intramuscular or subcutaneous methods. Reproduced from [24]. Licenced under creativecommons.org/licenses/by/3.0

nanoparticles to eliminate by phagocytes. This binds them to cell surface which is then engulfed into cells. The iron of the NP is stored in the blood cells, which is then eliminated through faeces and the coating is degraded and eliminated through kidneys [20–23]. Figure 6.4a shows various paths of internalization of IONP. Figure 6.5 shows electron microscopic images of the nanoparticles entering endothelial cells at various time points.

6.7.2 Cytotoxic Effects of Iron Oxide Nanoparticles

Bare iron oxide nanoparticles create reactive oxygen species (ROS) due to their high chemical reactivity, Fig. 6.7. This can lead to genotoxicity where damage to DNA occurs as a result of Fenton reaction. This owes to the small size of IONP, see Fig. 6.6. The accumulation of IONP can have effect on cell morphology like change in actin cytoskeleton due to cellular stress. This affects the cell migration, growth and differentiation. The toxicity can be reduced by coating the nanoparticles, which results in a smaller number of reactive oxidative sites [26, 27].

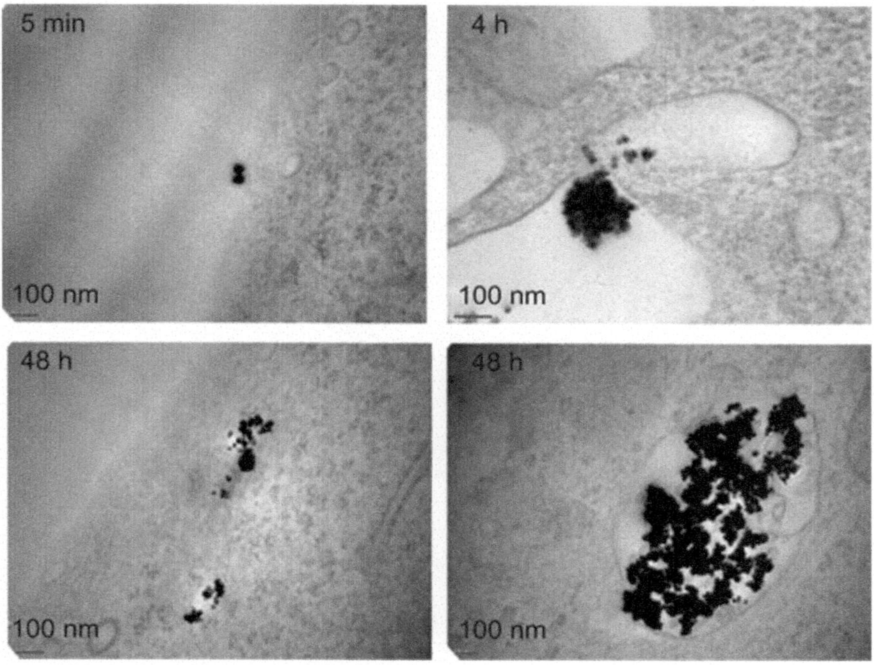

Fig. 6.5 Subcellular localization of nanoparticles. Transmission electron microscopy was performed on human umbilical vascular endothelial cells (HUVEC) exposed to nanoparticles for the indicated times. Reproduced from [25]

6.7.3 Biocompatibility Analysis Methods

The crucial features of the IONP that results in cytotoxicity are its small size, shape, structure and charge of the surface of the particles, solubility, composition and aggregation nature. A thorough understanding of the pharmacokinetics of the IONP can help in designing the contrast agents. IONP with size less than 10 nm will be removed quickly through kidneys. IONPs with size greater than 200 nm will be distributed to liver, spleen and bone marrow as part of the immune reactions. Super paramagnetic iron oxide nanoparticles are available in market as MRI contrast agents as they posseses the ability to enhance imaging due to lymphotropic properties and the ability to cause magnetic field gradient in tissues. They are used to image cancer, cardiovascular and cerebrovascular pathologies, autoimmune disorders, transplantation tracking, etc. [29, 30].

The first barrier to IONP when entering the body is endothelium. The interaction of IONP with endothelium can be studied using cell culture and it has been observed that within few hours of incubation, the IONP will be internalized in endothelial cells and aggregate in the cytoplasm even with zero coating or surface modifications

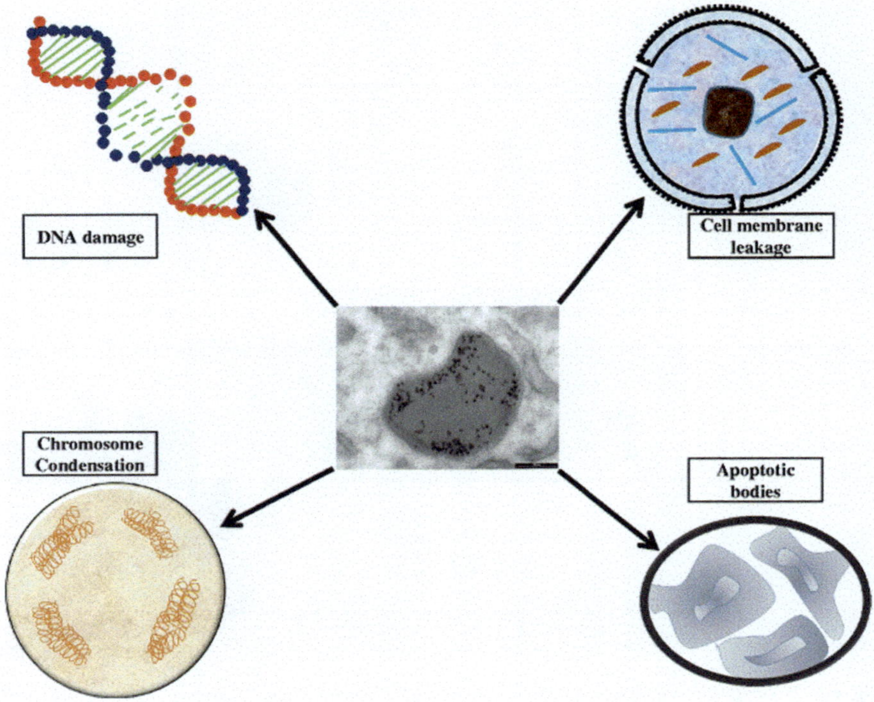

Fig. 6.6 Schematic representation of cytotoxicity of IONP. Reproduced from [28]

of the nanoparticles. The effect on cell functioning can be analysed using various methods [31].

6.7.3.1 MTT Assay

MTT assay measure the viability of cells when incubated with the nanoparticles. It has been observed that negligible toxicity was found within 24 h of incubation. But cell death occurs after 48 h of incubation which can be analysed using LDH leakage assay. After 24 h of incubation, toxicity increases due to necrosis. Necrotic mechanism was observed by LDH leakage into extracellular fluid and decrease in phosphorylation of Akt protein. Apart from this, ROS production was enhanced after 24 h of exposure which is an indication of oxidative stress. Thus, bare iron oxide nanoparticles can cause high level of cytotoxicity after 24 h of exposure as a result of necrosis and oxidative stress [32, 33].

IONP capped by biomolecules like aminoacid and coated with biopolymers like PEG shows zero toxicity in MTT assay. This proves that surface modification and coating can improve the biocompatibility of IONP [34].

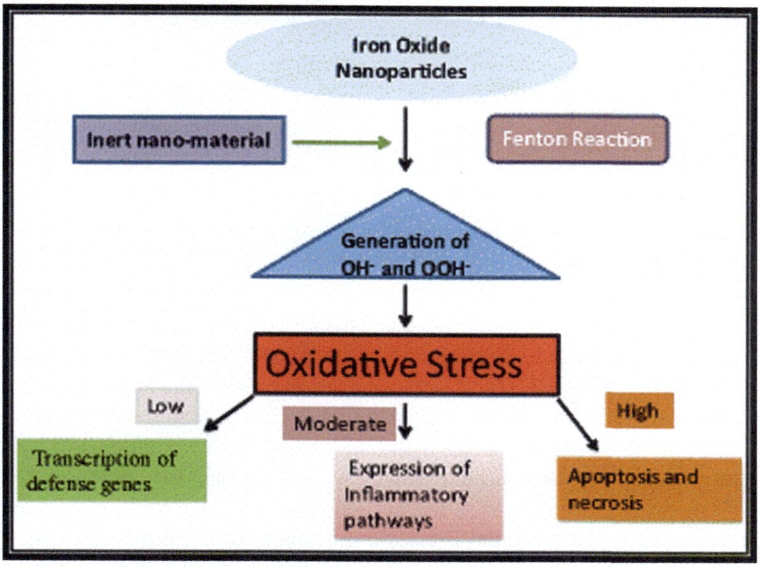

Fig. 6.7 Stages of ROS generation by Iron oxide nanoparticles and their stages. Reproduced from [28]

6.7.3.2 Hemolysis Assay

This evaluates the level of blood cell disruption by the material exposed to. IONP can cause red blood cell lysis but surface modified and coated IONP causes negligible hemolysis. Iron oxide nanoparticles normally show blood toxicity as per the literature. But organic coating can reduce the toxicity to a negligible level which can be considered as a mandatory design criterion for using IONP in vivo. In a study conducted for increasing the blood circulation time of magnetic nanoparticles, the particles were encapsulated into red blood cells by hypotonic dialysis method [35–37].

6.7.4 In Vivo Analysis

Blood analysis after injection of the nanoparticles helps to analyse the effects of nanoparticles in the physiological system. Blood analysis will show the percentage of elimination of nanoparticles from the circulatory system. It was observed that after 72 h, almost 75% of particles were eliminated and it can also be detected in the urine. Augmentation of white blood cells was observed while no effect on red blood cells. Inflammation was observed in lungs, liver and kidneys but zero effect on general health of the organism. Thus, bare IONP of small size causes long-term toxicity and accordingly surface coating or ligand attachment is necessary for its use in clinics [23].

6.7.5 *Histomorphometric Analysis*

Histomorphometry study is a quantitative analysis of the microscopic organization and structure of tissue by computer assisted analysis of images. This kind of analysis gives an in-depth knowledge of the interaction of nanoparticles with various organs. For prolonging the blood circulation time and to enhance colloidal stability, IONP can be attached to a biomolecule as capping agent and can be coated with biopolymers like polyethylene glycol (PEG). This increases the blood circulation time of the nanoparticles. Accumulation in liver can occur after 2 h of administration. When the particle size is large, accumulation in liver happens soon. Most of the modified IONP will be eliminated from blood stream after 24 h of administration, and the key organs responsible for this elimination are the kidneys Fig. 6.8. Degradation of IONP can occur in spleen [23, 38]. In short, IONP do not result in organ damage and will be eliminated from the body within 72 h of administration. But high dose (500 mg Fe/kg) can cause mild damage to liver and spleen [39].

6.8 Biodegradability

Superparamagnetic iron oxide nanoparticles (SPION) with appropriate coating have been used as MRI contrast agents. The coating should be biodegradable and should not hinder the magnetic resonance signal from the SPIONs for the purpose of imaging. Contrast agents with size around 50 to 200 nm will be easily cleared from body but still can linger around which can hinder with the quality of further scans. Large size particles of size around 1 μm can cause microvessel blockage [41].

Biodistribution of iron oxide nanoparticles need to be understood when new designs are implemented. Even though the primary organs where IONP accumulate are the liver, lymph nodes and spleen, various coatings and modifications can make them accumulate in brain, heart or lungs. Biodegradability of magnetic iron oxide nanoparticles is high when compared to others, which accounts for its low toxicity. The mechanism of biodegradation can be compared with ferritin metabolism where lysosomal enzymes aid in release of iron ions. This is the process of biodegradation of IONP happening inside the physiological system [42]. The steps in degradation of IONP inside the kidney are as follows:

(i) The nanoparticles reach glomeruli in nephrons (basic functional unit of kidney) through the renal hilum artery
(ii) Podocyte cells enclose the capillaries in the glomeruli

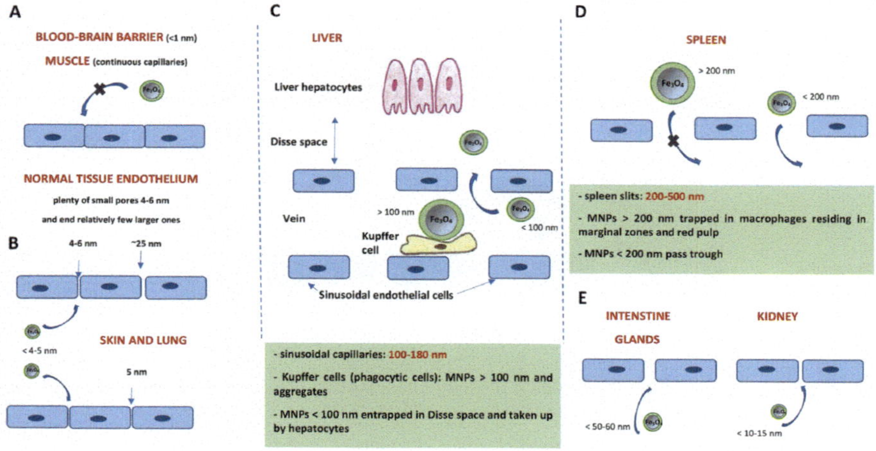

Fig. 6.8 Size-dependent physiological barriers against magnetic nanoparticles blood circulation. Even the smallest magnetic nanoparticles do not cross the blood–brain barrier and the blood vessel epithelium in the muscles, as long as the tissues are not cancerous (**a**). Nanoparticles with a diameter of about 5 nm or less are able to penetrate through small pores such as in the epithelium of the lungs and skin (**b**). Sinusoidal capillaries in the liver are fenestrated (100–180 nm) and lined with the Kupffer cells which quickly uptake large nanoparticles (> 100 nm) or agglomerates tagged with opsonins, whereas smaller nanoparticles (< 100 nm) are captured and hidden in the Disse space from where they can be collected by hepatocytes (**c**). Nanoparticles larger than about 200 nm get trapped in the marginal zones and the red pulp of the spleen, where they are absorbed by splenic macrophages (**d**). In the kidneys, nanoparticles with $d_H < 10$–15 nm in diameter are filtered out, whereas nanoparticles with $d_H < 50$–60 nm can penetrate through the pores in the intestines and glands (**e**). Reproduced from [40]

(iii) NPs escape through the openings or spaces in between the podocytes and reach the urine through ureter (Fig. 6.9) [42].

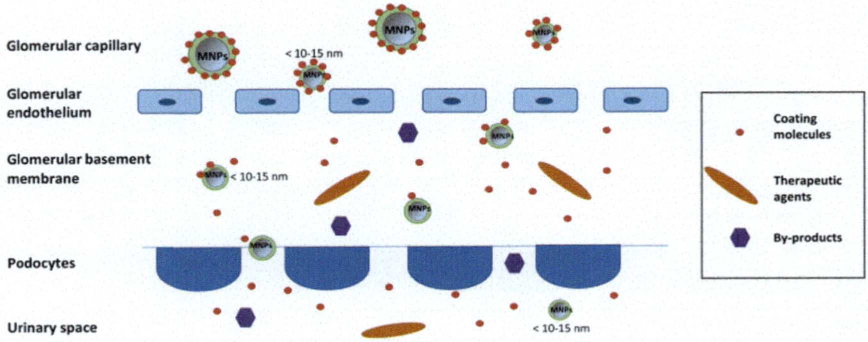

Fig. 6.9 Degradation pathway of coated magnetic nanoparticles in the kidney. Reproduced from [42] Licenced under creativecommons.org/licenses/by-nc/4.0/deed.en

6.9 Conclusion

Even though some superparamagnetic iron oxide nanoparticles are approved by FDA for clinical use, they were withdrawn from market as more efficient alternatives were available. Since MRI is a non-invasive technique for diagnosis of dreadful diseases like cancer, the type of contrast agent used determines the efficiency of the method. A single IONP design cannot meet all the needs, but efficient designing can create highly sensitive contrast agents. MRI technique is the most demanded diagnostic tool for early detection of cancer, eliminating the need for unnecessary expensive treatments and monitoring the treatment of tumours. Understanding the interaction of the IONP with physiological environment deeply is the greatest challenge in applying nanoparticles as contrast agents and in other biomedical applications. The correct understanding of the patient condition and disease etiology on pharmacokinetics of nanoparticles can help in designing contrast agents for imaging. The particle size and coating determine the pharmacokinetics of the IONP inside the body. The size of 15–100 nm would be the optimal size for in vivo usage. Deep studies on the influence of size of the designed nanoparticles on the way of clearance and pharmacokinetics should be performed.

References

1. R. Reda, A. Zanza, A. Mazzoni, A. Cicconetti, L. Testarelli, D. Di Nardo, An update of the possible applications of magnetic resonance imaging (MRI) in dentistry: a literature review. J. Imaging **7**(5), 75 (2021)
2. Y.D. Xiao, R. Paudel, J. Liu, C. Ma, Z.S. Zhang, S.K. Zhou, MRI contrast agents: classification and application. Int. J. Mol. Med. **38**(5), 1319–1326 (2016)

3. M. Salehipour, S. Rezaei, J. Mosafer, Z. Pakdin-Parizi, A. Motaharian, M. Mogharabi-Manzari, Recent advances in polymer-coated iron oxide nanoparticles as magnetic resonance imaging contrast agents. J. Nanopart. Res. **23**, 1–35 (2021)
4. M. Jeon, M.V. Halbert, Z.R. Stephen, M. Zhang, Iron oxide nanoparticles as T1 contrast agents for magnetic resonance imaging: fundamentals, challenges, applications, and prospectives. Adv. Mater. **33**(23), 1906539 (2021)
5. A. Avasthi, C. Caro, E. Pozo-Torres, M.P. Leal, M.L. García-Martín, Magnetic nanoparticles as MRI contrast agents, in *Surface-Modified Nanobiomaterials for Electrochemical and Biomedicine Applications* (2020), pp. 49–91
6. Z. Li, R. Bai, J. Yi, H. Zhou, J. Xian, C. Chen, Designing smart iron oxide nanoparticles for MR imaging of tumors. Chem. Biomed. Imaging **1**(4), 315–339 (2023)
7. T. Ahn, J.H. Kim, H.M. Yang, J.W. Lee, J.D. Kim, Formation pathways of magnetite nanoparticles by coprecipitation method. J. Phys. Chem. C **116**(10), 6069–6076 (2012)
8. M.O. Besenhard, A.P. LaGrow, A. Hodzic, M. Kriechbaum, L. Panariello, G. Bais, A. Gavriilidis et al., Co-precipitation synthesis of stable iron oxide nanoparticles with NaOH: new insights and continuous production via flow chemistry. Chem. Eng. J. **399**, 125740 (2020)
9. M. Unni, A.M. Uhl, S. Savliwala, B.H. Savitzky, R. Dhavalikar, N. Garraud, C. Rinaldi et al., Thermal decomposition synthesis of iron oxide nanoparticles with diminished magnetic dead layer by controlled addition of oxygen. ACS Nano **11**(2), 2284–2303 (2017)
10. R. Chalasani, S. Vasudevan, Form, content, and magnetism in iron oxide nanocrystals. J. Phys. Chem. C **115**(37), 18088–18093 (2011)
11. E. Wetterskog, C.W. Tai, J. Grins, L. Bergstrom, G. Salazar-Alvarez, Anomalous magnetic properties of nanoparticles arising from defect structures: topotaxial oxidation of Fe1−x O| Fe3− δO4 core| shell nanocubes to single-phase particles. ACS Nano **7**(8), 7132–7144 (2013)
12. M. Salvador, G. Gutiérrez, S. Noriega, A. Moyano, M.C. Blanco-López, M. Matos, Microemulsion synthesis of superparamagnetic nanoparticles for bioapplications. Int. J. Mol. Sci. **22**(1), 427 (2021)
13. W. Wu, Q. He, C. Jiang, Magnetic iron oxide nanoparticles: synthesis and surface functionalization strategies. Nanoscale Res. Lett. **3**, 397–415 (2008)
14. Y. Yuan, D. Rende, C.L. Altan, S. Bucak, R. Ozisik, D.A. Borca-Tasciuc, Effect of surface modification on magnetization of iron oxide nanoparticle colloids. Langmuir **28**(36), 13051–13059 (2012)
15. L. Wu, C. Wang, Y. Li, Iron oxide nanoparticle targeting mechanism and its application in tumor magnetic resonance imaging and therapy. Nanomedicine **17**(21), 1567–1583 (2022)
16. L. Zhu, Z. Zhou, H. Mao, L. Yang, Magnetic nanoparticles for precision oncology: theranostic magnetic iron oxide nanoparticles for image-guided and targeted cancer therapy. Nanomedicine **12**(1), 73–87 (2017)
17. X. Xu, X. Zhou, B. Xiao, H. Xu, D. Hu, Y. Qian, J. Tang et al., Glutathione-responsive magnetic nanoparticles for highly sensitive diagnosis of liver metastases. Nano Lett. **21**(5), 2199–2206 (2021)
18. Z. Zhao, M. Li, J. Zeng, L. Huo, K. Liu, R. Wei, J. Gao et al., Recent advances in engineering iron oxide nanoparticles for effective magnetic resonance imaging. Bioactive Mater. **12**, 214–245 (2022)
19. S.D. Iancu, C. Albu, L. Chiriac, R. Moldovan, A. Stefancu, V. Moisoiu, Z. Bálint et al., Assessment of gold-coated iron oxide nanoparticles as negative T2 contrast agent in small animal MRI studies. Int. J. Nanomed., 4811–4824 (2020)
20. N. Ajinkya, X. Yu, P. Kaithal, H. Luo, P. Somani, S. Ramakrishna, Magnetic iron oxide nanoparticle (IONP) synthesis to applications: present and future. Materials **13**(20), 4644 (2020)
21. A. Frtús, B. Smolková, M. Uzhytchak, M. Lunova, M. Jirsa, Š Kubinová, O. Lunov et al., Analyzing the mechanisms of iron oxide nanoparticles interactions with cells: a road from failure to success in clinical applications. J. Control. Release **328**, 59–77 (2020)
22. V. Mulens-Arias, J.M. Rojas, D.F. Barber, The intrinsic biological identities of iron oxide nanoparticles and their coatings: unexplored territory for combinatorial therapies. Nanomaterials **10**(5), 837 (2020)

23. U.S. Gaharwar, R. Meena, P. Rajamani, Biodistribution, clearance and morphological alterations of intravenously administered iron oxide nanoparticles in male Wistar rats. Int. J. Nanomed., 9677–9692 (2019)
24. A.L. Cortajarena, D. Ortega, S.M. Ocampo, A. Gonzalez-García, P. Couleaud, R. Miranda, A. Ayuso-Sacido et al., Engineering iron oxide nanoparticles for clinical settings. Nanobiomedicine **1**, 2 (2014)
25. A. Hanini, A. Schmitt, K. Kacem, F. Chau, S. Ammar, J. Gavard, Evaluation of iron oxide nanoparticle biocompatibility. Int. J. Nanomed., 787–794 (2011)
26. U.S. Gaharwar, R. Meena, P. Rajamani, Iron oxide nanoparticles induced cytotoxicity, oxidative stress and DNA damage in lymphocytes. J. Appl. Toxicol. **37**(10), 1232–1244 (2017)
27. B. Laffon, N. Fernández-Bertólez, C. Costa, F. Brandão, J.P. Teixeira, E. Pásaro, V. Valdiglesias, Cellular and molecular toxicity of iron oxide nanoparticles. Cell. Mol. Toxicol. Nanopart., 199–213 (2018)
28. Y. Javed, K. Akhtar, H. Anwar, Y. Jamil, MRI based on iron oxide nanoparticles contrast agents: effect of oxidation state and architecture. J. Nanopart. Res. **19**, 1–25 (2017)
29. H. Arami, A. Khandhar, D. Liggitt, K.M. Krishnan, In vivo delivery, pharmacokinetics, biodistribution and toxicity of iron oxide nanoparticles. Chem. Soc. Rev. **44**(23), 8576–8607 (2015)
30. M.N. Savari, A. Jabali, Pharmacokinetics of IONPs, in *Theranostic Iron-Oxide Based Nanoplatforms in Oncology: Synthesis, Metabolism, and Toxicity for Simultaneous Imaging and Therapy* (Springer Nature, Singapore, 2023), pp. 67–113
31. A. Orlando, M. Colombo, D. Prosperi, M. Gregori, A. Panariti, I. Rivolta, E. Cazzaniga et al., Iron oxide nanoparticles surface coating and cell uptake affect biocompatibility and inflammatory responses of endothelial cells and macrophages. J. Nanopart. Res. **17**, 1–13 (2015)
32. R. Hachani, M.A. Birchall, M.W. Lowdell, G. Kasparis, L.D. Tung, B.B. Manshian, N.T. Thanh et al., Assessing cell-nanoparticle interactions by high content imaging of biocompatible iron oxide nanoparticles as potential contrast agents for magnetic resonance imaging. Sci. Rep. **7**(1), 7850 (2017)
33. M.A. Abakumov, A.S. Semkina, A.S. Skorikov, D.A. Vishnevskiy, A.V. Ivanova, E. Mironova, V.P. Chekhonin et al., Toxicity of iron oxide nanoparticles: size and coating effects. J. Biochem. Mol. Toxicol. **32**(12), e22225 (2018)
34. H.L. Chee, C.R.R. Gan, M. Ng, L. Low, D.G. Fernig, K.K. Bhakoo, D. Paramelle, Biocompatible peptide-coated ultrasmall superparamagnetic iron oxide nanoparticles for in vivo contrast-enhanced magnetic resonance imaging. ACS Nano **12**(7), 6480–6491 (2018)
35. C.S. Tiwary, R.J. Mudakavi, S. Kishore, S. Kashyap, R. Elumalai, D. Chakravortty, K. Chattopadhyay et al., Magnetic iron nanoparticles for in vivo targeted delivery and as biocompatible contrast agents. RSC Adv. **6**(115), 114344–114352 (2016)
36. Y. Takeuchi, H. Suzuki, H. Sasahara, J. Ueda, I. Yabata, K. Itagaki, K. Murase et al., Encapsulation of iron oxide nanoparticles into red blood cells as a potential contrast agent for magnetic particle imaging. Adv. Biomed. Eng. **3**, 37–43 (2014)
37. A. Antonelli, M. Magnani, Red blood cells as carriers of iron oxide-based contrast agents for diagnostic applications. J. Biomed. Nanotechnol. **10**(9), 1732–1750 (2014)
38. H. Nosrati, M. Salehiabar, M. Fridoni, M.A. Abdollahifar, H. Kheiri Manjili, S. Davaran, H. Danafar, New insight about biocompatibility and biodegradability of iron oxide magnetic nanoparticles: stereological and in vivo MRI monitor. Sci. Rep. **9**(1), 7173 (2019)
39. Q. Zhou, Y. Wei, For better or worse, iron overload by superparamagnetic iron oxide nanoparticles as a MRI contrast agent for chronic liver diseases. Chem. Res. Toxicol. **30**(1), 73–80 (2017)
40. J. Nowak-Jary, B. Machnicka, Pharmacokinetics of magnetic iron oxide nanoparticles for medical applications. J. Nanobiotechnol. **20**(1), 305 (2022)
41. P.V. Hauser, H.M. Chang, N. Yanagawa, M. Hamon, Nanotechnology, nanomedicine, and the kidney. Appl. Sci. **11**(16), 7187 (2021)

42. J. Nowak-Jary, B. Machnicka, In vivo biodistribution and clearance of magnetic iron oxide nanoparticles for medical applications. Int. J. Nanomed., 4067–4100 (2023)
43. P. Mi, Stimuli-responsive nanocarriers for drug delivery, tumor imaging, therapy and theranostics. Theranostics **10**(10), 4557 (2020)

Chapter 7
Recent Advances in Biomedical Imaging for Cancer Diagnosis and Therapy

James C. L. Chow

Abstract With rapid advances of biomedical imaging, cancer diagnosis and therapy nowadays can be carried out less invasively, incorporating more sophisticated imaging and image-guided procedures. Advances of artificial intelligence/machine learning, nanotechnology, high-performance computing, and big data processing hugely benefit developments of imaging modalities such as ultrasonography, computed tomography, magnetic resonance imaging, and positron emission tomography. This results in many advanced and novel processes in cancer diagnosis and treatment, such as machine learning in tumour recognition, auto-contouring in radiation treatment planning, image fusion/integration, and image-guided cancer therapy. In this chapter, we will review some recent biomedical imaging techniques and how they are implemented and commissioned to the current cancer diagnosis and therapy. We will also provide the future prospectives regarding how the dramatic changes in biomedical imaging will affect the detection, recognition, and treatment of cancers.

7.1 Introduction

Medical imaging is a process to image the interior of a body for diagnosis, treatment, or visual representation of organ function. In disease diagnosis, medical imaging can form a database of normal anatomy and physiology to identify any abnormalities of a patient. In treatment, image-guided therapy can help achieve accurate and precise radiation dose delivery in radiotherapy, or medical intervention. Medical imaging

J. C. L. Chow (✉)
Radiation Medicine Program, Princess Margaret Cancer Centre, University Health Network, Toronto, ON M5G 1X6, Canada
e-mail: james.chow@uhn.ca

Department of Radiation Oncology, University of Toronto, Toronto, ON M5T 1P5, Canada

TECHNA Institute for the Advancement of Technology for Health, University Health Network, Toronto, ON M5G 2C4, Canada

N. Kalarikkal et al. (eds.), *Multimodal Biomedical Imaging Techniques*,
Biological and Medical Physics, Biomedical Engineering,
https://doi.org/10.1007/978-981-96-1124-9_7

is a part of biological imaging incorporating radiology. There are different biomedical imaging modalities based on various technologies such as x-ray radiography, computed tomography (CT), magnetic resonance imaging (MRI), and ultrasound (US). There are also different functional imaging techniques such as positron emission tomography (PET) and single-photon emission computed tomography (SPECT). Biomedical imaging is described as noninvasive because the procedure does not require introducing any instrument or probe into the patient's body. For example, x-ray radiography uses x-ray radiation with a source outside the patient's body. The x-ray beams penetrated into the body are absorbed at different rates due to the attenuation coefficients of human tissues such as lung, muscle, bone, and fat. Due to its noninvasive characteristic, biomedical imaging is popularly used for disease diagnosis and therapy. In 2010, there were five billion biomedical imaging studies conducted worldwide [1]. In 2006, over 50% of total ionizing radiation exposure was from biomedical imaging in the United States, raising concerns about radiation exposure in general diagnosis [2].

Cancer is a frightening and puzzling set of diseases that can develop anywhere in the human body. Generally, cancer begins with genetic changes that interfere with the body's normal processes. Cancer starts and grows from the body's cells that transform into pathological organisms or tumours. Unlike other diseases, the cancerous cells inside the tumour process genetic changes that cause them to surpass cell signals responding to cell differentiation. This causes programmed cell death and proliferation. This abnormal behavior allows the cancerous cells to grow uncontrollably and invade surrounding healthy cells. The cancerous cells can even spread through the body through lymphatic drainage and direct infiltration of the vascular system. In Canada, it is estimated that about 43% of Canadians will develop cancer in their lifetime and about 25% of these cancers will end in death [3]. In 2022, it was estimated that about 233,900 Canadians would be diagnosed with cancer. The five-year survival rate of cancer patients is about 64%, indicating a need to improve cancer prevention, diagnosis, and treatment [4]. Currently, there are many cancer treatment options such as radiotherapy, chemotherapy, immunotherapy, surgery, targeted therapy and so on. Among them, radiotherapy is very important because approximately over 50% of cancer patients will receive radiotherapy during their course of illness, with an estimation that radiotherapy contributes to about 40% of curative treatment [5]. In the radiotherapy chain depicting the whole treatment process, biomedical imaging plays a very significant role in the cancer diagnosis, treatment simulation, radiation treatment planning, image-guided patient setup and delivery. In this chapter, we will explore the recent developments of biomedical imaging in cancer diagnosis and therapy, in particular radiotherapy.

It is well known that developments of biomedical imaging have evolved through many novel technologies such as high-performance computing, big data processing/analytic, artificial intelligence (AI)/machine learning, and nanotechnology [6–9]. High-performance computing refers to using supercomputers or computer clusters to solve computational problems. Examples include cell processor and graphical processing unit computing borrowing the concepts of parallel, grid and cloud computing [10, 11]. With such a big computing enhancement, biomedical image processing, registration and deformation can be finished in a very short period of time. Moreover, more complicated computing algorithms such as machine learning can be run conveniently in biomedical imaging such as tumour recognition in cancer detection. Since machine learning requires training of the machine using a big dataset, a robust, secure and reliable image database is desired. With the recent evolution of big data processing such as big data cloud, image data from different institutions can be combined together, making deep learning process for cancer diagnosis possible [12]. Moreover, the recent progresses of data mining, data filtering, and data security makes the biomedical image processing more precise, reliable and efficient, taking advantage of AI and machine learning. Apart from the innovations in computing, the recent development of nanotechnology also benefits biomedical imaging. With more and more functionalized nanomaterials being designed and synthesized successfully, the nano-radiosensitizer can help to target the tumour and enhance the cancer cell kill in theranostics [13, 14]. These functionalized nanomaterials can act as both an image contract agent and dose enhancer in cancer therapy. To date, different cell, preclinical and clinical experiments are being conducted in biomedical imaging for cancer nanotheranostics [15, 16].

In this chapter, we will explore some recent advances in biomedical imaging modalities such as CT, MRI, and PET/SPECT. A brief review of the concept on the imaging modality will be given. For cancer diagnosis, we will introduce the recent progress in image contrast agent design and synthesis, tumour recognition based on machine learning, and auto-contouring of tumour and organs-at-risk in radiation treatment planning. In addition, we will review the recent developments of cancer therapies incorporating biomedical imaging such as image-guided radiotherapy using cone-beam CT (CBCT), MRI-guided radiotherapy and US-guided radiotherapy. Finally, we will discuss the future trends of work based on current findings and the future of the field in cancer diagnosis and therapy.

7.2 Recent Advances in Biomedical Imaging Modalities

Imaging modality is a medical imaging technique based on a physical mechanism to detect patient internal signals of anatomical structures and/or physiological events. The imaging techniques include CT, MRI and PET/SPECT. Images generated by kilovoltage (kV) x-ray such as from the CT scanner feature an image resolution of about 0.5 mm. CT image contains information of relative electron density that is useful in the calculation of absorbed dose considering heterogeneous correction in radiation treatment planning. MRI generates image through the exposure of hydrogen atoms to magnetic fields and radio waves. MRI has resolution of about 1.5 mm (magnetic field strength = 1.5/3 T) and depicts physiology and anatomy because the hydrogen atoms are related to the body chemistry rather than the organ structures. PET creates image utilizing radiation associated with altered glucose molecules to show glucose uptake. PET is important to check cancer metastasis as tumour uses to dominate the glucose uptake process. Moreover, though PET imaging has relatively low resolution of 4–5 mm, it can combine with CT to provide anatomic and physiologic images at one time for cancer diagnosis.

7.2.1 Computed Tomography

A CT scanner uses a rotating x-ray tube and a row of detectors located at the opposite direction in a gantry to measure x-ray attenuations, when the radiation passing through the patient's body. When the gantry rotates around the body, multiple x-ray measurements are taken from different beam angles. The measured attenuations are processed by a computer using tomography reconstruction algorithms to produce 3D images. CT image is essential in cancer diagnosis because it is usually employed to locate the tumour and its surrounding normal tissues. In cancer therapy, the CT image set can be input to a radiation treatment planning system, so that tumour and organs-at-risk can be contoured by the radiation oncologist or planner. Using the relative electron density information acquired from the CT, densities of various tissues such as lung, bone, and soft tissue can be identified in the body. Heterogeneous correction can therefore be carried out in radiation dose calculation to provide an accurate dose distribution in the treatment plan for delivery.

7.2.1.1 Photon-Counting CT

Photon-counting CT employed a photon-counting detector to detect the x-ray signal passing through the patient's body. Unlike typical CT scanner using energy integrating detectors, the photon-counting detector measures the photons as a voltage on a capacitor proportional to the x-ray detected. The detector registers the interactions of individual photons. By tracking the photon energy from each interaction, the detector can record an approximate energy spectrum. This makes the photon-counting CT a special or energy-resolved CT technique for diagnosis. The challenge of this new technique is that the detector is susceptible to noise. However, the photon-counting detector has a number of advantages such as improvement of signal-to-noise ratio, spatial resolution, reduction of radiation doses in irradiation, and capable of identifying multiple contrast agents under several energies. With the advance of detector technology, CT scanner equipped with photon-counting detector becomes available in the market recently. The first clinical photon-counting CT was approved by the Food and Drug Administration in September 2021 [17]. Kalender et al. [18] demonstrated that the photon-counting CT could provide a high spatial resolution of 0.1 mm under a very low glandular dose < 0.005 Gy for breast imaging. This resolution is sufficient for reliable detectability of calcifications of soft tissue in the breast, and soft tissue delineation.

7.2.1.2 Dual-Energy CT

Dual-energy CT is different from dual-source CT though both scanners have two x-ray sources in the gantry. Dual-source CT can have two x-ray sources with the same energy or not. If the latter (i.e. two sources with different energies) is used, the scanner becomes a dual-energy CT. It should be noted that dual-source CT using the same x-ray energy for the two sources can shorten the scanning time of the patient, because both sources are in action at the same time during the gantry rotation. For dual-energy CT, though the concept was proposed over 50 years ago, it is a relatively new technology in clinical practice, benefited by the recent advances of CT hardware and post-processing computer capabilities [19]. Dual-energy CT uses a normal x-ray for standard CT and an additional less powerful x-ray to generate image. The attenuation of x-ray inside the patient's body depends on the atomic number, density of the material and x-ray energy. The differential attenuation of the x-ray beam at different energy levels results in a separation of materials with different atomic numbers but similar CT attenuation. Therefore, dual-energy CT is good at detecting materials such as iodine, calcium or uric acid. For example, dual-energy CT can selectively increase the effect of some x-ray contrast agents such as iodine to generate better images of blood vessels for CT angiography and can improve the image quality if the patient has metal implant in the scanned area. In addition, only one single examination is needed to acquire images with and without contrast agents. Kruis et al. [20] found that implementation of dual-energy CT in cancer diagnosis and therapy could highly increase the accuracy of treatment delivery in particular radiotherapy. The application of dual-energy CT in radiotherapy could improve the dose calculation accuracy in treatment planning in the presence of contrast agent, reduce the metal artifact in the images, improve tumour/organs-at-risk delineation, and enhance visualization of the lymph nodes. Iodine, when employed as an intravenous contrast agent, can enhance the visibility of soft tissues in CT scans. Nevertheless, the enhancement in blood flow can sometimes be quite subtle, such as in the case of prostate tumours (Fig. 7.1) [20], or even appear less dense when compared to the surrounding tissue, as seen in pancreatic adenocarcinoma. This diminished contrast can lead to significant discrepancies in defining treatment boundaries in various medical contexts, including head and neck, gastric, pancreatic, prostate, bladder, and even pulmonary cancer cases.

7.2.1.3 Cone-Beam CT

CBCT generates a divergent x-rays from the source to form a cone of beam passing through the patient's body and then projected on a 2D detector. In adaptive radiotherapy, CBCT using kV x-rays is attached to a medical linear accelerator in radiation dose delivery. The first medical linear accelerator equipped with CBCT was developed in the late 1990s and early 2000s, and is now very popular in the latest generation of linear accelerator [21]. In the late 2010s, CBCT was also equipped on board with particle therapy delivery systems [22]. Unlike conventional CT utilizing a single row

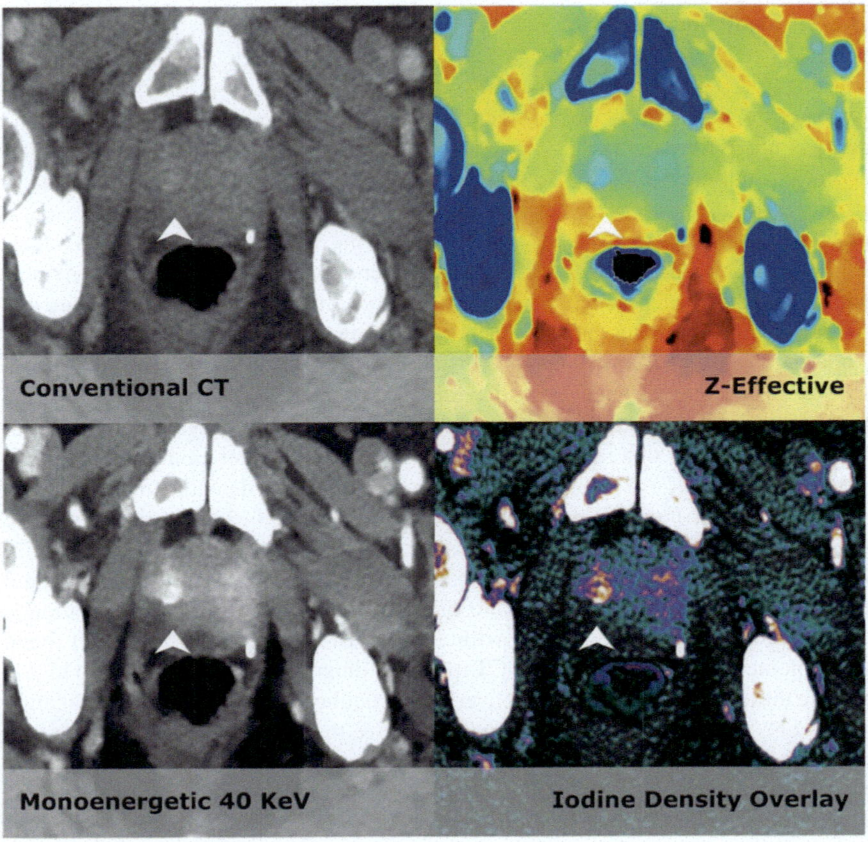

Fig. 7.1 In prostate radiotherapy, conventional CT struggles to detect tumours. MRI is added for precise tumour localization. Spectral CT, as shown in this image of an 82-year-old male with confirmed prostate cancer, offers improved tumour visibility, potentially enhancing treatment precision [20]

of detectors for fan beam, CBCT uses a 2D approach to obtain projection data from a cone beam. The scintillator in the CBCT detector converts x-ray to visible light, which is picked up by a charge-coupled device camera and recorded. The measured signals are then converted to images using CBCT reconstruction algorithms similar to typical tomographic reconstruction algorithms (e.g. filtered back-projection or iterative reconstruction). However, modification is needed using FDK algorithm as the reconstruction is 3D. In radiotherapy, kV-CBCT is used for patient setup in daily delivery.

7.2.2 *Magnetic Resonance Imaging*

MRI is a biomedical imaging modality to obtain information of anatomy and physiological processes from the patient's body. Unlike CT and PET/SPECT, MRI does not involve x-ray and therefore has no safety concern on ionizing radiations. MRI uses a strong magnetic field, magnetic field gradient and radio wave to generate images inside the body. The strong magnetic field forces protons (hydrogen atoms in water) to align with the field. A radiofrequency current is then pulsed through the patient to simulate the protons, making them spin out of equilibrium, straining against the pull of magnetic field. When the radiofrequency field is turned off, the MRI detector can detect the energy released, as the protons realign with the magnetic field. Since different tissues have different magnetic and chemical properties, the radiologist can identify the tissues based on the amount of energy released from the protons and the time taken for the protons to realign with the magnetic field.

7.2.2.1 0.55 T MRI

A low-field MRI system is recently developed with advantages of reduced costs, increased accessibility to patient's metal implant and some favorable physical properties [23]. The 0.55 T MRI (High-V MRI) system has some super diagnostic capabilities using a high-performance low magnetic field (0.55 T) to generate images. This new MRI system can improve metal implant imaging which has been an issue in conventional MRI (e.g. 1.5 T) due to artifacts. The 0.55 T MRI system can generate images with reduced metal distortions to improve cancer diagnosis (e.g. limb sarcoma) for implant imaging. The low-field MRI is also good at reducing susceptibility artifacts such as air-tissue interfaces of sinuses and orbits. The low 0.55 T magnetic field can reduce geometric distortions in diffusion imaging and hence improve the diagnostic quality. Furthermore, the MRI system performs very well in pulmonary imaging, which has air-tissue interfaces leading to fast signal decay. Bandettini et al. [24] compared the cine cardiovascular MR imaging between the 0.55 T and 1.5 T MRI. They found that there were no significant differences in the right and left ventricle end-diastolic volume between the two scanners. Bandettini et al. [24] concluded that the 0.55 T MRI system could offer good cardiovascular MR image quality, and quantification of biventricular volumes comparable to the 1.5 T MRI for diagnosis. Campbell-Washburn et al. [25] and Lavrova et al. [26] both explored the use of a 0.55 T MRI scanner for neuroimaging. Campbell-Washburn et al. applied the MR fingerprinting method to obtain 3D whole brain T1 and T2 maps at the lower MRI field. They simulated 100 pseudo-replicas of in vivo data to assess the measurement precision and compared the results with an MRI phantom. They found a strong correlation between the MR fingerprinting and the phantom measurements, and concluded that the method is feasible for T1 and T2 mapping at 0.55 T. Lavrova et al. evaluated the quality of brain images obtained by various pulse sequences at 0.55 T as shown in Fig. 7.2 and the suitability of this scanner for clinical use. They had radiologists review the brain images and rate their diagnostic quality (R1 and R2 in Fig. 7.2). They found that the brain images were acceptable for clinical purposes and that the scanner could be integrated into the clinical workflow.

7.2.2.2 11.7 T MRI

A very powerful MRI scanner with magnetic field strength up to 11.7 T was developed and commissioned in 2020 [27]. The scanner used a magnet of 132 tons, with dimensions of 5 m external diameter, 0.9 m internal diameter and 5 m length. With such an extraordinary dimension of magnet, a normal magnetic field strength of 11.7 T was produced compared to the typical strength of 1.5/3 T for conventional MRI used in hospital. The big magnet was powered by a current of 1.5 kA with conductor coils cooled by liquid helium at a temperature of −271.35 °C. Under such a strong magnetic field, a very high image resolution of $0.08 \times 0.08 \times 0.2$ mm^3 was achieved. This image resolution was much higher than the conventional MRI of $1.5 \times 1.5 \times 4$ mm^3 measured by Moats et al. [28].

7.2.2.3 Artificial Intelligence

AI is a branch of computer science that aims to create machines and systems capable of performing tasks that typically require human intelligence. This encompasses a wide range of applications, from problem-solving and decision-making to speech recognition, natural language understanding, and biomedical imaging. AI plays a crucial role in the field of biomedical imaging, which involves the use of advanced technologies to visualize and analyse various aspects of the human body. AI in this context encompasses the development of computer algorithms and systems that can replicate human-like intelligence to interpret and process medical images, such as x-rays, MRIs, and CT scans. Machine learning is a subset of AI that, when applied to biomedical imaging, leverages data (e.g. inter-institutional data) to train models for tasks like disease detection, organ segmentation, and treatment planning. Deep learning is a specialized branch of machine learning that employs neural networks with multiple layers to process complex biomedical images, enabling more accurate and detailed analysis. The relationship between machine learning and deep learning in biomedical imaging lies in the fact that deep learning techniques, particularly convolutional neural networks, have become increasingly effective in extracting intricate patterns from medical images, enhancing the accuracy and efficiency of diagnostics and research in this crucial domain.

AI in biomedical imaging has great potential to improve diagnosis, prognosis, and treatment of various diseases and conditions. However, it also faces many challenges and limitations that need to be addressed and overcome. Some of these challenges include: (1) the availability, quality, and privacy of the data that are used to train and test the AI algorithms; (2) the validation, generalization, and adaptation of the AI algorithms to different populations, settings, and imaging modalities; (3) the

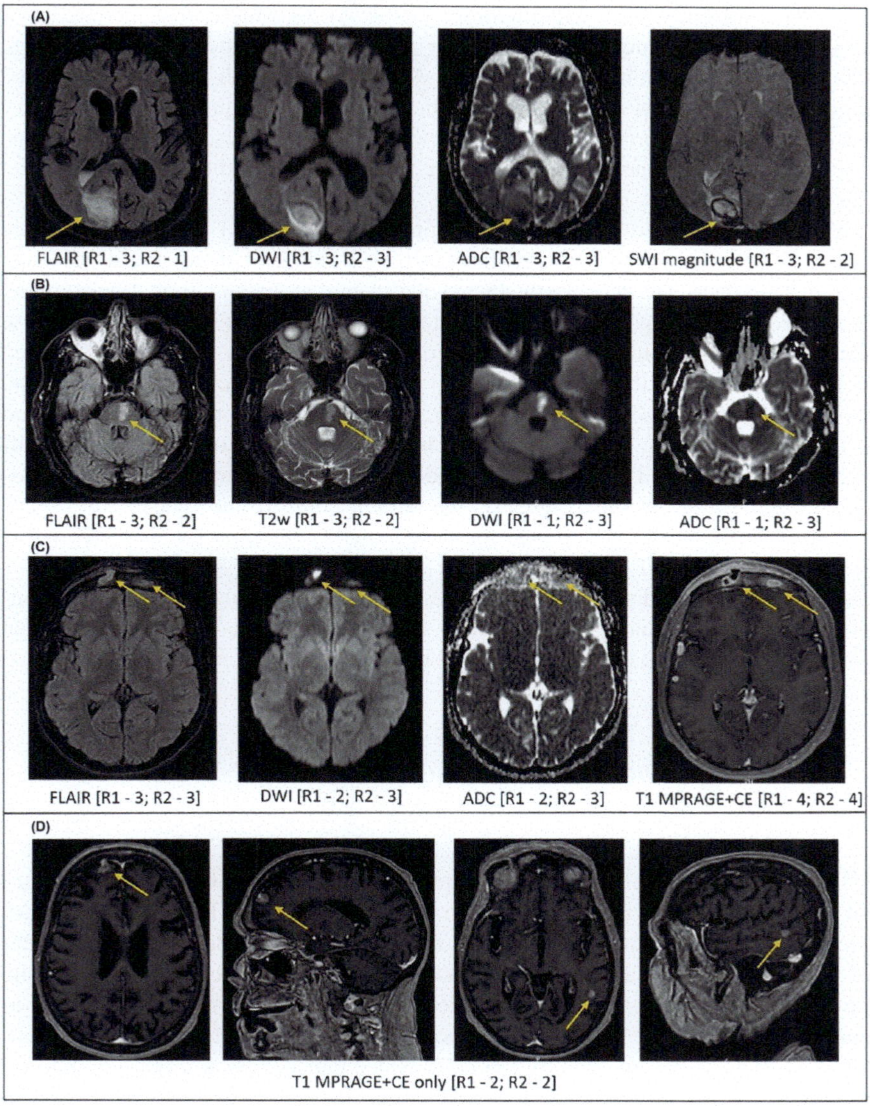

Fig. 7.2 Images obtained from patients scanned at 0.55 T, illustrating diverse clinical cases: **a** The first image shows a right occipital intraparenchymal hemorrhage extending into the right occipital horn of the lateral ventricle, characterized by hyperintensity on FLAIR/T2, restricted diffusion, and hypointensity on SWI. **b** The second image depicts an acute pontine infarct in the left paramedian pons, represented as a hyperintense lesion on FLAIR/T2 with diffusion restriction. **c** The third image displays Pott's puffy tumour, where frontal sinus soft tissue extends to subgaleal soft tissues, showing isointense FLAIR signal, diffusion restriction, and contrast enhancement. **d** The fourth image reveals metastatic disease, featuring contrast-enhancing lesions in the right frontal and left temporal lobes. Each overall quality of image was rated by two readers (R1 and R2), with scores ranging from 1 (unacceptable for diagnostic use) to 4 (excellent) [26]

reliability, robustness, and explainability of the AI algorithms and their outputs, especially in the presence of errors, biases, or uncertainties; and (4) the integration, collaboration, and communication of the AI algorithms with the human experts, clinicians, and patients who use them. These challenges require interdisciplinary and collaborative efforts from AI developers, biomedical researchers, clinicians, ethicists, regulators, and patients to ensure that AI in biomedical imaging is safe, effective, and ethical.

AI-powered workflows are recently implemented to MRI process to produce better clinical result. This includes the machine training of deep neural networks through a big dataset to improve the image quality and processing time. For example, AI is used in the targeted reconstruction method to increase the signal-to-noise ratio of the image using the iterative reconstruction process. The AI-based deep learning can leverage the raw data to deep learning reconstruction technology, and therefore shorten the scan time and enhance the image resolution. Some AI-based post-processing image analysis software is also developed to assist clinical staff in disease diagnosis. This software includes sites of brain and prostate. The AI-powered workflow not only increases the patient comfort and reduces the image acquisition time, but also increases the image quality. Zhang et al. [29] discussed the applications of AI in brain MRI for different diseases such as Alzheimer's disease, Parkinson's disease, major depressive disorder, schizophrenia, attention-deficit/hyperactivity disorder, and autism spectrum disorder. They concluded that using AI to assist disease diagnosis would be the technical trend in the future.

7.2.3 Positron Emission Tomography and Single-Photon Emission Computerized Tomography

PET is a functional imaging technique uses radiotracers to generate images and measure physiological processes. In cancer diagnosis, ^{18}F-FDG (FDG: fluorodeoxyglucose) is commonly used as radiotracer to detect cancer metastasis. FDG is a glucose analog taken up by glucose-using cells (e.g. cancer cells). The metabolic trapping of the radioactive glucose molecule in the cell allows the PET scan to generate image signals. The tissue metabolic activity is proportional to the concentration of FDG tracer uptaken by the cells. Since cancer cell is more dominant in the glucose uptake compared to its surrounding normal cells, ^{18}F-FDG PET scan is used to detect cancer metastasis for diagnosis. The spatial resolution of PET is relatively low (4–5 mm) but it can combine with a CT scanner with higher resolution of 1 mm, to provide a useful imaging tool for accessing functional and anatomical information based on a single fused image. SPECT is a nuclear medicine tomographic imaging technique based on gamma rays. The gamma-emitting radioisotope (e.g. isotope of gallium (III)) is delivered to the patient through injection into the bloodstream. SPECT imaging is created using a gamma camera to acquire multiple image projections through different angles. A 3D image set is then constructed by a tomographic reconstruction algorithm using computer. SPECT is similar to PET because both modalities use radiotracers and detect gamma rays. However, radiotracers in SPECT emit gamma radiation that is measured directly, while PET tracers emit positrons that finally result in two photons emitted in opposite directions.

7.2.3.1 PET-CT

In most clinical applications, PET is combined with a CT system to provide anatomical information essential for interpreting PET signals accurately. The concept of PET-CT was proposed in 1987 [30]. For cancer diagnosis, PET-CT is not performed alone without previous diagnosis due to its lack of sensitivity. For example, PET-CT should not be used in early-stage breast cancer without symptoms as per the National Comprehensive Cancer Network. However, PET-CT is recommended to identify unsuspected regional nodal disease with diagnostic CT, particularly when the standard staging process is suspicious. In radiotherapy, PET-CT is mainly used as part of the diagnostic procedures to evaluate the treatment results. With recent advances of personalized radiotherapy/medicine, radiobiological/volumetric characteristics of tumour can be integrated into the routine radiation planning process. In radiation treatment planning, delineating of the tumour can be enhanced by the information on tumour biology acquired by the PET-CT. Jelercic et al. [31] reviewed the role of PET-CT in radiation treatment planning of solid tumours. From evidences of clinical trials of lung cancer, they found that PET-CT would be an integral part in the treatment planning procedure in the future.

7.2.3.2 PET-MRI

PET-MRI provides functional and anatomical information of a specific organs or body system at molecular level, and this novel imaging modality becomes a powerful tool to study the molecular biology details, pharmacokinetics and cancer mechanisms. Although the first PET-MRI experiment was carried out in 1990s, clinical application was only accomplished in recent years due to a few technical challenges to integrate PET and MRI into a single system [32]. One of the challenges in combining PET and MRI is that a PET detector capable of operating in a strong magnetic field generated by the MRI needs to be developed. Moreover, the radio frequency shield of MRI needs to be designed to provide magnetic field homogeneity in the presence of PET detector. Recently, PET-MRI is used for breast cancer patients because the system can acquire both metabolic data and high-contrast morphological images in one single exam. PET-MRI is recommended to improve diagnostic accuracy of breast cancer and decrease false-positive rates over those from MRI. Pinker et al. [33] found that for 78 indeterminate or suspicious lesions, half of the unnecessary biopsies could be performed using PET-MRI compared to conducting MRI alone.

7.2.3.3 SPECT-CT

Similar to PET, SPECT has been a nuclear medicine imaging modality for several decades. Recently, SPECT is integrated with CT to combine the functional (SPECT) and anatomic (CT) imaging for cancer diagnosis and therapy. Nowadays, with advances of image processing and fusion, a SPECT-CT system can perform both modalities on a single gantry, and provide fused functional and anatomic image in one single exam. Apart from tumour delineation based on nuclear imaging findings, SPECT-CT allows accurate and rapid attenuation correction of SPECT studies. This feature has been proved to be useful in many cardiac, neurologic and oncologic applications. Israel et al. [34] found that the improved diagnostic accuracy by using SPECT-CT could be associated with greater diagnostic confidence and better inter-specialty communication. They concluded that SPECT-CT should be used as a routine imaging technique and be integrated into the clinical decision-making process in cancer therapy.

7.3 Cancer Diagnosis

Cancer can be diagnosed by biopsy, lab and imaging tests. Lab tests detect levels of substances (e.g. biomarkers) in the patient body reflecting sign of cancer. Those substances can be found in blood, urine, or other body fluids from the patient. In addition, some lab tests are based on testing blood or tissue samples for tumour markers, which are produced by cancer cells or other cells showing response to cancer. Biopsy is a medical procedure to remove a tissue sample, which is investigated by the pathologist using a microscope. The biopsy sample can be obtained through a needle, with endoscopy (e.g. colonoscopy and bronchoscopy) or with surgery. Imaging tests use biomedical images to help oncologist to visualize the presence of tumour in the patient's body. The imaging modalities include CT, MRI, PET, US, and x-ray. In this chapter, we focus on some recent advances of cancer diagnosis using biomedical imaging. These include the development of image contrast agent such as gold nanoparticles in nanoparticle-enhanced radiotherapy. Moreover, we explore the applications of AI technology in cancer diagnosis. These consist of machine learning in biomedical imaging for tumour recognition and auto-contouring in radiation treatment planning.

7.3.1 Image Contrast Agent

Image contrast agent can increase the contrast of the target or tumour in biomedical imaging so that radiation staff can highlight and contour the tumour easily and accurately in treatment planning. With the recent advance in nanotechnology, nanoparticles are used as a contrast agent for cancer diagnosis [35, 36]. For example, magnetic nanoparticles such as gadolinium and iron oxide are used as MRI contrast agents [37], and heavy-atom nanoparticles such as iodine, bismuth, and gold nanoparticles are used as CT contrast agents [38]. When gold nanoparticles are added to the tumour, the compositional atomic number of the tumour is increased. This increases the attenuation of the radiation beam passing through the tumour due to photoelectric enhancement, where the photoelectric absorption is approximately proportional to Z^3/E^3, where Z is the atomic number and E is the radiation beam energy [39]. Therefore, gold nanoparticles are an effective contrast agent in medical imaging when using kV x-rays such as CT/CBCT. A study was carried out to investigate the contrast ratio of different nanoparticles using various particle concentrations and beam energies for CT [40]. The contrast ratio was calculated using the equation:

$$C = \frac{\textit{target value} - \textit{background value}}{\textit{background value}} \tag{7.1}$$

In Eq. (7.1), C is equal to

$$\frac{I_t - I_b}{I_b} \tag{7.2}$$

I_t is the transmitted x-ray intensity and I_b is the transmitted background intensity. Monte Carlo simulation was used to calculate these intensities based on the Beer-Lambert Law [41], where I_o is the incident intensity and x is the thickness of the matter:

$$I_t = I_o e^{(-\mu x)} \tag{7.3}$$

The mass attenuation coefficient, μ, is the sum of the three interactions (photoelectric effect, Compton effect, and pair production) between the x-ray and matter in the proper energy range [42]. Dependences of contrast ratio on the nanoparticle type (Fe_3O_4, Ag, I, Pt, and Au), nanoparticle concentration (3–40 mg/mL) and radiation beam energy (100–140 kVp) are shown in Fig. 7.3.

In Fig. 7.3, it is seen that the contrast ratio increased with an increase in nanoparticle concentration from 3 to 40 mg/mL. This range of concentration was selected as per the usual range used in preclinical experiment for nanoparticle-enhanced radiotherapy [43]. The contrast ratio increased with higher particle concentration because more nanoparticles accumulated in the tumour, leading to a higher attenuation of photon beam, compared to the surrounding normal tissues. Moreover, it is seen that gold nanoparticles had the highest contrast ratio compared to other nanoparticles. This is because gold has the highest atomic number among other elements. Since photoelectric enhancement is proportional to the compositional atomic number (Z^3) of the tumour under irradiation, tumour added with gold nanoparticles has the highest attenuation (i.e. contrast) [36]. For dependence of contrast ratio on the photon beam energy, Fig. 7.3 shows that the lower the beam energy, the higher the contrast ratio. This is shown in Fig. 7.3a that the maximum contrast ratio of gold nanoparticles for

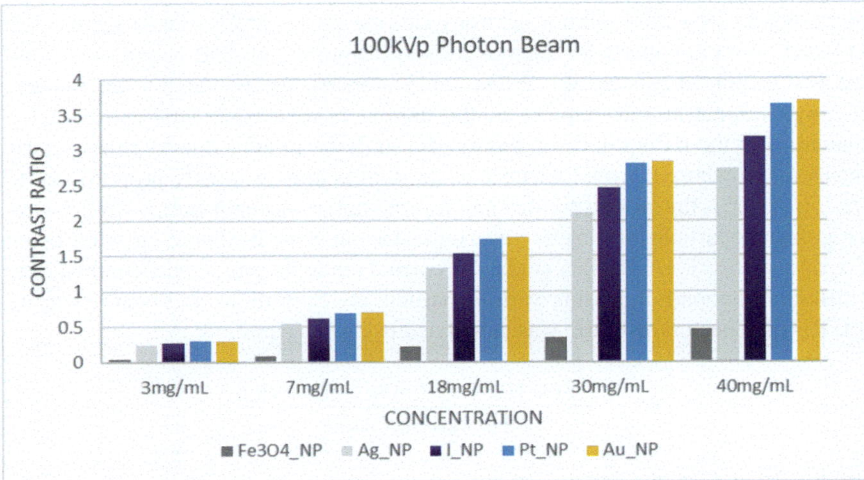

(A)

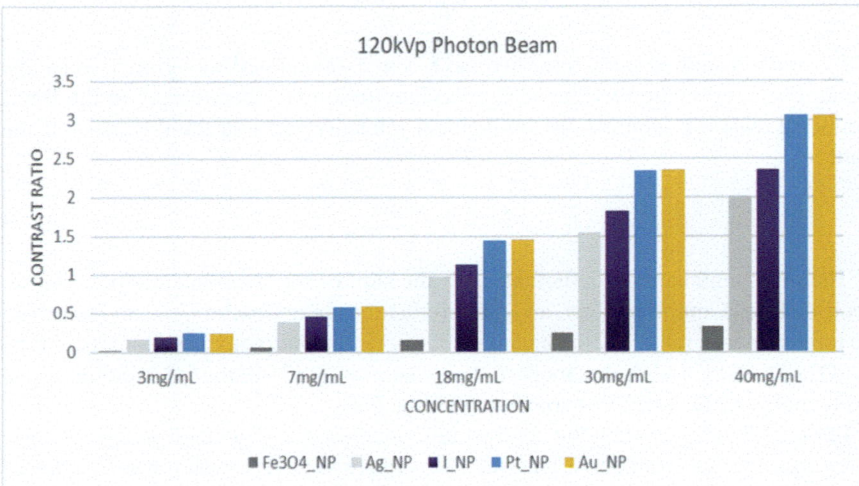

(B)

Fig. 7.3 Relationship between contrast ratio and nanoparticle concentration (mg/mL) with the delivery of **a** 100 kVp, **b** 120 kVp, **c** 130 kVp, and **d** 140 kVp photon beams for the iron oxide (Fe_3O_4), silver (Ag), iodine (I), platinum (Pt), and gold (Au) nanoparticles (NP) [40]

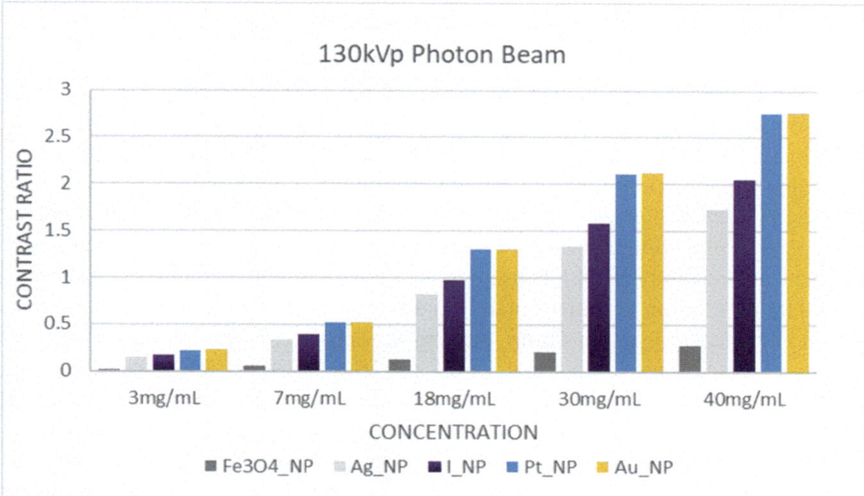

(C)

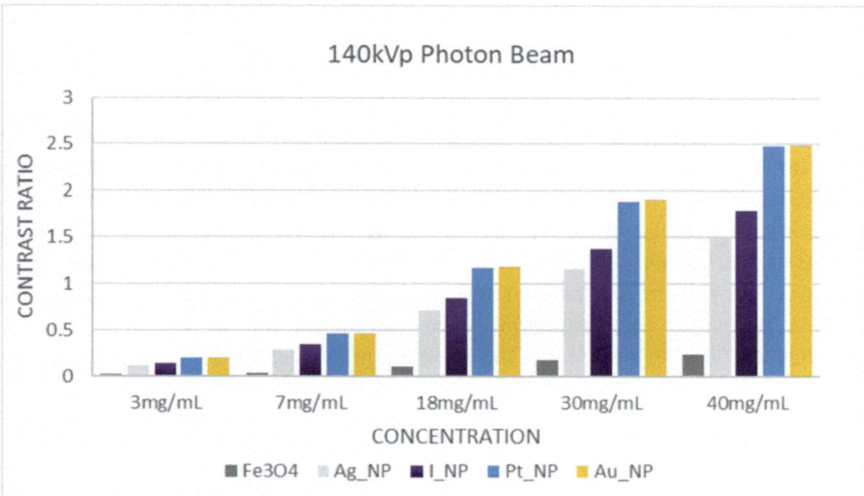

(D)

Fig. 7.3 (continued)

40 mg/mL was 3.71 using the 100 kVp photon beam. When the energy was increased to 120 kVp (Fig. 7.3b), the maximum contrast ratio was equal to 3.07. The maximum contrast ratios further decreased to 2.77 and 2.49 for the 130 and 140 kVp photon beams as shown in Fig. 7.3c and d, respectively. This is because the photoelectric cross-section is inversely proportional to the radiation beam energy ($1/E^3$) [44].

7.3.2 Machine Learning in Biomedical Imaging

Since biomedical imaging became available in digital format, computerized image processing and visualization techniques have been used to achieve more accurate and robust image analysis, leading to more reliable and faster clinical decision making. Practical applications combined advances in computer calculation, machine learning, and imaging technology as mentioned in Sect. 7.2. Machine learning is an AI algorithm to perform a computer task after training with a dataset [45]. In radiotherapy, machine learning has been widely used in radiation treatment planning. For example, to evaluate a radiation treatment plan based on the dose distribution in the patient [46]. In cancer diagnosis, machine learning is used to recognize the tumour from biomedical imaging modalities such as CT and MRI based on image classification.

Classification between objects such as tumour is easy for human but machine due to complicated machine training using various algorithms such as artificial neural network, convolutional neural network, decision tree, support vector machines and K-nearest neighbor [47]. To perform machine learning in tumour recognition, image pre-processing is the first step to prepare a dataset to be trained. A selected machine learning algorithm is used to predict the results based on the dataset, and the training algorithm keeps on comparing the real and trained result to determine the deviation. Machine training is carried out continuously until the deviation is non-significant enough to be acceptable. In this state, the machine learning algorithm is validated and can be implemented to the imaging system.

For recent advances of machine learning application in cancer diagnosis, Nayak et al. [48] developed an autonomous brain tumour segmentation method based on deep learning. They used a unique convolutional neural network architecture to recognize brain tumour cells, which were very difficult to identify because of their heterogeneous nature. In the proposed model, they used python code simulated on a Kaggle platform using TensorFlow, and the optimized "Adam" was selected for classifying the brain tumours in the simulation. After machine training of dataset from brain MRI, Nayak et al. [46] tested that the accuracy percentage in the simulation of the model was about 100% based on nine runs. On the other hand, Rasool et al.. [49] proposed a hybrid deep learning model based on MRI for brain tumour classification. In their work, they used a new hybrid convolutional neural network-based architecture to classify three brain tumour types, namely, glioma, meningioma, and pituitary tumour using MRI. Two algorithms were used in the study. The first algorithm combined a per-trained Google-Net model of the convolutional neural network algorithm for feature extraction with support vector machine for pattern classification. The second algorithm integrated a finely tuned Google-Net with a soft-max classifier. There were 1,426 glioma images, 708 meningioma images, 930 pituitary tumour images, and 396 normal brain images used in machine training and evaluation. Rasool et al. [49] demonstrated that their deep learning method had an accuracy of 93.1% from the finely tuned Google-Net model, while accuracy could be further improved to 98.1%, when the Google-Net as a feature extractor, was combined with a support vector machine classifier.

For breast cancer diagnosis using machine learning, Rasool et al. [50] evaluated different machine-based predictive models for breast cancer using the data exploratory techniques. They evaluated the breast cancer diagnostic accuracy based on four models, namely, support vector machine, linear regression, K-nearest neighbor and ensemble classifier. All models were implemented on the Wisconsin Diagnostic Breast Cancer and Breast Cancer Coimbra Dataset. Each classifier's efficiency was evaluated by the standard performance metrics and training time. Results showed that accuracies of models were 99.3%, 98.06%, 97.35%, and 97.61% for support vector machine, linear regression, K-nearest neighbor and ensemble classifier, respectively. On the other hand, Bacha et al. [51] proposed a novel machine learning method for breast cancer diagnosis. The method was based on the differential evolution algorithm of a radial-based function kernel extreme learning machine. Through tuning two parameters for the model: (1) penalty parameter and (2) RBF-kernel parameter, the differential evolution algorithm was validated with machine training using the Mammographic Image Analysis Society and Wisconsin Breast Cancer Database. Bacha et al. [52] also developed an automated computer-assisted medical diagnosis system to detect breast cancer using machine learning based on digital mammogram. The machine learning model was verified using two available datasets, namely, Mammographic Image Analysis Society Dataset and the INbreast Dataset. From the experimental verification, the proposed system resulted in an accuracy of 100% for the former dataset. For the malignant-benign classification using the latter dataset, however, the system gave an accuracy of 99.93%.

For lung cancer diagnosis based on biomedical imaging using machine learning, Sun et al. [53] compared the performance of deep learning with the traditional computer-aided diagnosis system. Three machine learning algorithms, namely, convolutional neural network, deep belief network and stacked denoising autoencoder were used in this study. The machine learning models were trained using CT cases from Lung Image Database Consortium. The accuracies of the convolutional neural network, deep belief network and stacked denoising autoencoder were 0.7976, 0.8119 and 0.7929, respectively. The accuracy of the traditional computer-aided diagnosis system was 0.7940. It is seen that the deep learning algorithms of convolutional neural network and deep belief network were slightly better than the computer-aided diagnosis system. To further improve the lung cancer diagnosis using machine learning, Hussain et al. [54] suggested an image enhancement method to improve the lung cancer image quality before machine learning. The enhancement methods included image adjustment, gamma correction, and contrast stretching. Gray-level co-occurrence matrix features of the enhanced image were extracted, and machine learning was then carried out using methods such as decision tree, support vector machine, radial base function, and polynomial. Hussain et al. [54] found that without the image enhancement, the highest performance of machine learning resulted in an accuracy of 99.89%. While with the suggested image enhancement, accuracy of 100% was achieved.

7.3.3 *Machine Learning in Auto-contouring*

In the radiotherapy chain, radiation treatment planning is a very important step to create a plan to the patient for radiation dose delivery. This process is carried out by the radiation oncologist, medical physicist, and radiotherapist using a computerized treatment planning system. Typically, CT image set of patient acquired from the CT-simulator is input to the planning system, and the radiation oncologist identifies the target (tumour) and organs-at-risk in the patient's body. Then, the radiation oncologist or planner contour and define the volumes-of-interest for the target and critical organs in the treatment plan. The target and organ contouring are indeed very time-consuming and labor-intensive, because the planner needs to work on many volumes-of-interest in every CT slide in order to outline the targets and critical organs in the treatment plan. Auto-contouring can therefore save time for the planner in creating the treatment plan. However, based on the early version of computer-aided system, the auto-contouring process was not perfect and manual correction/adjustment was usually required for every volume-of-interest. With the recent advance of AI, different machine learning algorithms were applied in learning the feedback from the planner regarding contour corrections. This AI-assisted auto-contouring system has become more and more robust based on the dataset from previously successful treatment plans.

For auto-contouring in cervical cancer, Ma et al. [55] evaluated the deep learning-based auto-segmentation algorithm using a CT dataset containing 535 cervical cancer cases. The deep learning tool automatically delineated the target volumes of the pelvic lymph drainage and parametrial area. With the training, validation and test number for the pelvic lymph drainage area equal to 157, 20, and 23 (definitive radiotherapy group), and 272, 30, and 33 (postoperative radiotherapy group), dice similarity coefficient, mean surface distance, and Hausdorff distance were determined to evaluate the contouring accuracy. Results showed that the mean dice similarity coefficient, mean surface distance and Hausdorff distance for the deep-learning tool were 0.88, 1.32 mm, and 21.6 mm for the pelvic lymph drainage area, and 0.86, 1.15 mm, and 20.78 mm for the parametrial area. The contouring accuracy for the deep learning tool was found comparable to that of senior radiation oncologists and was better than that of junior/intermediate radiation oncologists. Ma et al. [55] therefore concluded that their deep learning contouring tool could improve target contouring accuracy, reduce contouring time and improve the clinical efficiency in cervical cancer therapy.

For head-and-neck radiotherapy, a novel deep learning auto-segmentation algorithm was developed by Koo et al. [56], and the performance of the algorithm was compared with other commercial algorithms. In this study, 864 head-and-neck cancer cases were used to train and evaluate the deep learning algorithm, which was constructed using fully convolutional network with combined U-Net and V-Net. Machine training was based on the dice loss plus cross-entropy loss function with Adam optimizer using 75 validation cases. The auto-segmented structures were evaluated with dice similarity coefficient, Hausdorff distance and voxel-penalty metric. Results showed that the average values of dice similarity coefficient, Hausdorff distance, and voxel-penalty metric were equal to 0.81, 84.1, 1.6 mm; 0.74, 62.8, 3.2 mm; and 0.66, 46.8, 3.3 mm for the *A*, *B*, and *C* structures. 93% of model *A* structures evaluated was clinically useful. Koo et al. [56] therefore concluded that their deep learning contouring tool had high performance in radiation treatment planning.

Mao et al. [57] conducted a retrospective study to evaluate the performance of auto-contouring for lung cancer therapy. The study included 10 patients with 297 fractions of treatment. Two types of auto-contouring processes were used: (1) automatic contours without modification (A-ART) and (2) automatic contours were modified manually by radiation oncologists and physicists (S-ART). Results showed that by comparing the target volumes between A-ART and S-ART, the dice similarity coefficient, mean contour distance and Hausdorff distance were 0.93 ± 0.05, 1.5 ± 1.2 mm and 4 ± 2.3 mm. The average target coverages were improved from 0.96 ± 0.04 to 1 ± 0.02 and 1.02 ± 0.04 for A-ART and S-ART, respectively. Statistical tests showed that both A-ART and S-ART hugely improved the cumulative target dose. Mao et al. [55] concluded that the online adaptive radiotherapy with automatic contouring could improve target coverage and spare organs-at-risk for lung cancer.

For auto-contouring in MRI-guided radiotherapy using machine learning, Ding et al. [56] implemented and tested an automatic contour refinement process based on deep learning using 80 abdominal MRI cases. They evaluated the deep learning process using four accuracy metrics, namely, dice similarity coefficient, mean distance to agreement, surface dice similarity coefficient and added path length. Based on the good results from the test, Ding et al. [58] concluded that the automatic contour refinement process could quickly correct some inaccurate contours generated from the MRI deep learning tool. They added that the process might be integrated into the auto-segmentation step to enhance MRI-guided radiotherapy.

7.4 Image-Guided Cancer Therapy

Image-guided cancer therapy uses imaging technologies such as CT, MRI, and US to more accurately treat the patient. Advantages of imaged-guided cancer therapy include allowing higher dose to kill cancer cells, avoiding harm to the healthy tissue and keeping patient's side effect to a minimum. In this chapter, we will focus on image-guided radiotherapy as it is the most popular option in cancer therapy.

7.4.1 CBCT Image-Guided Radiotherapy

In radiotherapy, kV-CBCT is used in image-guided radiotherapy for patient setup in daily treatment delivery. The CBCT hardware is attached on the gantry of the medical linear accelerator and the CBCT image acquired before the treatment is compared to the CT image used in treatment planning. Setting up the patient based on the CBCT image can ensure the irradiation geometry close to that created in the treatment plan. For the advance of nanotechnology applied in biomedical imaging, heavy–atom nanoparticles such as gold nanoparticles are used as radiosensitizer in cancer therapy [59]. Addition of nanoparticles to the tumour can result in an increase in image contrast of the target, and an enhancement of cancer cell kill during irradiation. Gold nanoparticles are now popular as a contrast agent in gold nanoparticle-enhanced image-guided radiotherapy. However, one concern is that the addition of gold nanoparticles to the tumour would increase the imaging dose, which leads to various side effects and secondary cancers [60]. Recently, the increase in CBCT imaging dose due to addition of gold nanoparticles was investigated using Monte Carlo simulation [61].

Monte Carlo simulation is a mathematical algorithm based on random sampling, and is particularly good at determining numerical solution from a problem which is very difficult to find out the absolute solution. Since the accuracy of Monte Carlo result depends on the number of Monte Carlo experiments, a very long computing time is required to reduce the simulation uncertainty. This computing time issue is recently solved by the advance of high-performance computing [62]. To determine the imaging dose enhancement with gold nanoparticles added to the tumour, Monte Carlo simulation was carried out based on a Rando-type phantom using phase-space kV photon beams from a CBCT. The increase in imaging dose can be calculated by the imaging dose enhancement ratio (IDER) [61]:

$$IDER = \frac{D_{(water+nanoparticles)}}{D_{(water)}} \tag{7.4}$$

In Eq. (7.4), $D_{(water+nanoparticles)}$ is the imaging dose of the phantom with addition of nanoparticles to the patient's body, and $D_{(water)}$ is the imaging dose of the phantom without adding nanoparticles. Monte Carlo simulation based on the EGSnrc computing code [63] was carried out using CBCT beam energies ranging from 120 to 140 kVp. The projected area of the CBCT was 41×41 cm^2 at a distance of 153.6 cm from the source based on the general treatment practice. Apart from gold nanoparticles, other nanoparticles, namely, platinum, iodine, silver, and iron oxide were used to determine the IDER varying with concentration from 3 to 40 mg/mL.

Figure 7.4 shows the relationship between the IDER and photon beam energy for different types of nanoparticles [61]. The nanoparticle concentration was equal to 18 mg/mL. It is seen from Fig. 7.4 that all values of IDER were greater than one, showing the addition of nanoparticles increased the imaging dose in image-guided radiotherapy. Furthermore, IDER is found decreased with a decrease in CBCT beam energy. This is due to the probability of photoelectric absorption is inversely proportional to the photon beam energy ($1/E^3$) [39]. Among different types of nanoparticles, gold and platinum performed the best compared to others. This is because they have the highest atomic numbers ($Z_{gold} = 79$ and $Z_{platinum} = 78$) among other nanoparticles in the study, and the photoelectric enhancement is proportional to the atomic number (Z^3) [64].

Figure 7.5a and b show dependences of IDER on the concentration and photon beam energy for gold nanoparticles. In Fig. 7.5a, it is seen that the imaging dose

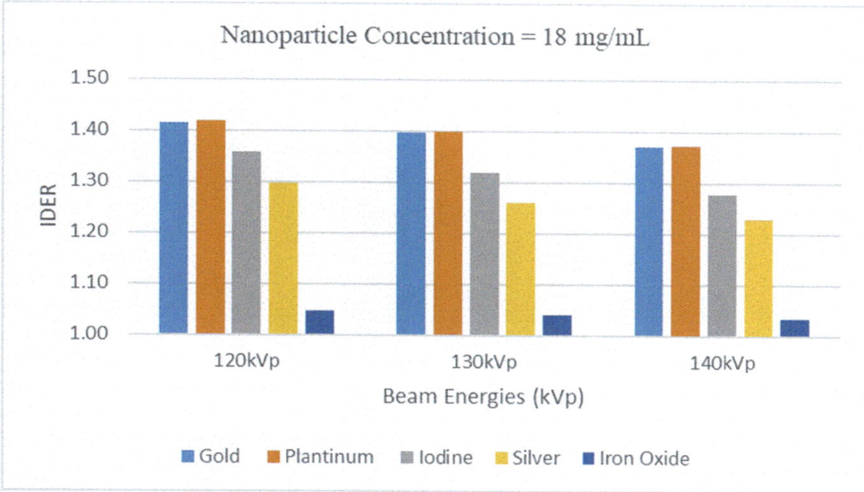

Fig. 7.4 Relationship between imaging dose enhancement ratio (IDER) and CBCT beam energy for different types of nanoparticles (gold, platinum, iodine, silver, and iron oxide). The nanoparticle concentration is equal to 18 mg/mL [61]

increased with the nanoparticle concentration, because more nanoparticles in the patient's body could result in larger photoelectric enhancement. For the variation of IDER on the photon beam energy, gold nanoparticles were found increased in IDER with a decrease in CBCT photon beam energy.

Considering the typical prescription dose per fraction of 2 Gy and the maximum IDER of 1.6 for platinum nanoparticles with concentration equal to 40 mg/mL using the 120 kVp photon beam in the study, the increase in imaging dose was only about 0.63% of the prescription dose. This was within the dose delivery uncertainty of $\pm$ 5% in radiotherapy, and therefore the increase in imaging dose due to addition of nanoparticles was not significant in affecting the radiation dose delivery [60].

7.4.2 MRI-Guided Radiotherapy

MRI-guided radiotherapy is used to carry out using a MRI-linear accelerator system, which is a combination of MRI and radiotherapy to treat cancers. Since MRI is integrated into the linear accelerator, the system can deliver radiation dose and monitor the target volume at the same time. MRI-guided radiotherapy can help radiation staff to have better control over the radiation dose delivery, because they can view the internal anatomy and tumour of the patient in real time when the radiation beam is on. This allows the radiation oncologist to fine-tune and personalize the treatment plan which cannot be arrived in other types of radiotherapy. One concern of MRI-guided radiotherapy is that a strong magnetic field is present during the treatment.

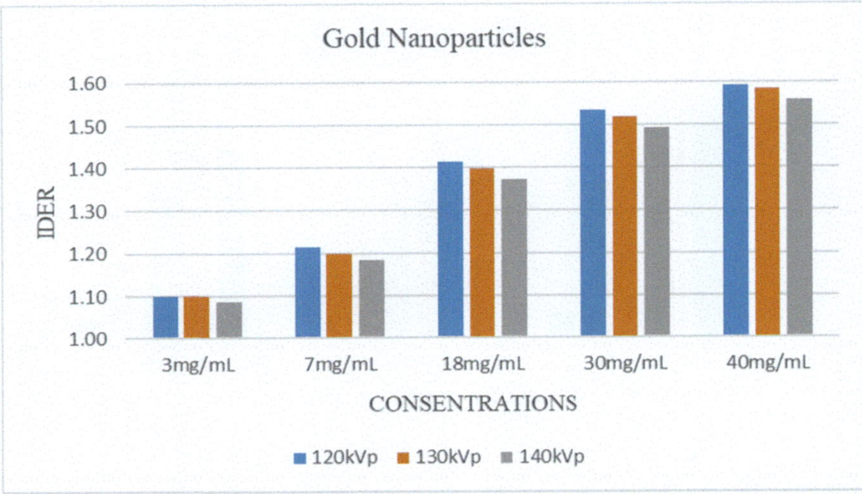

(A)

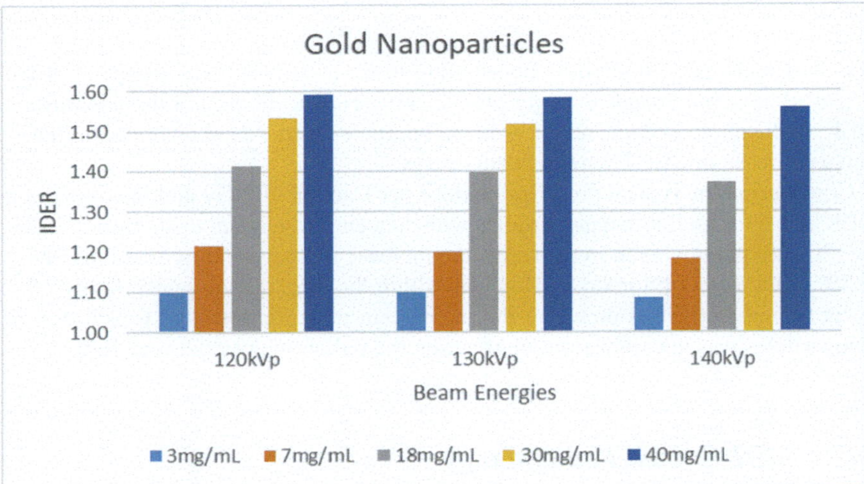

(B)

Fig. 7.5 Relationship between imaging dose enhancement ratio (IDER) and **a** concentration and **b** CBCT photon beam energy for gold nanoparticles [61]

As dose delivery depends on the energy deposition of secondary electrons generated from the radiation beam and electron is a charged particle, the magnetic field from the MRI may influence the dose distribution in the patient [65]. In nanoparticle-enhanced MRI-guided radiotherapy, imaging agents such as gold nanoparticles are added to the patient as radiosensitizer. Since the addition of nanoparticles to the tumour increases the secondary electron yield, it is worthwhile to investigate if the presence of nanoparticles will affect the dose enhancement at the deoxyribonucleic acid (DNA). It is because DNA dosimetry or damage [66] is related to the cancer cell kill or treatment outcome in cancer therapy.

Monte Carlo simulation was conducted to determine the DNA dosimetry when a magnetic field from the MRI was present in the patient [67]. The Geant4-DNA Monte Carlo code, which is specifically designed for nanodosimetry, was employed to predict the dose enhancement at the DNA [68]. For a single gold nanoparticle model containing a DNA irradiated by a photon beam, the dose enhancement ratio (DER) is calculated as:

$$DER = \frac{Dose\ in\ the\ DNA\ with\ gold\ nanoparticle\ addition}{Dose\ in\ the\ DNA\ without\ gold\ nanoparticle\ addition} \quad (7.5)$$

When there was no gold nanoparticle added to the patient, the simulation model had a water nanoparticle to replace the gold nanoparticle. This mimicked the situation of a homogenous water equivalent tumour, and the DER was equal to one. Usually, DER was greater than one with gold nanoparticles added to the patient [69].

Figure 7.6 shows the gold nanoparticle irradiated by a photon beam (50 keV) in the simulation model and the electron tracks generated from the gold nanoparticle were seen in red. When the secondary electrons (i.e. yellow dots) were generated along the electron paths, energy depositions happened. DNA damage could be observed when energy depositions occurred in the DNA to produce lethal damages such as double-strand breaks.

Figure 7.7 shows the relationship between the DER and nanoparticle size (radius = 15, 25 and 30 nm), with different orientations (i.e. parallel to the x, y, and z-axes) of magnetic fields with strengths equal to 0, 1 and 2 T [67]. The strength of magnetic field was to mimic the typical field strength of a conventional MRI (i.e. 1–2 T). The photon beam energies were varied as 50 (Fig. 7.7a), 100 (Fig. 7.7b)), and 150 keV (Fig. 7.7c) to irradiate the gold nanoparticle. Considering the same nanoparticle size with different photon beam energies, it is seen in Fig. 7.7 that the DER decreased when the beam energy increased. For nanoparticle size equal to 25 nm radius, the DER was found decreased from 6.09 to 4.55 when the photon beam energy increased from 50 to 150 keV. This is because the photoelectric enhancement decreased with an increase in kV beam energy. The trends of DER variation on nanoparticle size and photon energy agreed well with results from other works using similar simulation model [70, 71]. However, considering the largest gold nanoparticle (radius = 50 nm) with the lowest beam energy (50 keV) in the presence of magnetic field, the DER only increased by about 2.5–5% for magnetic field strength equal to 1 and 2 T, parallel

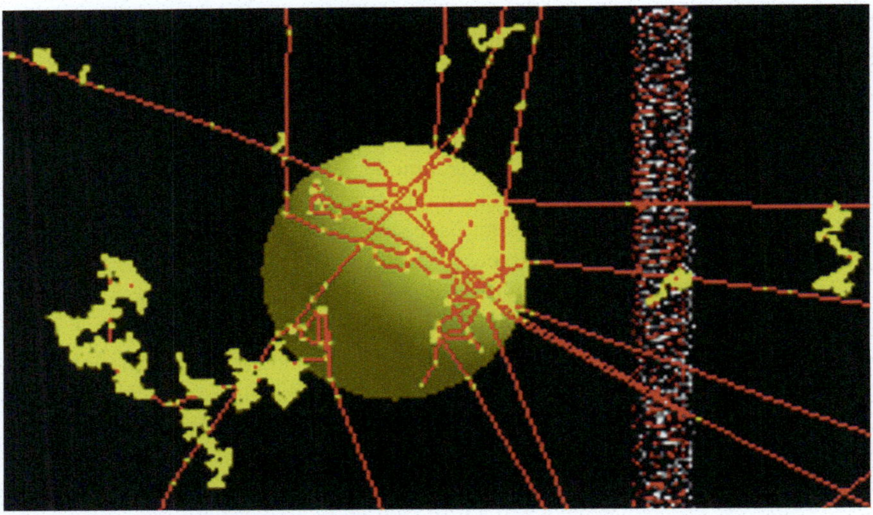

Fig. 7.6 Schematic diagram showing the Monte Carlo simulation model of the irradiated gold nanoparticle with the DNA. Secondary electrons (yellow dots) were generated on the electron path (red) from the particle when photons interacted with the gold. The photon beam energy is equal to 50 keV. It can be seen that some secondary electrons were generated inside the DNA and energy depositions would happen leading to DNA damages [67]

to the x, y, and z-axes. The increase in DER was found even smaller in smaller nanoparticles using higher beam energies.

7.4.3 US-Guided Radiotherapy

Ultrasound provides good contrast and real time images of soft tissue without radiation dose and with relatively low cost. Moreover, in radiotherapy, US imaging system can be designed for special use in the radiation treatment room for patient setup. Nowadays, US imaging can provide daily soft-tissue imaging to determine locations of target tumour and surrounding tissues in treatment delivery. Highly conformal radiotherapy planning techniques such as intensity modulated radiotherapy and volumetric modulated arc therapy are therefore possible with the help of US imaging to verify the patient setup. Many studies are therefore carried out or in progress concerning the development of US-guided radiotherapy for different cancers [72–74].

For free-breathing US-guided radiotherapy, Mezheritsky et al. [75] developed a novel deep learning-based motion modelling framework for US image-guided radiotherapy. The framework took advantage of US imaging such as relatively inexpensive, portal, and non-ionizing. The machine learning algorithm included an image similarity-based rigid alignment model with a deep deformable motion model. Machine training was carried out on a variety of deformations and anatomies of the

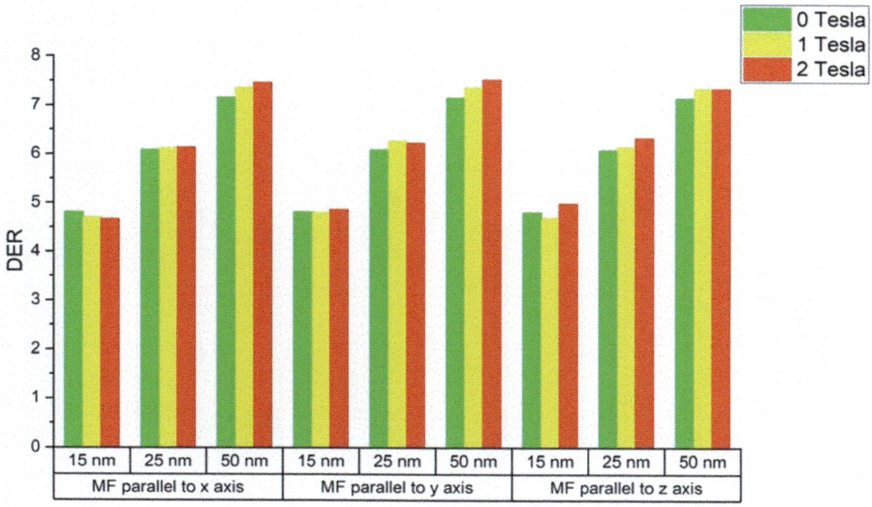

(A)

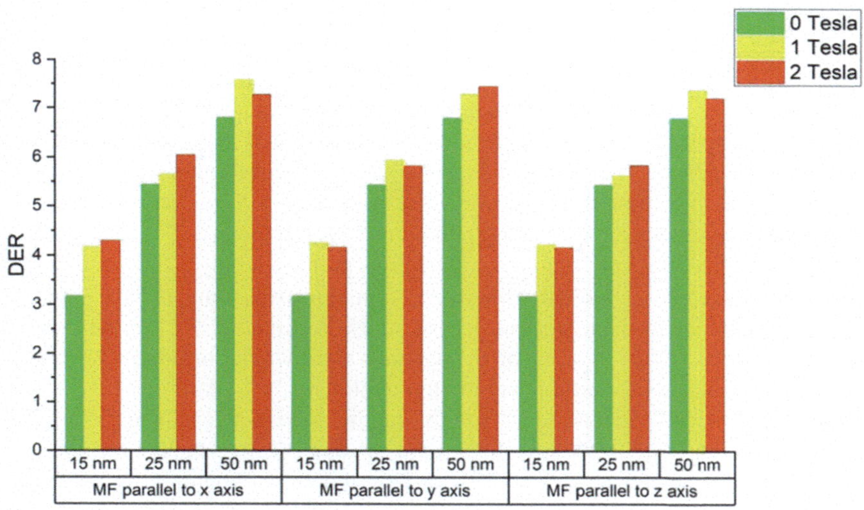

(B)

Fig. 7.7 Relationships between DER and nanoparticle size with variations of magnetic field strength and orientation using photon beam energies of **a** 50, **b** 100, and **c** 150 keV [67]

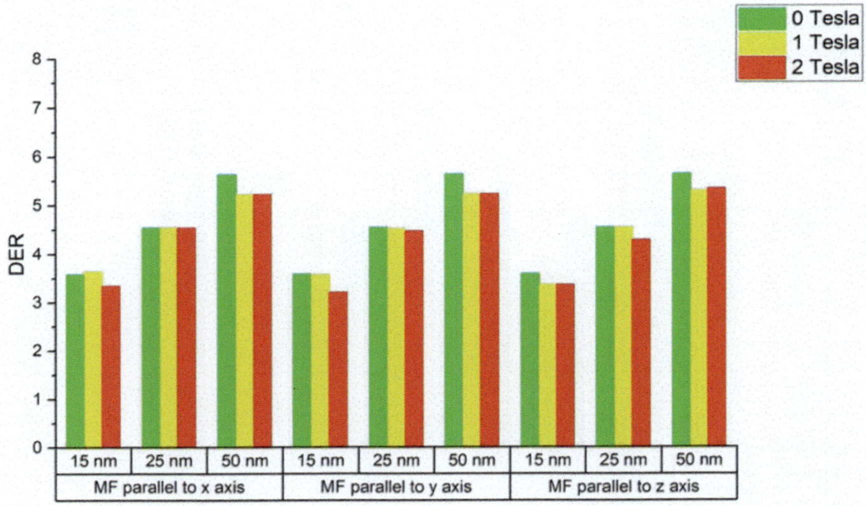

(C)

Fig. 7.7 (continued)

liver site, and the model was verified on a 3D + t US dataset of 20 volunteers. Mezheritsky et al. [75] found that the mean tracking error was 3.5 ± 2.4 mm for the surpass comparable methodologies, which demonstrated that the technique was clinically applicable to radiotherapy. Perrin et al. [76] carried out a treatment planning study to demonstrate the feasibility of transthoracic US guidance to facilitate ventricular tachycardia ablation using photon and proton therapy. For non-invasive treatment to ablate the arrhythmogenic structures, proton therapy was a good option because of its steep dose gradient. Moreover, US imaging was found to have potential to provide guidance in the treatment. In this study, Perrin et al. [76] found that the physical presence of the US probe on the patient's chest would introduce constraints on usable beam angles for proton and photon treatments.

Matrosic et al. [77] performed a study for quality assurance test on an US image-guided radiotherapy delivery system, through dosimetrically benchmarking a gel dosimetry measurement, based on a deformable abdominal phantom. In the study, the dynamically deformable phantom was dosimetrically benchmarked by delivering a liver volumetric modulated arc therapy plan and measuring the dose distribution. The measured dose was carried out and compared to the plan generated from the treatment planning system, and measurements using radiochromic film and ionization chamber. From the results, Matrosic et al. [77] found a good qualitative matching of dose distributions and improvements of the gamma-analysis results for the US-gated dosimeter. To investigate the factors affecting the accuracy and precision in transperineal US motion, Grimwood et al. [78] conducted a clinical study to assess the prostate motion in US-guided radiotherapy. The study required patients imaged using transperineal US. The prostate motion of the patient was measured online at the patient setup and offline by experienced observers. CBCT with markers was used as a comparator and the influence of different clinical factors was examined. From 330 fractions in 22 patients, it is found that random errors increased in fractions with poor image quality, which were associated with inconsistent probe placement, large anatomical changes and unfavorable imaging conditions. Grimwood et al. [78] concluded that though US can provide unique potential to assess inter- and intrafraction organ motion on linear accelerator, inconsistent image quality, inexperienced operator and pressures of clinical environment would degrade the precision and accuracy in treatment delivery.

For US-guided transrectal implantation of fiducial markers without local anesthesia, Sen et al. [79] conducted a study to evaluate the procedure-associated pain and side effects in patients with early-stage prostate cancer. There were 46 patients in this study. Before radiotherapy, all patients were implanted with three intraprostatic fiducial markers using the US-guided transrectal method without local anesthesia. The levels of pain and side effects from the patients were graded and recorded using a patient-based scoring system. From the results, Sen et al. [79] concluded that image-guided radiotherapy accompanied by the US-guided implantation of fiducial markers without local anesthesia was safe with acceptable levels of pain and low side effect. Bednarz et al. [80] performed the first-in-human imaging using the MR compatible e4D US probe for respiration-induced positional change in radiotherapy. These motion management strategies used an electronic 4D MR-compatible US probe designed for hands-free operation in the treatment room. In the study, the US hardware was tested for MR compatibility, and an electromagnetic interference shielding was used to enclose the probe in the operation. Results showed that the presence of US probe only made very minor artifacts in the MR images at a depth smaller than 1 cm. Moreover, the system demonstrated good concordance between the MR and US datasets. Bednarz et al. [80] concluded that their US probe could acquire images during MR acquisition without impacting either the MR or US image quality. Such system could also be applied in interventional procedures of biopsies and drug delivery.

7.5 Portable CT and MRI Scanners in Oncologic Emergencies

Portable scanners (CT and MRI) are emerging technologies that have the potential to improve the diagnosis and management of oncologic emergencies, especially in situations where shifting the patient might not be feasible. Oncologic emergencies are life-threatening conditions that can result from cancer or its treatment, such as spinal cord compression, brain metastases, tumour lysis syndrome, or superior vena cava syndrome [81]. These conditions require prompt and accurate imaging to assess the extent of the problem and guide the appropriate intervention. Portable scanners offer several advantages over conventional scanners, such as they can be brought to the patient's bedside, reducing the time, cost, and risk of transporting the patient to another location [81, 82]. They can also provide fast and reliable results, enabling timely and informed decision-making by the clinicians [83]. Moreover, they can be used in remote or low-resource settings, where access to conventional scanners may be limited or unavailable [82]. However, portable scanners also have some limitations and challenges. For example, they have lower image quality and resolution than conventional scanners, due to the use of lower magnetic fields or smaller detectors [83]. They may not be able to perform some advanced imaging techniques or applications, such as contrast-enhanced studies, functional imaging, or molecular imaging (Fig. 7.8). In addition, they may require additional training, validation, and regulation to ensure their safety, efficacy, and ethical use [83]. Therefore, portable scanners are not meant to replace conventional scanners, but rather to complement and augment them, especially in situations where conventional scanners are not feasible or accessible. Portable scanners may have a significant impact on the diagnosis and management of oncologic emergencies, as well as other clinical scenarios, by providing rapid and convenient imaging at the point of care.

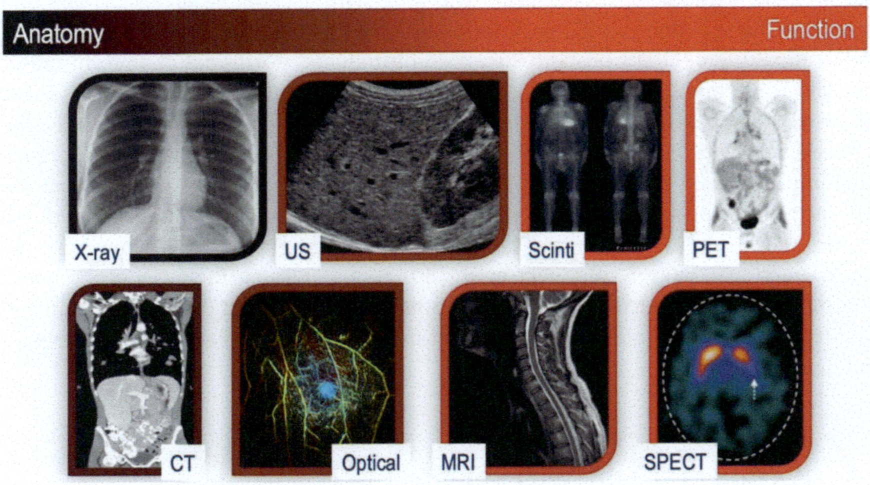

Fig. 7.8 Different imaging techniques, in conjunction with their color-coded capacity to represent both anatomical and functional information, can be illustrated. For instance, x-ray imaging primarily conveys anatomical data, PET imaging offers functional and molecular insights, and MRI and optical imaging can supply a combination of anatomical and functional information depending on the chosen protocol or operational mode [83]

7.6 Theranostics

Theranostics is a term combined with therapeutics and diagnostics. It is used to describe using one radioactive drug to identify the tumour and another radioactive drug to treat the tumour. One example of theranostics is to treat the neuroendocrine tumour with the SSTR2 proteins on the cancer cell membranes as the somatostatin receptor as suggested by Lopez et al. [84]. The receptor is proteins served as a target for cancer drugs. First, the Ga-68 DOTATOC radioactive drug is used to target the SSTR2. The Ga-68 DOTATOC is injected into the patient's body throughout the bloodstream to all organs. When the Ga-68 drug finds the tumour with SSTR2 on the cell membranes, the drug will bind to the cell, acting as a contrast agent for the PET scan. Second, once the neuroendocrine cancer is identified using the Ga-68-DOTATOC in PET scan, the Ga-68 can be replaced with another radionuclide such as Lu-177 or Y-90 to kill the cancer cells that have SSTR2. The therapeutic Y-90-DOTATOC and Lu-177-DOTATATE can both be injected into the patient and travel to tumour having the SSTRS proteins. These two drugs will kill the cancer cells through damaging the cells' DNA. Normal cells surrounding the tumour, which have no SSTR2 proteins, are not affected by the drugs. To date, many studies of theranostics on biomedical imaging are being carried out and in progress [85–87].

7.7 Future Prospective and Conclusions

In this chapter, it is seen that there were many recent rapid advances of biomedical imaging, which benefited cancer diagnosis and therapy. Such advances allow cancer treatment such as radiotherapy and can be delivered at higher dose and more conformal to the target. The future trend of work can be seen using new imaging agents and biomarkers into different levels of treatment. For example, applying biomarkers in radiation treatment planning to optimize the plan based on biological dose in image-guided radiotherapy. These studies can be boosted by other innovative technologies such as AI/machine learning [88], high-performance computing, nanotechnology [89], device engineering, and big data processing. The future of precision cancer diagnosis and therapy will incorporate with evolutions of functional imaging, molecular therapy, AI-assisted adaptive planning, image-guided therapy, nanotheranostics, and advanced dose delivery techniques such as stereotactic body radiotherapy and proton therapy. Some innovations in biomedical imaging such as virtual reality and 3D imaging incorporated into MRI and CT scan can permit clinician to look at a full and more realistic view of the patient's body. It is concluded that the future of biomedical imaging rests in both the innovations of computational speed and technological improvements, as well as developments of imaging agents and biomarkers for cancer diagnosis and therapy.

References

1. C.A. Roobottom, G. Mitchell, G. Morgan-Hughes, Radiation-reduction strategies in cardiac computed tomographic angiography. Clin. Radiol. **65**(11), 859–867 (2010)
2. N.E. Bolus, NCRP report 160 and what it means for medical imaging and nuclear medicine. J. Nucl. Med. Technol. **41**(4), 255–260 (2013)
3. D.R. Brenner, A. Poirier, R.R. Woods, L.F. Ellison, J.M. Billette, A.A. Demers, E. Holmes et al., Projected estimates of cancer in Canada in 2022. CMAJ **194**(17), E601–E607 (2022)
4. S. Lee, Cancer statistics at a glance. Can. Cancer Soc. (2021)
5. R. Baskar, K.A. Lee, R. Yeo, K.W. Yeoh, Cancer and radiation therapy: current advances and future directions. Int. J. Med. Sci. **9**(3), 193 (2012)
6. J.C.L. Chow, Internet-based computer technology on radiotherapy. Rep. Pract. Oncol. Radiother. **22**(6), 455–462 (2017)
7. S. Siddique, J.C.L. Chow, Application of nanomaterials in biomedical imaging and cancer therapy. Nanomaterials **10**(9), 1700 (2020)
8. S. Siddique, J.C.L. Chow, Recent advances in functionalized nanoparticles in cancer theranostics. Nanomaterials **12**(16), 2826 (2022)
9. S. Siddique, J.C.L. Chow, Artificial intelligence in radiotherapy. Rep. Pract. Oncol. Radiother. **25**(4), 656–666 (2020)
10. J.C.L. Chow, A performance evaluation on Monte Carlo simulation for radiation dosimetry using cell processor. J. Comput. Methods Sci. Eng. **11**(1–2), 1–12 (2011)
11. J.C.L. Chow, P. Lam, D.A. Jaffray, Monte Carlo dose calculation using a cell processor based PlayStation 3 system. J. Phys. Conf. Ser. **341**(1), 012028 (IOP Publishing, 2012)
12. Y.C. Chen, D.J.K. Hong, C.W. Wu, M. Mupparapu, The use of deep convolutional neural networks in biomedical imaging: a review. J. Orofac. Sci. **11**(1), 3 (2019)
13. J.C.L. Chow, Synthesis and applications of functionalized nanoparticles in biomedicine and radiotherapy, in *Additive Manufacturing with Functionalized Nanomaterials* (Elsevier, 2021), pp. 193–218
14. J.C.L. Chow, Application of nanomaterials in biomedical imaging and cancer therapy. Nanomaterials **12**(5), 726 (2022)
15. J.C.L. Chow, Computer method and modeling: medical biophysics applications in cancer therapy, medical imaging and drug delivery. AIMS Biophys. **8**(3), 233–235 (2021)
16. I.S. Dennahy, Z. Han, W.M. MacCuaig, H.M. Chalfant, A. Condacse, J.M. Hagood, L.R. McNally et al., Nanotheranostics for image-guided cancer treatment. Pharmaceutics **14**(5), 917 (2022)
17. V. Mergen, A. Euler, K. Higashigaito, M. Eberhard, H. Alkadhi, Photon-counting CT: initial clinical experience, in *Spectral Imaging* (Springer, Cham, 2022), pp. 363–376
18. W.A. Kalender, D. Kolditz, C. Steiding, V. Ruth, F. Lück, A.C. Rößler, E. Wenkel, Technical feasibility proof for high-resolution low-dose photon-counting CT of the breast. Eur. Radiol. **27**(3), 1081–1086 (2017)
19. D. Odedra, S. Narayanasamy, S. Sabongui, S. Priya, S. Krishna, A. Sheikh, Dual energy CT physics—a primer for the emergency radiologist. Front. Radiol. **2**, 820430 (2022)
20. M.F. Kruis, Improving radiation physics, tumor visualisation, and treatment quantification in radiotherapy with spectral or dual-energy CT. J. Appl. Clin. Med. Phys. **23**(1), e13468 (2022)
21. P. Alaei, E. Spezi, Imaging dose from cone beam computed tomography in radiation therapy. Phys. Med. **31**(7), 647–658 (2015)
22. G. Landry, C.H. Hua, Current state and future applications of radiological image guidance for particle therapy. Med. Phys. **45**(11), e1086–e1095 (2018)
23. A.E. Campbell-Washburn, R. Ramasawmy, M.C. Restivo, I. Bhattacharya, B. Basar, D.A. Herzka, R.S. Balaban et al., Opportunities in interventional and diagnostic imaging by using high-performance low-field-strength MRI. Radiology **293**(2), 384 (2019)
24. W.P. Bandettini, S.M. Shanbhag, C. Mancini, D.R. McGuirt, P. Kellman, H. Xue, A.E. Campbell-Washburn et al., A comparison of cine CMR imaging at 0.55 T and 1.5 T. J. Cardiovasc. Magn. Reson. **22**(1), 1–10 (2020)

25. A.E. Campbell-Washburn, Y. Jiang, G. Körzdörfer, M. Nittka, M.A. Griswold, Feasibility of MR fingerprinting using a high-performance 0.55 T MRI system. Magn. Reson. Imaging **81**, 88–93 (2021)
26. A. Lavrova, S. Mishra, J. Richardson, M. Masotti, R. Kurokawa, M. Kurokawa, P. Itriago-Leon, V. Gulani, B. McCracken, K. Wright, H. Hussain, T. Moritani, N. Seiberlich, Quality assessment of routine brain imaging at 0.55 T: initial experience in a clinical workflow. NMR Biomed. e5017 (2023)
27. L. Quettier, G. Aubert, J. Belorgey, C. Berriaud, P. Bredy, G. Dilasser, P. Vedrine et al., Commissioning completion of the Iseult whole body 11.7 T MRI system. IEEE Trans. Appl. Supercond. **30**(4), 1–5 (2020)
28. R.A. Moats, S. Velan-Mullan, R. Jacobs, I. Gonzalez-Gomez, D.J. Dubowitz, T. Taga, W.E. Laug et al., Micro-MRI at 11.7 T of a murine brain tumor model using delayed contrast enhancement. Mol. Imaging **2**(3), 15353500200303112 (2003)
29. Z. Zhang, G. Li, Y. Xu, X. Tang, Application of artificial intelligence in the MRI classification task of human brain neurological and psychiatric diseases: a scoping review. Diagnostics **11**(8), 1402 (2021)
30. D.W. Townsend, Combined PET/CT: the historical perspective, in *Seminars in Ultrasound, CT, and MR*, vol. 29, No. 4 (NIH Public Access, 2008), p. 232
31. S. Jelercic, M. Rajer, The role of PET-CT in radiotherapy planning of solid tumours. Radiol. Oncol. **49**(1), 1–9 (2015)
32. J.H. Jung, Y. Choi, K.C. Im, PET/MRI: technical challenges and recent advances. Nucl. Med. Mol. Imaging **50**(1), 3–12 (2016)
33. K. Pinker, W. Bogner, P. Baltzer, G. Karanikas, H. Magometschnigg, P. Brader, T.H. Helbich et al., Improved differentiation of benign and malignant breast tumors with multiparametric ^{18}fluorodeoxyglucose positron emission tomography magnetic resonance imaging: a feasibility study ^{18}FDG PET-MRI of breast tumors: feasibility. Clin. Cancer Res. **20**(13), 3540–3549 (2014)
34. O. Israel, O. Pellet, L. Biassoni, D. De Palma, E. Estrada-Lobato, G. Gnanasegaran, F. Giammarile et al., Two decades of SPECT/CT–the coming of age of a technology: an updated review of literature evidence. Eur. J. Nucl. Med. Mol. Imaging **46**(10), 1990–2012 (2019)
35. S. Siddique, J.C.L. Chow, Gold nanoparticles for drug delivery and cancer therapy. Appl. Sci. **10**(11), 3824 (2020)
36. J.A. Moore, J.C.L. Chow, Recent progress and applications of gold nanotechnology in medical biophysics using artificial intelligence and mathematical modeling. Nano Express **2**(2), 022001 (2021)
37. J.C.L. Chow, Magnetic nanoparticles as contrast agents in magnetic resonance imaging and radiosensitizers in radiotherapy, in *Fundamentals and Industrial Applications of Magnetic Nanoparticles* (2022), pp. 291–316
38. J.C.L. Chow, Recent progress of gold nanomaterials in cancer therapy, in *Handbook of Nanomaterials and Nanocomposites for Energy and Environmental Applications* (2020), pp. 1–30
39. J.C.L. Chow, Photon and electron interactions with gold nanoparticles: a Monte Carlo study on gold nanoparticle-enhanced radiotherapy. Nanobiomater. Med. Imaging, 45–70 (2016)
40. F. Albayedh, J.C.L. Chow, Monte Carlo simulation on the imaging contrast enhancement in nanoparticle-enhanced radiotherapy. J. Med. Phys. **43**(3), 195 (2018)
41. D.F. Swinehart, The beer-lambert law. J. Chem. Educ. **39**(7), 333 (1962)
42. J.C.L. Chow, Dose enhancement effect in radiotherapy: adding gold nanoparticles to tumor in cancer treatment, in *Nanostructures for Cancer Therapy* (Elsevier, 2017), pp. 383–403
43. J.F. Hainfeld, H.M. Smilowitz, M.J. O'Connor, F.A. Dilmanian, D.N. Slatkin, Gold nanoparticle imaging and radiotherapy of brain tumors in mice. Nanomedicine **8**(10), 1601–1609 (2013)
44. J.C.L. Chow, Characteristics of secondary electrons from irradiated gold nanoparticle in radiotherapy, in *Handbook of Nanoparticle* (Springer, Switzerland, 2015), pp. 1–19
45. S. Siddique, J.C.L. Chow, Machine learning in healthcare communication. Encyclopedia **1**(1), 220–239 (2021)

46. F. Ng, R. Jiang, J.C.L. Chow, Predicting radiation treatment planning evaluation parameter using artificial intelligence and machine learning. IOP SciNotes **1**(1), 014003 (2020)
47. B. Mahesh, Machine learning algorithms-a review. Int. J. Sci. Res. (IJSR) **9**, 381–386 (2020)
48. D.R. Nayak, N. Padhy, P.K. Mallick, D.K. Bagal, S. Kumar, Brain tumour classification using noble deep learning approach with parametric optimization through metaheuristics approaches. Computers **11**(1), 10 (2022)
49. M. Rasool, N.A. Ismail, W. Boulila, A. Ammar, H. Samma, W. Yafooz, A.H.M. Emara, A hybrid deep learning model for brain tumour classification. Entropy **24**(6), 799 (2022)
50. A. Rasool, C. Bunterngchit, L. Tiejian, M.R. Islam, Q. Qu, Q. Jiang, Improved machine learning-based predictive models for breast cancer diagnosis. Int. J. Environ. Res. Public Health **19**(6), 3211 (2022)
51. S. Bacha, O. Taouali, A novel machine learning approach for breast cancer diagnosis. Measurement **187**, 110233 (2022)
52. S. Bacha, K. Ben Abdellafou, A. Aljuhani, O. Taouali, N. Liouane, Early detection of digital mammogram using kernel extreme learning machine. Concur. Comput. Pract. Exp. e6971 (2022)
53. W. Sun, B. Zheng, W. Qian, Computer aided lung cancer diagnosis with deep learning algorithms, in *Medical imaging 2016: computer-aided diagnosis*, vol. 9785 (SPIE, 2016), pp. 241–248
54. L. Hussain, H. Alsolai, S.B.H. Hassine, M.K. Nour, M.A. Duhayyim, A.M. Hilal, M. Rizwanullah et al., Lung cancer prediction using robust machine learning and image enhancement methods on extracted gray-level co-occurrence matrix features. Appl. Sci. **12**(13), 6517 (2022)
55. C.Y. Ma, J.Y. Zhou, X.T. Xu, J. Guo, M.F. Han, Y.Z. Gao, J.S. Maltz et al., Deep learning-based auto-segmentation of clinical target volumes for radiotherapy treatment of cervical cancer. J. Appl. Clin. Med. Phys. **23**(2), e13470 (2022)
56. J. Koo, J.J. Caudell, K. Latifi, P. Jordan, S. Shen, P.M. Adamson, V. Feygelman et al., Comparative evaluation of a prototype deep learning algorithm for autosegmentation of normal tissues in head and neck radiotherapy. Radiother. Oncol. (2022)
57. W. Mao, J. Riess, J. Kim, S. Vance, I.J. Chetty, B. Movsas, A. Kretzler, Evaluation of auto-contouring and dose distributions for online adaptive radiation therapy of patients with locally advanced lung cancers. Pract. Radiat. Oncol. (2022)
58. J. Ding, Y. Zhang, A. Amjad, J. Xu, D. Thill, X.A. Li, Automatic contour refinement for deep learning auto-segmentation of complex organs in MRI-guided adaptive radiotherapy. Adv. Radiat. Oncol. 100968 (2022)
59. J.C.L. Chow, Application of nanoparticle materials in radiation therapy, in *Handbook of Ecomaterials*, ed. by L.M.T. Martinez, O.V. Kharissova, B.I. Kharisov, pp. 3661–3681
60. J.C.L. Chow, Evaluation of the risk and benefit of using functionalized nanomaterials as contrast agents in image-guided radiotherapy: a Monte Carlo study on the imaging dose and contrast enhancement, in *Handbook of Functionalized Nanomaterials* (Elsevier, 2021), pp. 281–308
61. D. Mututantri-Bastiyange, J.C.L. Chow, Imaging dose of cone-beam computed tomography in nanoparticle-enhanced image-guided radiotherapy: a Monte Carlo phantom study. AIMS Bioeng. **7**(1), 1–11 (2020)
62. J.C.L. Chow, Application of cloud computing in preclinical radiation treatment planning. Int. J. Comput. Res. **22**(3), 209 (2015)
63. I. Kawrakow, Accurate condensed history Monte Carlo simulation of electron transport I EGSnrc, the new EGS4 version. Med. Phys. **27**(3), 485–498 (2000)
64. J.C. Chow, C.A. Santiago, DNA damage of iron-gold nanoparticle heterojunction irradiated by kV photon beams: a Monte Carlo study. Appl. Sci. **13**(15), 8942 (2023)
65. X.J. Zheng, J.C.L. Chow, Radiation dose enhancement in skin therapy with nanoparticle addition: a Monte Carlo study on kilovoltage photon and megavoltage electron beams. World J. Radiol. **9**(2), 63 (2017)
66. A.J. Raaijmakers, B.W. Raaymakers, J.J. Lagendijk, Magnetic-field-induced dose effects in MR-guided radiotherapy systems: dependence on the magnetic field strength. Phys. Med. Biol. **53**(4), 909 (2008)

67. M. Jabeen, J.C.L. Chow, Gold nanoparticle DNA damage by photon beam in a magnetic field: a Monte Carlo study. Nanomaterials **11**(7), 1751 (2021)
68. S. Incerti, G. Baldacchino, M. Bernal, R. Capra, C. Champion, Z. Francis, C. Zacharatou et al., The geant4-dna project. Int. J. Model. Simul. Sci. Comput. **1**(02), 157–178 (2010)
69. A. Sadiq, J.C. Chow, Evaluation of dosimetric effect of bone scatter on nanoparticle-enhanced orthovoltage radiotherapy: a Monte Carlo phantom study. Nanomaterials **12**(17), 2991 (2022)
70. C. He, J.C.L. Chow, Gold nanoparticle DNA damage in radiotherapy: a Monte Carlo study. AIMS Bioeng. **3**(3), 352–361 (2016)
71. N.H. Huynh, J.C.L. Chow, DNA dosimetry with gold nanoparticle irradiated by proton beams: a Monte Carlo study on dose enhancement. Appl. Sci. **11**(22), 10856 (2021)
72. M. Van Herk, Different styles of image-guided radiotherapy, in *Seminars in Radiation Oncology,* Vol. 17, No. 4. (WB Saunders, 2007), pp. 258–267
73. L.A. Dawson, M.B. Sharpe, Image-guided radiotherapy: rationale, benefits, and limitations. Lancet Oncol. **7**(10), 848–858 (2006)
74. D. Verellen, M.D. Ridder, N. Linthout, K. Tournel, G. Soete, G. Storme, Innovations in image-guided radiotherapy. Nat. Rev. Cancer **7**(12), 949–960 (2007)
75. T. Mezheritsky, L.V. Romaguera, W. Le, S. Kadoury, Population-based 3D respiratory motion modelling from convolutional autoencoders for 2D ultrasound-guided radiotherapy. Med. Image Anal. **75**, 102260 (2022)
76. R. PPerrin, P. Maguire, A. Garonna, G. Weidlich, S. Bulling, M. Fargier-Voiron, A. Mirandola et al., Case report: treatment planning study to demonstrate feasibility of transthoracic ultrasound guidance to facilitate ventricular tachycardia ablation with protons. Front. Cardiovasc. Med. **9** (2022)
77. C.K. Matrosic, W. Culberson, A. Shepard, S. Jupitz, B. Bednarz, 3D dosimetric validation of ultrasound-guided radiotherapy with a dynamically deformable abdominal phantom. Phys. Med. **84**, 159–167 (2021)
78. A. Grimwood, K. Thomas, S. Kember, G. Aldis, R. Lawes, B. Brigden, H.A. McNair et al., Factors affecting accuracy and precision in ultrasound guided radiotherapy. Phys. Imaging Radiat. Oncol. **18**, 68–77 (2021)
79. C.A. Sen, Ultrasound-guided transrectal implantation of fiducial markers for image-guided radiotherapy of prostate cancer without local anesthesia: Patient-reported gastrointestinal–genitourinary system complications and pain. J. Cancer Res. Ther. **17**(2), 353 (2021)
80. B.P. Bednarz, S. Jupitz, W. Lee, D. Mills, H. Chan, T. Fiorillo, T.K. Foo et al., First-in-human imaging using a MR-compatible e4D ultrasound probe for motion management of radiotherapy. Phys. Med. **88**, 104–110 (2021)
81. J. Turpin, P. Unadkat, J. Thomas, N. Kleiner, S. Khazanehdari, S. Wanchoo, M. Schulder et al., Portable magnetic resonance imaging for ICU patients. Crit. Care Explor. **2**(12) (2020)
82. T.C. Arnold, C.W. Freeman, B. Litt, J.M. Stein, Low-field MRI: clinical promise and challenges. J. Magn. Reson. Imaging **57**(1), 25–44 (2023)
83. T. Beyer, L. Bidaut, J. Dickson, M. Kachelriess, F. Kiessling, R. Leitgeb, O. Mawlawi et al., What scans we will read: imaging instrumentation trends in clinical oncology. Cancer Imaging **20**(1), 1–38 (2020)
84. P.J.P. López, M.S. Pinera, A.M. Fusco, B.D. Brasero, P.P. Muntané, E.R., Clinical codiasimpact of Ga68-DOTATOC PET/CT on neuroendocrine tumors: a preliminary experience. Endocrinología, Diabetes y Nutrici6n (Engl. ed.) **67**(10), 636–642 (2020)
85. Z. Zhang, M. Kang, H. Tan, N. Song, M. Li, P. Xiao, B.Z. Tang et al., The fast-growing field of photo-driven theranostics based on aggregation-induced emission. Chem. Soc. Rev. (2022)
86. L. Filippi, I. Palumbo, O. Bagni, O. Schillaci, C. Aristei, B. Palumbo, Somatostatin receptor targeted PET-imaging for diagnosis, radiotherapy planning and theranostics of meningiomas: a systematic review of the literature. Diagnostics **12**(7), 1666 (2022)
87. S. Debnath, N. Zhou, M. McLaughlin, S. Rice, A.K. Pillai, G. Hao, X. Sun, PSMA-targeting imaging and theranostic agents—current status and future perspective. Int. J. Mol. Sci. **23**(3), 1158 (2022)

88. J.C.L. Chow, Artificial intelligence in radiotherapy and patient care, in *Artificial Intelligence in Medicine.* (Springer International Publishing, Cham, 2022), pp.1275–1286
89. A. Spina, J.C.L. Chow, Dosimetric impact on the flattening filter and addition of gold nanoparticles in radiotherapy: a Monte Carlo study on depth dose using the 6 and 10 MV FFF photon beams. Materials **15**(20), 7194 (2022)

Chapter 8
Bioimaging Applications of Metal and Metal Oxide Nanoparticles

Abinash Das, Sampad Ghosh, and Nabakumar Pramanik

Abstract The early detection of diseases and health disorders is crucial for maintaining good health. To achieve this, there is a pressing need to advance imaging techniques and contrast agents. There is an urgent need for advancements in imaging techniques and contrast agents to achieve fast and detailed imaging of tissue microstructures and accurately characterize lesions. Nanoparticle technology offers a promising solution in this regard, thanks to its unique properties. Metal nanohybrids have garnered significant attention due to their exceptional properties compared to larger individual nanoparticles. Metal and metal oxide nanoparticles represent a novel class of hybrid nanomaterials with a wide range of intriguing properties, finding applications in both nanoscience and biomedical research. Understanding how nanoparticles interact with animal and plant cells is crucial for establishing their significant role and enhancing their effectiveness in health and medical applications. This chapter explores the role of different types of metal-supported nanomaterials and their complex hybrid structures in various biomedical imaging modalities, including antibacterial, antifungal, anticancer applications, and more. Metal and metal oxide nanohybrids are discussed in contexts such as gene delivery, theranostics, and catalysis. Due to their versatility, nanohybrids can be employed in tasks like tumor detection, cell tracking, and visualizing specific disease regions. Future research will focus on developing multifunctional nanoplatforms to ensure accurate, safe, effective, and versatile imaging, which will be essential for future applications in healthcare.

A. Das · N. Pramanik (✉)
Department of Chemistry, National Institute of Technology, Arunachal Pradesh, Jote 791113, Arunachal Pradesh, India
e-mail: pramaniknaba@gmail.com

S. Ghosh
Department of Students Welfare, Maulana Abul Kalam Azad University of Technology, Haringhata, Nadia 741249, West Bengal, India

N. Kalarikkal et al. (eds.), *Multimodal Biomedical Imaging Techniques*,
Biological and Medical Physics, Biomedical Engineering,
https://doi.org/10.1007/978-981-96-1124-9_8

8.1 Introduction

Biomedical imaging probes based on inorganic nanoparticles have been the subject of substantial research as a possible replacement for traditional molecular imaging probes. "Bioimaging" describes capturing images of living things or biological processes. Bioimaging at the subcellular or molecular level has emerged due to recent advancements in the creation of imaging probes. As a result of their physical and physiological features, organic molecules and metalorganic compounds, which comprise the vast bulk of existing imaging probes, have limited clinical usefulness [1]. Fluorescent dyes employed in optical imaging are susceptible to photobleaching, while magnetic resonance imaging (MRI) contrast agents composed of Gd^{3+} chelates have a lacklustre impact owing to their low magnetic moment. Poor targeting efficiency and inadequate imaging enhancement result from the short in vivo half-life of these small molecule-based probes [2, 3].

Nanotechnology has made it possible to create unprecedented, high-performance imaging probes. Due to their novel physical and chemical characteristics that emerge at the nanoscale, inorganic nanoparticles are one of the most investigated materials in this field [4]. As a result, they provide a wealth of opportunities for developing contrast agents boasting high-performance and multimodal capabilities [5]. Because their surfaces may be readily changed to better adhere to the chosen ligands, these nanoparticles are regarded as dependable drug carriers with a robust, tailored delivery of therapeutic compounds. The four primary phases of a drug delivery system (DDS) are retention, targeting, evasion, and release. It also keeps its components in place by encasing, connecting, or releasing them from the body's defenses [6].

Metal and metal oxide nanoparticles and nanomaterials come in different shapes and sizes, including zero-dimensional nanoparticles, one-dimensional nanomaterials, two-dimensional nanomaterials, and three-dimensional nanomaterials. Several nanostructures, such as polymers, liposomes, dendrimers, carbon, and silicon-based nanoparticles, have been used effectively for drug delivery and other therapeutic applications [7, 8].

Metal and metal oxide nanoparticles have been used successfully in the biomedical area for diagnostics and therapies for many years. Magnesium (Mg), gold (Au), silver (Ag), aluminium (Al), palladium (Pd), cerium (Ce), copper (Cu), iron (Fe), selenium (Se), platinum (Pt), and titanium (Ti) are the most attractive metallic nanoparticles because of their exceptional and unique characteristics [9–11].

Metal oxides are widely used in sensor and biosensor technologies. Copper (II) oxide (CuO), zinc oxide (ZnO), gold oxide (AuO), titanium dioxide (TiO_2), iron oxide (Fe_2O_3), graphene oxide (GO), etc. are all good options for sensors because they have desirable properties such as high sensitivity, fast response/recovery (res/rec) time, outstanding reproducibility and stability, and low cost at the expense of easy fabrication [12].

While numerous researchers have extensively examined the characteristics, uses, and synthesis methods of metal and metal oxide nanoparticles, this chapter uniquely spotlights the biomedical dimension of these nanoparticles. It delves deep into their role in diagnostics and therapeutic applications, placing a special emphasis on their significance in the field of therapeutic and diagnostics in the field of health care.

8.2 Emerging Metal and Metal Complex for Bioimaging

8.2.1 Zinc Metal

Compared to conventional organic fluorophores, fluorescent metal complexes have become an essential class of imaging agents because of their photophysical characteristics, which include high fluorescence, a significant Stokes shift, long fluorescence lifetimes, good photostability, and a large two-photon absorption cross-section [13]. For organic molecules, metal–organic coordination is a commonly used method because it facilitates the incorporation of counter-anions and metal ions [14, 15]. Fluorescent metal complexes have been developed as sensors and labels for in vitro and in vivo bioimaging throughout the last two decades, with a focus on complexes based on second- and third-row transition metals and lanthanides.

Comparing Zn(II) to other heavy metal complexes like Ir(III), Ru (II), and Pt(II), we can see that it has a low cytotoxicity and a modest molecular weight. Because of its closed-shell electronic form, which produces fluorescence qualities when bound to a ligand, Zn(II) is also a biocompatible metal seen often in metalloenzymes. Zinc is the body's second most prevalent transition metal and an essential trace element involved in metabolism, gene expression, and signal transduction. Proteins (which make up around 10% of the human body) contain high concentrations of zinc. Zn(II) is a metal ion that plays a crucial role in human development and growth, as well as in the neurological, immunological, and reproductive systems. Zn(II) is considered an active signalling ion in cell biology, creating a balance either within or outside the cell [16–18].

In recent studies, a set of novel terpyridine Zn(II) derivatives (**complexes 1** and **2**) with high two-photon absorption cross-sections was synthesized and characterized [19]. It was discovered that the photophysical characteristics of the related complexes may be significantly altered by changing the halide ions involved. Both the nuclear and plasma membranes may be traversed by the complex (**1&2**), which also has a strong interaction with ribonucleic acid (RNA) in nucleoli and the cytoplasm. A possible dye for intracellular fluorescence imaging and nucleic acid labeling, complex (1&2) preferentially binds to RNA rather than deoxyribonucleic acid (DNA) in cells, as shown by two-photon fluorescence microscopy (Fig. 8.1).

Many proteins involved in controlling transcription include what are called zinc fingers, which are particular specific DNA-binding domains. Regulating gene expression and apoptosis using small-molecule Zn(II) complexes is a topic that receives little attention in the literature. To improve the specific function of zinc finger proteins, scientists have recently synthesized a series of chiral Zn-Salen complexes (Fig. 8.2, **complexes 3–5**). The metal complexes seemed to pass through the cell membrane in

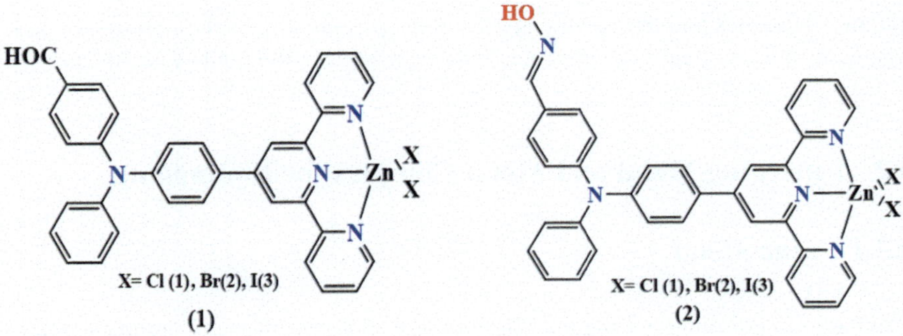

Fig. 8.1 Molecular structure of complex 1 and 2

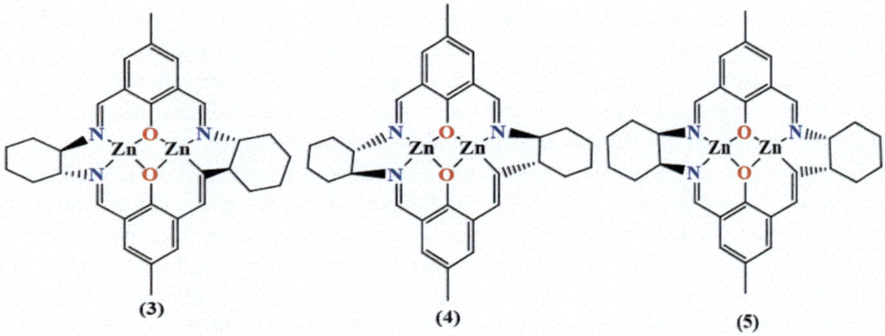

Fig. 8.2 Molecular structure of complex 3, 4 and 5

confocal microscopy images. The well-known complex [the meso-Zn(II) complex] regulates cell cycle and anti-apoptosis genes via binding to bcl-2 and bcl-xL. This research opens up a new avenue for exploring Zn(II) compounds as therapeutic agents [20].

Hydrogen peroxide (H_2O_2) is a major reactive oxygen species (ROS) that contributes to various bodily functions in several ways. Aging and neurodegenerative disorders like Parkinson's and Alzheimer's may be caused by cellular oxidative stress, a biological state brought on by the disorganized control of H_2O_2. DNA damage and chromosomal instability may occur when H_2O_2 levels are not adequately controlled. Monitoring H_2O_2 in living systems is crucial because of its role as a secondary messenger in cell signaling pathways. In addition to mitochondria, endoplasmic reticulum, and cell membranes, previously reported literature suggests that lysosomes play a role in regulating H_2O_2 flux, which defines the ultimate biological outcome of ROS and the cross-talk among subcellular regions [21].

The original fluorescent sensors for biological zinc used quinoline moieties, and despite their ultraviolet (UV) excitation energy, small quantum yields, and typically poor water solubility, they are still used in designs with strong selectivity for zinc

Fig. 8.3 Molecular structure of zinc sensor complex 6

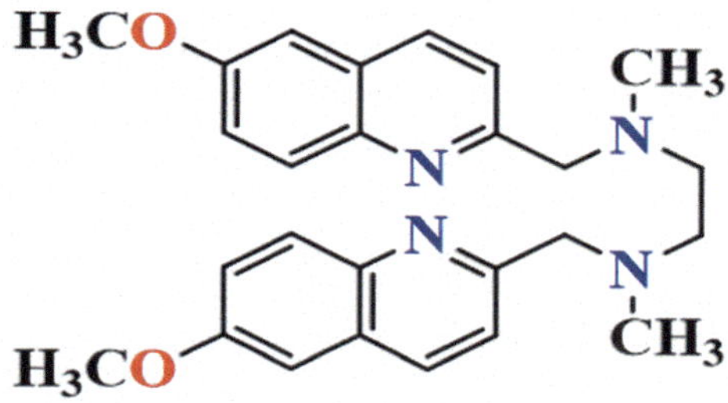

6-MeOBQDMEN

(6)

[22]. These moieties were used in the first fluorescent sensors for biological zinc. Yeast has been shown to be sensitive to two different water-soluble systems: a carboxamidoquinoline derivative and a β-cyclodextrin conjugation. A bisquinoline-based zinc sensor (**complex-6** in Fig. 8.3) demonstrated pH insensitivity and a 40-fold increase in fluorescence emission. This technique was able to detect exogenous zinc in mammalian cells despite zinc's poor solubility in water. Recent reports and proof-of-concept testing in biological systems have shown that sensors based on additional UV-exciting fluorophores are feasible.

Recent methods have used protein-based systems to probe Zn(II) ion distribution, building on previous publications that demonstrated the feasibility of FRET-based biosensors for on-the-go ion detection. A zinc finger domain is used in a FRET-based biosensor to bind zinc with micromolar affinity. This sensor indicated that mobile zinc accumulates and is released in the mitochondria of hippocampus neurons in response to a variety of stimuli. Pancreatic β -cell cytosol and insulin-containing granule Zn(II) concentrations were detected using ratiometric biosensors with affinities ranging from the pico-nanomolar range [23].

8.2.2 Iron Metal

Iron is a crucial element in the form of a trace metal. Iron's ability to transmit electrons and cycle between oxidation states makes it essential in a wide variety of metabolic activities, such as oxygen transport, DNA replication, and a wide variety of redox reactions [24–26]. Many redox enzymes, like cytochrome P450, rely on iron complexes as their catalytic center to conduct redox reactions that are important for energy generation and intermediate metabolism [27]. The extensive valence state transitions and good photomagnetic characteristics, notably for optical and MR imaging, have led to a surge in interest in the use of iron complexes as sensors in the past few years.

Some iron complex–based probes have been created for in vivo detection and have been widely employed in MRI research. The trivalent state of Fe(III) complexes is predicted to be maintained in the biological environment for routine MRI studies [28]. By altering ligand structures, the Fe(III) ion may be stabilized in solution by lowering the Fe(III)/Fe(II) redox potentials. As MRI agents, a series of Fe(III) macrocyclic complexes were described by Morrow's team [29].

Four Fe(III) macrocyclic complexes (Fig. 8.4,**complex-1–4**), including alcohol donor groups have recently been produced as efficient T_1 MRI CAs. Preparation of **Complexes 1–4** was accomplished using Fe(II) salts, whereas in neutral solution, stabilized high-spin Fe(III) exhibited modest redox potentials of less than – 200 mV. It is believed that the solid second-sphere contacts of water-mediated mediated by alcohol or hydroxide ligands or by a hydrogen bonding network are responsible for the high $r_1$relaxivity values of these Fe(III) complexes, which lack a fast exchanging inner-sphere water. Complex-1 exhibits T1 relaxivity in serum phantoms at 4.7 T, which is comparable to that of the Gd(III) agent used in clinical use. In mice, **complex-1** improves kidney selectivity and increases T_1 relaxation rates 30 min after injection, as shown in magnetic resonance imaging studies [30–32].

Recent studies have shown that **complex-5** has a positive influence on T_1 MRI contrast in addition to its high stability and biocompatibility. It has also been shown that **complex-5** may be employed as a photo-thermal agent for the efficient removal of tumors [33] (Fig. 8.5)

Complex-6, shown in Fig. 8.6, is an MR imaging probe based on Fe(III) that changes its relaxivity in response to pH via pH-mediated regulation of monomer and dimer shape. Complex-6 was created to self-associate with Ka at pH ≤ 7.4, and it can rapidly interconvert between a low-relaxivity antiferromagnetically coupled dimeric complex and the high-relaxivity monomeric S ¼ 5/2 Fe(III) complex in

Fig. 8.4 Molecular structure of iron complex (1–4) applicable as bioimaging factor

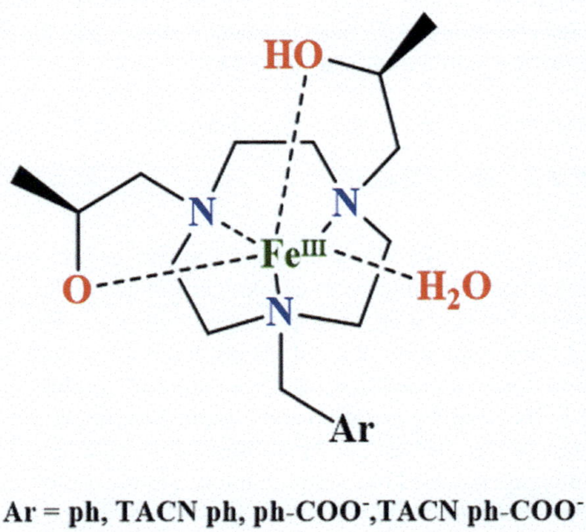

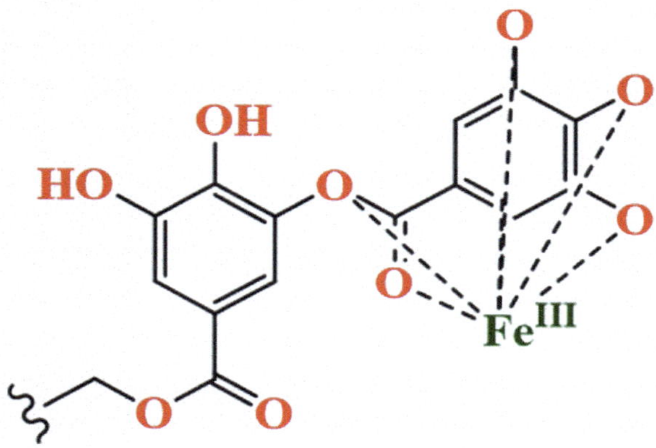

Fig. 8.5 Chemical structure of Iron complex-5 applicable as bioimaging factor

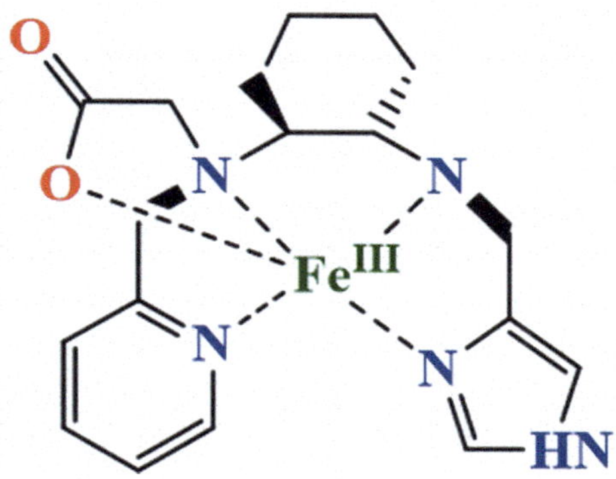

Fig. 8.6 Chemical structure of Iron complex -6 applicable as bioimaging factor

response to pH changes relevant to human pathophysiology (6.0–7.4), making it an ideal candidate for in vivo MRI [34].

In order to detect tissue inflammation in vivo, the scientists synthesized a redox-active Fe complex (Fig. 8.7**, complex-7**), which acts as a biochemically responsive MRI CA. Rapid oxidation of **complex-7** from Fe(II) to Fe(III) using hydrogen peroxide (H_2O_2) results in an order-of-magnitude shift in relaxivity. In reaction to L-Cys and H_2O_2, complex-7 may quickly switch between the Fe(II) and Fe(III) oxidation states. When injected into a mouse model of pancreatitis (the caerulein/LPS model), **complex-7**-Fe(II) causes a robust and selective tissue contrast enhancement, while having no effect on normal pancreatic tissue (saline-treated mice) [35].

8.2.3 Platinum

Applications of cyclometalatedplatinum(II) complexes in biology, such as bio-labeling, cytotoxicity, phototoxicity, and their ability to bind nucleic acids and

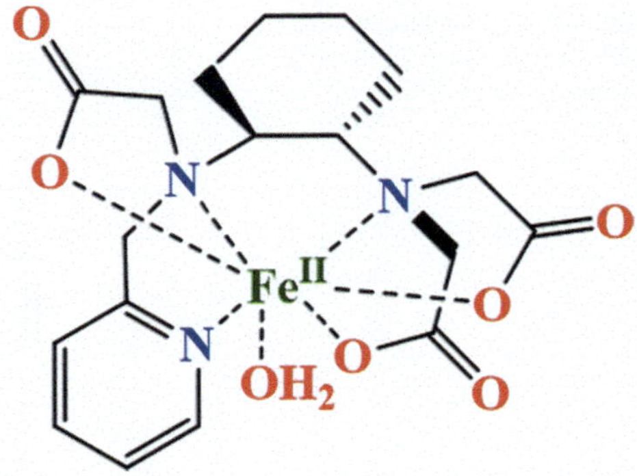

Fig. 8.7 Iron complex -7 molecular structure, applicable as bioimaging factor

proteins, have remained one of the main reasons for very fast growing of research in the field of TMCs.

Metal complexes with biological applications are practically never studied as aggregates but rather as individual monomeric species. Although aggregation might shield metal complexes from environmental interactions and pave the way for the development of species more resistant to chemical degradation, this is not the case.

In the instance of platinum(II) complexes, self-assembled structures lead to intriguing changes in characteristics that might be employed to enhance their usage as labels in either organic solvents or in the solid state. Platinum complexes with a square planar geometry are known to encourage metallophilic interactions, which drives the complexes' aggregate-forming abilities.

Figure 8.8 depicts the structures of two cyclometalatedplatinum(II) compounds presented in the article: [PtII(ppy)(bisNHC^{2C6})]OTf, complex-1 (where ppy is 2-phenylpyridine and bis-NHC is a bis-Nheterocyclic carbene), and [PtII(thpy)(bisNHC^{2C6})]OTf, **complex-2**. As is the case with most phosphorescent substances, the scientists demonstrated that the buffer used substantially quenches the reported green and orange emissive platinum complexes. However, the emission intensity is amplified if bovine serum albumin (BSA) is added to protect the complex from oxygen quenching. These two platinum(II) complexes were also found to be taken up by cells and to be poisonous to them. The authors claim that their Pt(II) systems selectively stain the endoplasmic reticulum (ER) (**complex-2**) after being internalized into HeLa cells. Moreover, 3-(4,5-dimethylthiazol-2-yl)-2,5- diphenyltetrazolium bromide (MTT) assay experiments revealed very high toxicity activity for both complexes towards several cancer cell lines such as nasopharyngeal carcinoma (HONE1, SUNE1), breast cancer (MCF-7), lung carcinoma (HCC827, H1975), hepatocellular carcinoma (HepG2), and a non-tumorigenic liver cell line (MIHA). They found that the toxicity of complex 3 was notably higher than that of cisplatin, a common treatment used to treat human tumors, by a factor of 5.3 to 60. They also described how ER stress controlled the toxicity and ultimately caused cell death [36].

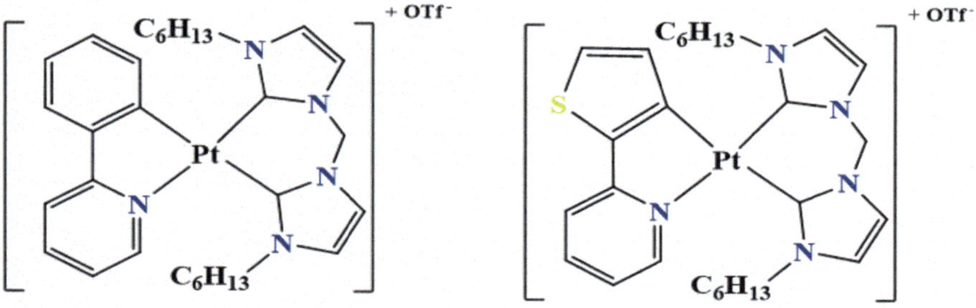

(1)

(2)

Fig. 8.8 Chemical structures of the bidentate platinum(II) complexes

8.2.4 Copper Complexes

Copper, the third most prevalent necessary trace metal, is often found in the bound form of proteins, where it is called a metalloprotein. Some metalloenzymes, such as copper and zinc-superoxide dismutase (Cu, Zn-SOD), have copper complexes at their catalytic centers, similar to the two metal ions outlined above. Copper is essential for life's metabolism because of its function in gene expression, electron transport, and redox reactions [37–39]. In recent years, optical and MR imaging investigations of copper complexes have received a great deal of interest.

Copper complexes may undergo a shift in their photophysical characteristics and fluorescence processes when the metal valence changes. Using a Cu(II)-BCA based color production system, the Liu lab developed a colorimetric alkaline phosphatase (ALP)-linked immunoassay for the sensitive measurement of rabbit IgG and prostate specific antigen (PSA) (Fig. 8.9) [40].

Fig. 8.9 Chemical structures of copper complex used as bioimaging agent

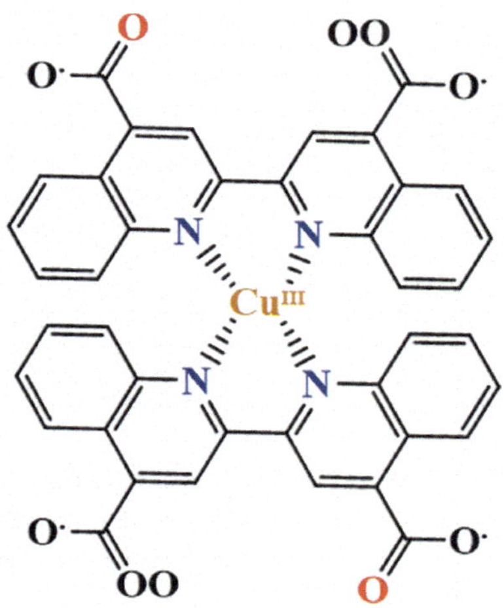

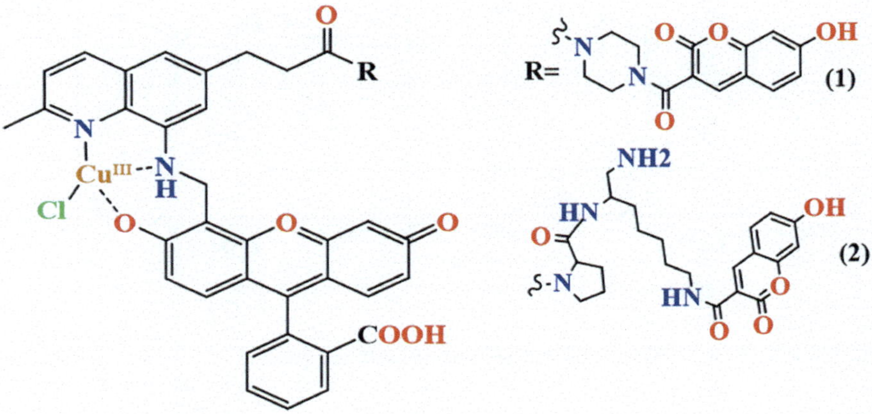

Fig. 8.10 Molecular structures of copper complexes used as bioimaging agent

Fluorescence and the free ligands, which likely interacted further with the bioactive molecules, are both restored when Cu(II) is reduced to Cu(I), which may displace the complexes. In one of a series of experiments involving the direct imaging of NO generation in live cells via turn-on Cu(II)-based fluorescence, the Lippard lab presented the first two Cu(II)-based fluorescent sensors (Fig. 8.10, R1 and R2) for direct ratiometric detection of nitric oxide (NO) [41].

8.2.5 Gadolinium

Nuclear magnetic resonance (NMR) is the foundation of MRI, which uses NMR signals to create tissue contrast to create pictures. The T_1 and T_2 relaxation times of protons within the tissues are shortened by gadolinium-based MRI contrast agents, which improves image contrast. Using contrast agents (CAs) improves contrast imaging by increasing the difference in signal strength between the tumor and surrounding normal tissue.

Mostof currently available clinical CAs for MRI employs Gd complexes, which are paramagnetic. Magnevist, a chelate of Gadolinium (Gd) (III), is a common MRI contrast agent. However, there may be biotoxicity concerns. It is constrained by somewhat reduced relaxivity and moderately shortened blood circulation time. The exceptional properties of gadolinium (III) NPs (Gd-NPs) have led to their use in biological contexts. Because of their favorable properties in terms of molecular size and functionalization, relaxivity, water solubility, and in vivo stability, as well as their low toxicity and thermodynamic and kinetic stability, these NPs are used to formulatethe formulation of improved CAs.

Gd has the most significant thermal neutron capture cross-section (s) of all stable radioisotopes, the most considerable spin magnetic moment of any element, and a

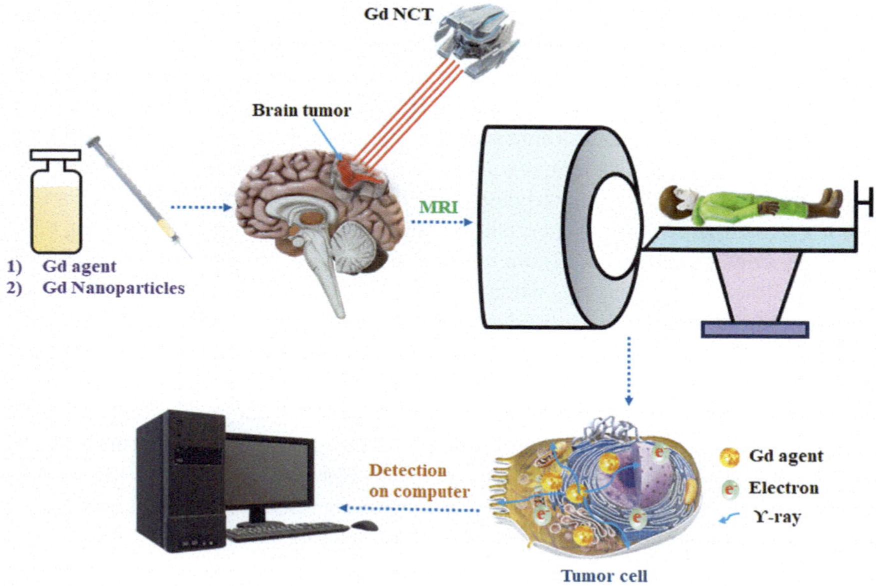

Fig. 8.11 Outline of Gadolinium theragnosis of malignant tumors (a brain tumor is illustrated)

high X-ray attenuation coefficient (m). For positive (T1) MRI, the first quality permits Gd agents to have a high longitudinal water proton relaxivity (r^1). The neutron capture treatment (NCT) of tumors is aided by the third characteristic, whereas CT benefits from the second. Based on these characteristics, Gd drugs may be employed for the theragnosis of malignant tumors (i.e., diagnosingof a malignant tumor by T_1 MRI or CT followed by treatment using NCT, as shown in Fig. 8.11 [42].

8.2.6 Gold (Au) Metal Structures

Conventionally, gold (Au), recognized for its low reactivity and chemical inertness in bulk form, has been deemed a secure material for biomedical purposes. Extensive research has delved into the properties of Au, revealing its distinctive chemical, electronic, and optical attributes that manifest at the nanoscale. As the dimensions decrease to nanometers, Au undergoes unique transformations, making it a focal point for exploration. Numerous investigations have been dedicated to crafting nanostructured Au with diverse morphologies, leveraging the innovative chemical effects at the nanoscale. These endeavors extend beyond biomedical applications, encompassing catalysis, energy, and environmental technologies, reflecting the versatile utility of nanostructured Au in various domains [43–45].

Recent studies have indicated that noble metal nanoparticles, including gold, silver, and platinum, exhibit significant potential in real-time activation and monitoring owing to their light absorption capabilities in biological tissue, particularly at near-infrared wavelengths. The inherent ability of common noble metallic nanoparticles to strongly absorb near-infrared radiation has positioned them as active contrast agents in in vivo imaging therapies. Despite their utility as standalone contrasting agents, noble metallic nanomaterials possess the versatility to amalgamate various imaging modalities, offering synergistic advantages over individual bioimaging techniques. A contemporary development involves the use of computed tomography for three-dimensional imaging, where a series of plane cross-sectional images are computationally interwoven to produce clear and comprehensive three-dimensional results. Notably, composite nanoparticles, such as the FeO/Si core and Au shell composite, have emerged as promising candidates for in vivo applications. These composites serve as dual contrast agents for computed tomography and magnetic resonance imaging, demonstrating excellent computed tomography attenuation and enhanced magnetic resonance imaging results in hepatoma. The versatility of Gd-Au nanocomposites has been validated for dynamic multimodal imaging as contrast agents, highlighting their potential in advancing imaging technologies [46].

8.2.7 Silver (Ag) Metal Structures

Silver (Ag) stands out as a noble metal characterized by its adjustable and distinctive plasmonic attributes, making it highly desirable for diverse applications. Compared to gold, silver exhibits superior thermal and electrical conductivities, along with enhanced efficiency in electron transfer, featuring sharper extinction bands. Furthermore, modified silver nanoparticles demonstrate heightened stability in both water and air. These intrinsic qualities render silver an appealing tool across various domains, including diagnostics, drug delivery, environmental applications, electronics, and antimicrobial interventions. Notably, the utilization of silver nanoparticles has gained prominence, particularly in the realms of biosensors and imaging, showcasing their efficacy in elevating the specific detection of clinical markers, as substantiated by multiple studies.

Silver clusters, garnering considerable attention for biomedical applications, are prized for their low toxicity, size-dependent characteristics, and emission properties. Despite silver's modest photo-oxidative stability, innovative ligands have been demonstrated to induce robust fluorescence signals during the formation process. A recent investigation employed cationic polyelectrolytes to modify fluorescent DNA-silver nanoclusters (NCs), leveraging electrostatic interactions between the positive polymer backbone and the negatively charged phosphate groups of the DNA strand. The outcomes revealed a remarkable threefold enhancement in fluorescence emission from silver nanoclusters, facilitating swift cellular imaging, heightened stability in the intracellular milieu, and improved cellular uptake of DNA-silver nanoclusters. In a parallel study, silver clusters with near-red fluorescence were synthesized using polymethacrylates as a template, serving as a glutathione sensing and bioimaging probe. Another noteworthy advancement involved stabilizing silver clusters (BSA-Ag) with bovine serum albumin, endowing them with near-infrared electrochemiluminescence properties, featuring a robust anodic spectral peak at 904 nm in aqueous media and exceptional bioimaging capabilities. An alternative effective strategy, as evidenced by recent research, entails the development of peptide-capped gold-silver clusters for lysosome-targeted imaging of hypochlorite, exhibiting a high fluorescence quantum yield. This integration approach proves advantageous in constructing heterogeneous composites, surpassing single metals in terms of electronic, optical, and catalytic properties [47].

8.2.8 Manganese Complexes

Numerous reported imaging agents employing manganese complexes are MRI probes because of the multiple oxidation states of manganese, and certain Mn complexes (such MnDPDP) have already received approval as clinical MRI CAs [48].

In mice, biodistribution showed that (Fig. 8.12, **Complex-1**) can be taken in by the liver with the help of both the hepatobiliary and kidney clearing routes. The amphiphilic anion of **Complex-1** increases hepatic targeting mediated by organic anion transporting polypeptides (OATPs) generated by functioning hepatocytes, as shown by imaging, biodistribution, and cellular uptake experiments using bromosulfophthalein (BSP). Cancers with changed OATP expression may now be more easily diagnosed with MRI thanks to the new CA **Complex-1** [49].

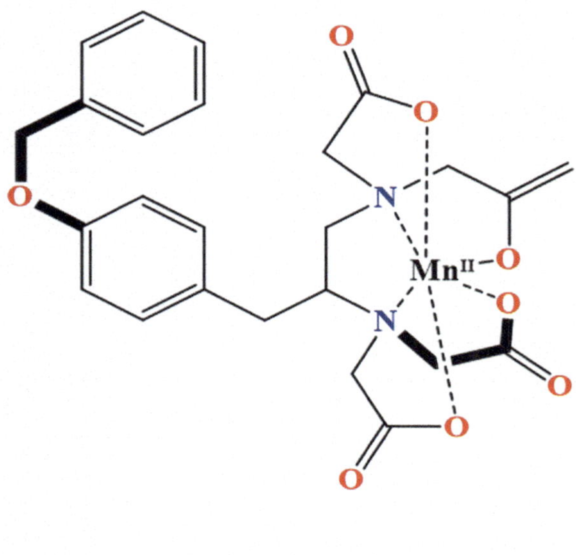

Fig. 8.12 Molecular structures of manganese complex used as bioimaging agent (Complex-1)

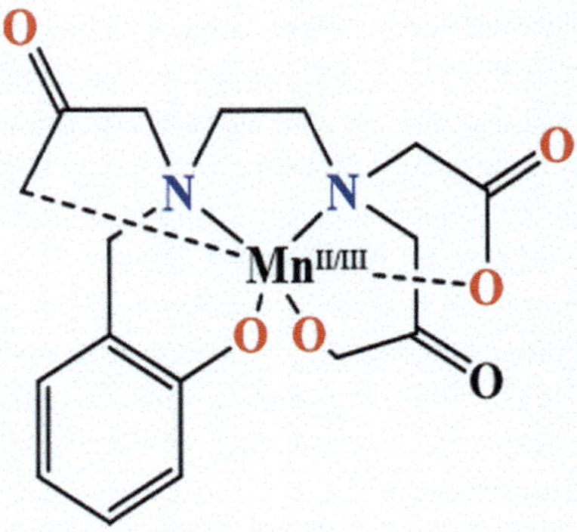

Fig. 8.13 Molecular structures of manganese complex used as bioimaging agent (Complex-2)

Relaxivity changes were seen with the addition of glutathione (GSH) and hydrogen peroxide (H_2O_2) to a simple Mn coordination complex (Fig. 8.13, **Complex-2**) described in 2013 by Caravan's group as a redoxactivated MR CA.

A Mn(II) complex (Fig. 8.14, **Complex-3**) containing quinol-containing ligands was created by Goldsmith's group to act as an MRI CA sensor to H_2O_2. The r^1 response of this compound may be disabled by oxidising its metal center with H_2O_2 to the less paramagnetic Mn(III). H_2O_2 may oxidize the ligand after altering it, leading to a reverse change in relaxivity.

The ligand with two quinols was used to create another redox-responsive mononuclear Mn(II) complex, which was reported by the same team as an MRI CA (Fig. 8.15, **Complex-4**). **Complex-4's** relaxivity response to H_2O is enhanced and its cytotoxicity is mitigated by the addition of the second quinol. Because para-quinones are unable to deprotonate to an anionic form, the oxidation of the ligand allows more H_2O to align with the metal ion, which increases relaxivity, as discovered by the researchers [50–52].

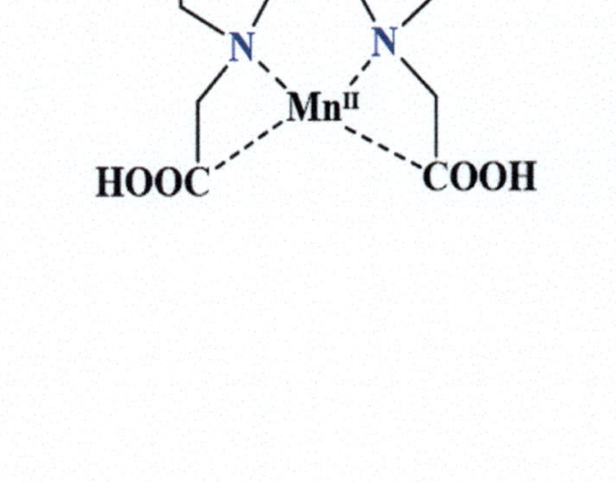

Fig. 8.14 Molecular structures of manganese complex used as bioimaging agent (Complex-3)

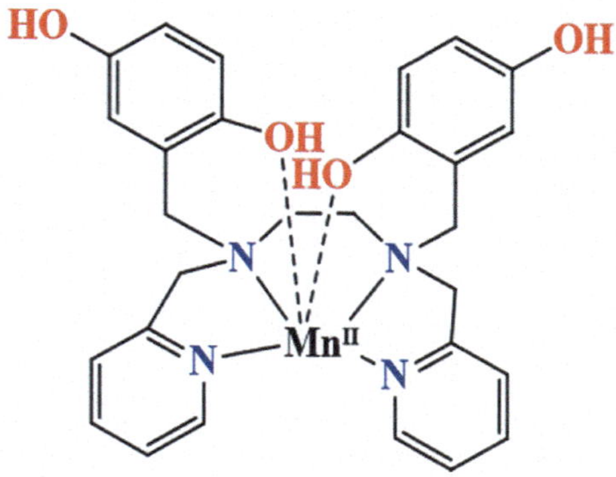

Fig. 8.15 Molecular structures of manganese complex used as bioimaging agent (Complex-4)

8.2.9 Rare Earth Complexes

Rare earth (RE) or lanthanide elements find extensive application in the doping of ZnO nanomaterials, allowing for precise modification of the electronic band structure. Ongoing research underscores that ZnO exhibits rapid recombination of photogenerated charge carriers, leading to diminished quantum efficiency when employed as a photocatalyst. Leveraging lanthanides, with their distinctive 4f configuration, holds promise for extending the recombination time of electron/hole pairs in the semiconductor, thereby enhancing the efficiency of the photocatalytic process and influencing antimicrobial activity. Cerium, for instance, has demonstrated the capability to augment the photocatalytic properties of ZnO nanorods through this mechanism, surpassing the performance of the commonly used reference material, titanium dioxide (TiO_2). The alterations in the band gap and the creation of trap states render RE doping intriguing. Additionally, from a luminescence perspective, this opens up possibilities for utilizing ZnO as an imaging agent [53].

Rare earth (RE) nanoparticles are ceramic nanoparticles composed of oxides, sulfides, halides, or oxysalts containing lanthanide group ions, typically with a + 3 charge. These nanoparticles exhibit distinctive optical properties owing to their 4f or 4f-5d orbital electrons. The unique spectroscopic and magnetic characteristics arise from the shielding effect of electrons in the 4f and 5d orbitals on electrons in the 5 s and 5p orbitals. Notably, Gd^{3+} with its seven electrons evenly distributed in the 4f orbital stands out, displaying superior performance in increasing signal intensity. Gd nanoparticles, like Gd_2O_3, are known to reduce T1 time and enhance T1-weighted images, particularly brightening vascular structures and indicating blood–brain barrier breakdown. GdF_3 nanoparticles serve as more suitable host matrices for luminescent centers. Introducing different coatings, such as 2-aminoethyl phosphate and poly (maleic anhydride-alt-1-octadicene), enhances the performance of Gd nanoparticles in decreasing T1 as a rare-earth contrast agent. Moreover, Gd nanoparticles, depending on the coating, can also be employed to decrease T2, with coatings playing a pivotal role. Functionalizing Gd nanoparticles with aspartic-dextran polymers or coating them with D-glucuronic acid results in a significant reduction in T2 time, showcasing the versatile applications of Gd nanoparticles in medical imaging [54, 55].

8.3 Metal Oxide Nanoparticles and Derived Composites

8.3.1 Zinc Oxide (ZnO)

Semiconductor nanoparticles provide substantial benefits in biological applications including bioimaging and medication administration because of their unique physical and chemical characteristics. ZnO nanoparticles' cheap cost and high biocompatibility make them an attractive option for use in bioimaging and medication delivery. ZnO nanoparticles are a viable new option since they are inexpensive, stable in air and sunlight, do not generate pollution, and have acceptable safety characteristics. Despite its potential, luminous ZnO nanoparticles with high stability in water have proven challenging to synthesize. ZnO's surface flaws are the primary source of its visible fluorescence, and water molecules may eliminate this fluorescence [56–58].

ZnO fluorescence must be excited by UV light since the bandgap of ZnO (3.37 eV at room temperature) resides in this range. However, as UV radiation can only penetrate the skin by a few millimeters, it is inappropriate for most in vivo research. To enable MRI methods to identify the labeled tissues deep inside the animal body, it was proposed to dope ZnO nanoparticles with magnetic elements to generate novel binary probes with fluorescence and magnetic functionalities. On the other hand, numerous ZnO nanoparticles doped with common magnetic elements including Fe, Co, Ni, and Mn exhibited essentially slight luminescence in the visible area. Human cervical carcinoma HeLa cells were fluorescently labeled with Gd^{3+}-doped ZnO nanoparticles using a sol–gel preparation method and then identified using magnetic resonance imaging (MRI) and fluorescence microscopy (FM). Since synthetic processes do not harm ZnO surface flaws at or near room temperature, sol–gel methods are favored as the best way to alter ZnO QDs. The use of wet chemical methods has progressed significantly. Polyvinylpyrrolidone (PVP), oleic acid (OA) and diethanolamine (DEA), polyethylene glycol methyl ether (PEGME), and polymethylmethacrylate (PMMA) are only a few examples of the reported capping groups for ZnO nanoparticles [56, 59].

Researchers have recently synthesized Fe_3O_4@ZnO core–shell nanoparticles to transport carcinoembryonic antigen (CEA) into dendritic cells (DCs) to realize immunotherapy for mouse cancer. Nanocarriers for efficient antigen delivery, these Fe_3O_4@ZnO core–shell nanoparticles also served as an imaging agent, detectable in vitro using confocal laser scanning microscopy (CLSM) and in vivo with magnetic resonance imaging (MRI). ZnO NPs have wide-ranging applications because of their efficient blue and near-UV emissions, which contain green or yellow luminescence associated with oxygen vacancies. The study's first-ever stable aqueous ZnO@polymer core–shell nanoparticles (ZnO@poly(MAA-co-PEGMEMA)) were synthesized using a straightforward sol–gel process. In aqueous liquids, the ZnO@polymer core–shell nanoparticles showed a high quantum yield and a stable, wide photoluminescence. ZnO-1 (made from LiOH) fluorescence green in human hepatoma cells, whereas ZnO-2 (made from NaOH) fluoresced yellow. Both ZnOs had average sizes of 3 and 4 nm. At doses lower than 0.2 mg/mL, these nanoparticles exhibited no toxicity towards human hepatoma cells. The cells were still alive after 45 min of treatment, and the luminescence was very stable throughout cell culture [60, 61].

Recent studies have used chemical precipitation to make ZnO NPs. There were blue, green, yellow, and orange flashes of light coming off of it. Changing the pH of the precipitation solutions affects the color of the emission. They produced ZnO@silica core–shell nanostructures by encapsulating ZnO NPs in silica to make them more stable in water. The produced ZnO@silica core–shell nanoparticles were highly water-resistant while still emitting light in the visible spectrum. Attaching to the surface of NIH/3T3 cells, it showed a spectrum of fluorescent colors, each emitted light at a slightly different wavelength [62].

In the quest to enhance the photocatalytic efficacy of ZnO, researchers have dedicated their efforts to address two pivotal challenges. Firstly, ZnO exhibits a wide band gap of 3.7 eV, rendering it inefficient in harnessing visible light and leading to a limited spectral range, consequently yielding a small apparent rate constant for photocatalysis. Secondly, the rapid recombination rate of photogenerated carriers diminishes the production of reactive oxygen species (ROS), thereby reducing overall efficiency (Fig. 8.16). Researchers have undertaken various strategies to overcome these hurdles, involving the modification of ZnO through processes such as doping, deposition of noble metals, or combining it with materials featuring narrower band gaps. These modifications serve to broaden the photoresponse spectrum, enhance the efficiency of charge separation, and augment ROS generation, ultimately resulting in superior photocatalytic performance [63].

8.3.2 Iron Oxide

Due to the magnetic features of the core and the small size of the particles, iron oxide nanoparticles (IONPs) are superparamagnetic and can only become magnetized in the presence of a magnetic field. As a result of their one-of-a-kind characteristics, IONPs have proven helpful as contrast agents in MRI scans, where they generate hypointense (dark) signals in T_2-weighted MR images. In nanomedicines authorized by the Food and Drug Administration (FDA), iron oxide has been the subject of the most significant research. Magnetic iron oxide NPs (MNPs), composed of magnetite (Fe_3O_4) or maghemite (Fe_2O_3), with specific sizes (from 15 nm and no more than 100 nm), have shown to be efficient contrast agents, drug delivery vehicles, and thermal-based therapies. The diagnostic, imaging, and photothermal therapeutic capabilities of this material are unparalleled. Biocompatibility is helping to

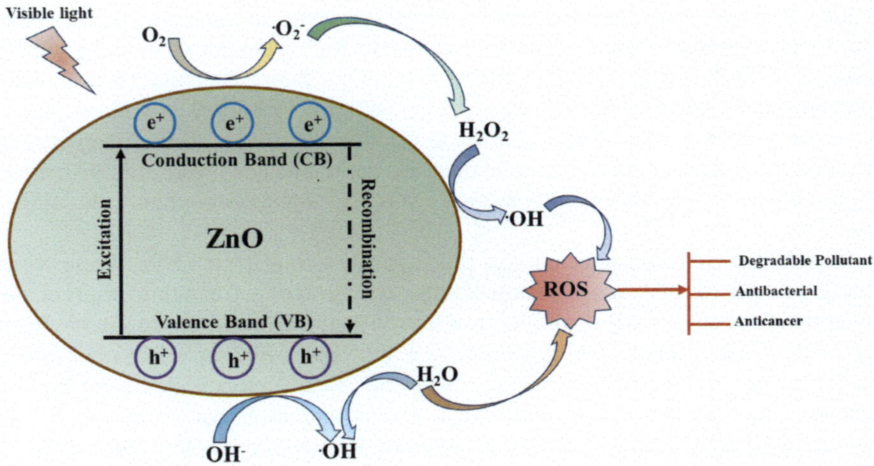

Fig. 8.16 Graphical representation of the photocatalytic mechanism of ZnO

meet a growing need in fields where organic materials cannot provide the necessary qualities. The use of contrast agents in imaging diagnostics and hyperthermia therapy for cancer, as well as medication administration using superparamagnetic iron oxide nanoparticles. The environmental applications of magnetic nanoparticles are many. Ferumoxsil is an IONP-based contrast agent authorized by the FDA in 1996 for imaging of the gastrointestinal (GI) tract but was subsequently removed from the market due to safety concerns. The majority of IONP-based imaging agents, however, need intravenous administration. The contrast agents ferumoxide and ferucarbotran were explicitly developed for liver imaging and gained approval in the United States and Europe. Both were taken off the market in 2008/2009 owing to various adverse effects, leaving just Resovist® (albeit only in certain countries) until this day. Ferumoxtran-10 was explicitly developed for imaging lymph nodes in metastatic prostate cancer [64, 65].

8.3.3 *Titanium Dioxide* (TiO_2)

In recent years, scientists have been more interested in TiO_2 because of its distinct chemical active surface and its wide range of possible uses. Nanotechnology has efficiently synthesized for the efficient synthesis of TiO_2 nanoparticles, nanotubes, and nanoprobes marked with fluorescent dyes or magnetic resonance contrast agents for use in fluorescence analysis or magnetic resonance imaging (MRI) of cells. The future of a cell may be seen via imaging and observation. Due to its energy gap (3.0–3.2 eV), TiO_2 can only be excited by UV light. It is challenging to use TiO_2 in biological applications since the nanoparticles are often insoluble in aqueous solution. High-temperature non-hydrolytic fabrication of water-soluble TiO_2 nanoparticles was shown in a study. With a diameter of 3.5 nm and a length of 10.4 nm, these nanorods are shorter and more rod-like than the commercially available Degussa P-25 nanoparticles, and they are more hazardous to human melanoma cells (A375) when exposed to UV light.

Mesoporous titania nanoparticles (MTNs) with outstanding biocompatibility (LC50 E 400 μg mL^{-1}) and a large surface area (ca. 237.3 m^2 g^{-1}) have been developed in the in a recent work, and they were further functionalized with a fluorescent molecule that contains phosphate (flavin mononucleotide; FMN). Green fluorescence from FMN-MTNs was seen in the cytoplasm after 4 h of incubation with BT-20 cells, suggesting that the FMN molecules were mostly intact inside the MTNs. As an intracellular bioimaging agent, FMN-MTNs performed well. Cancer cells may be tracked with pinpoint accuracy by conjugating the particular antibody (CEA) produced by cancer cells with these fluorescent contrast-modified TiO_2 particles.

Similar findings were reported in another work that compared the performance of TSAlClPc to that of undoped and nitrogen-doped TiO_2 NPs and their conjugates. For the HeLa tumor cell analysis, photosensitizers were used at doses between 5 and 20 mg/mL. Visible light irradiation of various locations between 420 and 800 nm (15.9 J/cm^2) and 420 and 575 nm (7.5 J/cm^2) was used to analyze the absorbance and photokilling impact on HeLa cells [66, 67].

8.4 Quantum Dots for Bioimaging

Numerous biological uses, including imaging, sensing, drug transport, and photodynamic treatment, have piqued interest in fluorescent nanodots in the twenty-first century. Bioimaging, which uses fluorescence microscopy to examine living cells, is a potent tool in the wide world of biomedicine. It enables the immediate observation of events taking place in living organisms. The use of organic and inorganic nanomaterials as fluorophores for bioimaging applications has been the subject of several publications. During the last decade, metal-based chalcogenide inorganic semiconductor nanocrystals such as quantum dots (QDs) like CdX (X = Se, Te) capped with different long-chain alkyl thiols were widely employed as fluorescent probes as an alternative to organic fluorophores. Semiconductor QDs have several advantages over organic fluorophores, including their tunability in terms of optical properties (absorption, emission wavelengths, and emission quantum yield) and their quantum confinement effect (i.e., change in particle of QDs leads to changes in optical properties; for example, an increase in particle size will shift the absorption in the longer wavelength due to the reduction in bandgap energy). As a result of their core–shell structure capped with ligands, quantum dots exhibit superior photostability versus organic fluorophores. However, most quantum dots (QDs) include hazardous heavy metals, including Cd, Pb, and Hg, which is a significant problem from the biocompatibility perspective. Toxic metals encased with a surface covering inside the cell but could still leak out into the medium and injure living cells might have been released during the imaging study [68].

8.4.1 ZnO Quantum Dots for Bioimaging

The ZnO quantum dot is an eco-friendly material with promising future uses in biomedical imaging and cancer detection. The ZnO QD is a promising material for use in biomedicine because of its low material cost and high biocompatibility. ZnO QDs show promise as non-toxic, thermally, and chemically stable alternatives for diagnostics and imaging, but their poor performance in water has severely constrained their usage. Applications for highly luminous ZnO quantum dots include cell imaging, LEDs, and anticounterfeiting measures. It has been reported that Gd-doped ZnO QDs may be created by modifying the surface using N-(2-aminoethyl) aminopropyltrimethoxysilane. The MTT cell proliferation experiment demonstrated that Gd-doped ZnO QDs were nontoxic up to 1 mM, allowing their use in fluorescence imaging of HeLa cells.

Covalent attachment of PAMAM (poly(amidoamine)) dendrons with the siloxane group as the central focus resulted in the synthesis of highly luminous water-soluble ZnO quantum dots. Because of its branching structure, strong water affinity, and multivalent surface groups, PAMAM dendrimers have been widely employed as carriers for improving the bioavailability of medicines and nanoparticles. After 20 days, ZnO@PAMAM's PL QY reached 59%, most likely owing to a surface-ordering of the siloxane-capping that yields ZnO QDs with more effectively shielded luminous centers from water assault. These highly luminescent ZnO QDs were successfully used for imaging of the gram-positive bacteria staphylococcus aureus, and their biocompatibility was significantly improved in comparison with Cd-based ones, as growth inhibition tests demonstrated that they could be used at concentrations up to 1 mM without affecting cell growth. Recent research declared that ZnO nanocrystals may be successfully encapsulated inside the nonpolar core of phospholipid micelles. Micelles serve as stabilizers of nanoparticles and enable them to perform site-specific drug delivery thanks to their unique qualities, such as a low critical micelle concentration and strong water stability. Micelle-encapsulated zinc oxide (ZnO) optical probes are appropriate for bioimaging applications because they can accurately pick the input wavelength, resistance to photobleaching, and lack of heat dissipation into the cells. ZnO-FA nanoparticle internalization in living cells demonstrates the promising future of micelle-encapsulated ZnO nanoprobes for intracellular photochemical interactions and photodynamic treatment [69].

8.4.2 Carbon Quantum Dots for Bioimaging

The term "carbon dot" is often used to describe a wide range of tiny carbon nanoparticles in aqueous or other solutions. The discovery that single- and multi-walled carbon nanotube surface defects became brightly emissive under certain conditions provided the first clue that nanoscale carbon particles could be photoluminescent. This was especially true when the nanotubes were functionalized for dispersion at the individual nanotube level without bundling or aggregation.

Employing their visible excitation and emission wavelengths, their intense fluorescence at the individual dot level, and their excellent photostability, carbon dots are shown to be helpful in the fluorescence imaging of cells. PEGylated carbon dots were incubated with Escherichia coli for fluorescent labeling of the bacterium cells, and oligomeric amino polymer-functionalized carbon dots were treated with Caco-2 cells for internalization, both in the pioneering investigation of carbon dots by current research. Confocal microscope imaging showed that most of the endocytosed carbon dots remained in the cytoplasm, with just a tiny percentage entering the nucleus.

The current research uses carbon dots produced from graphene oxide precursors to visualize MG-63 cells. Carbon dots were synthesized by hydrothermal carbonization of chitosan, and their bright fluorescence emissions from the cytoplasm were seen at 405 and 488 nm excitations. The carbon dots were used to create copies of A549 cells. When examined under a microscope, fluorescent spots were only visible in the cytoplasm and cell membrane, with considerably weaker signals in the nucleus. Carbon dots created by caramelizing PEG oligomers in the microwave were used to UV-label cells in a study. However, the imaging magnification was insufficient to pinpoint the precise positions of the dots inside the cell [70].

8.5 Applications of Metallic and Metal Oxide Nanoparticles

8.5.1 Biosensing and Nanoelectronic Applications

Nanotechnological intervention in electronics results in the creation of electronic components with reduced size and altered display, power consumption, and research advancement. Since nano-sized sensors have been developed, their use has expanded beyond the house to the battlefield. Materials created for nanoelectronics have several applications in biology, including detecting illness biomarkers and infectious organisms [71].

Biosensors are now common due to their wide variety of uses outside of the traditional biological sphere, including but not limited to monitoring remedy and disease progress, drug development, food regulation, environmental invigilation, and the biomedical research sector [72]. Numerous biosensors, such as the cellulose strip sensor constructed of NHs of silver-gold (Ag-Au) core–shell NPs with surface-enhanced Raman scattering (SERS), have been produced. Semiconductors, such as tungsten oxide (WO_3), titanium oxide (TiO_2), tin oxide (SnO_2), and molybdenum oxide (MoOx), as well as carbon-based NPs doped with metallic ions, were employed in the manufacture of biosensor NHs [73, 74].

Biosensor applications for TiO_2 NPs cover a range from inorganic-inorganic to inorganic–organic NHs [75]. Nanoparticles (NPs) have served as both ionization matrices and capturing probes in many applications; for example, Au-NPs control cellular absorption in laser desorption/ionization mass spectrometry (LDI-MS), and cationic moieties are used to functionalize the NPs' surfaces. The results of this study demonstrate that Au-NPs are persistent in the internal cellular milieu without using a matrix and that these cells may be employed as a probe for drug administration, protein capture, treatment, and enrichment [76].

8.5.2 Bioimaging

Biomedical imaging employs various nanoparticles, including solid lipid nanoparticles, nanotubes, metallic nanoparticles, quantum dots, dendrimers, polymeric nanoparticles, and liposomes. Nanoparticles are promising probes for disease detection because of their unique features, which include surface chemistry, magnetic properties, absorption properties, and emission qualities [77]. For example, L-glutathione was used to synthesize red fluorescent Au nanoclusters, which were then combined with porphyrin derivatives [tetrasodium pyrophosphate (TSPP)] for in vivo and in vitro imaging to identify cancer and noncancer cells. The bioimaging method is simplified since the synthesized nanoclusters are expelled without causing any harm to the live cells being imaged. As a result, this approach may be investigated by scientists as they create novel nanomaterials based on metals for use in bioimaging. Similarly, Bengal rose extract was used to synthesize dextran-coated gadolinium oxide nanoparticles (Gd_2O_3 NPs), which showed good paramagnetic behavior and might be used for bioimaging and tracking. In another study, a biodegradable nanoparticle was used to create a superparamagnetic iron oxide nanoparticle's core–shell structure, which was then used as an MRI contrast agent. This nanoparticle had a hydrophobic poly(lactic-co-glycolic acid) [PLGA] core and a positively charged glycol chitosan shell [78–80]. In a similar way, methoxy[poly(ethylene glycol)]trimethoxy-silane (PEG-silane) and bismuth oxide (Bi_2O_3) were used to make PEGylated bismuth (PEGBi) nanoparticles. PEG-Bi nanoparticles synthesized in this study performed very well in in vivo X-ray computed tomography imaging and photothermal cancer treatment. Nanostructured bismuth oxide was used as a radio sensitizer and gadolinium nanoparticles were used as a contrasting agent in an experiment to detect glucose [81, 82].

8.5.3 Anti-diabetic Activity

Diabetes is a metabolic illness caused by sugar buildup in the blood due to the body's inability to produce enough insulin to enable the cells to respond to insulin. Consequently, there are two types of diabetes: insulin-dependent and insulin-independent. Two of the most important regulators of this condition are the enzymes a-glucosidase and amylase. The enzyme secretion in diabetes is reduced with bio-synthesized nanoparticles of metal oxides, some of which are discussed below. Lonicera japonica was used in the biological synthesis of Ag NPs due to the plant's wide range of therapeutic qualities. These include antiviral, anti-inflammatory, and anti-diabetic effects. Diabetic carbohydrate-digesting enzymes, a-amylase and a-glucosidase, were effectively inhibited by the synthesized nanoparticle's antidiabetic action [83]. Similarly, green-synthesized silver nanoparticles made from the leaf extracts of Pouteria sapota stopped a-amylase from working, and the non-enzymatic glycosylation of hemoglobin proved that the particles were anti-diabetic. In a similar vein, selenium nanoparticles are shown to be effective in treating diabetes because they inhibit hypoglycemia action while also protecting cells from oxidative stress [84]. In addition, the anti-diabetic benefits of chitosan-stabilized selenium nanoparticles (CsSeNPs) were discovered in a rat model [85]. It has also been shown that when selenium was mixed with the cerium oxide nanoparticle, ROS levels were lowered with increased insulin secretion, thereby controlling the production of insulin and minimizing the oxidative stress [86–88]. Similarly, ZnO NPs with potent antidiabetic properties were synthesized using a microwave-assisted technique and an extract of Vaccinium arctostaphylos L. fruits. They were effective in lowering fasting blood glucose (FBS) and keeping certain lipid levels under control in diabetic rats. According to the experiment results, green-synthesized ZnO NPs are superior to traditional ZnO NPs in their anti-diabetic characteristics [89]. Anti-diabetic uses of biogenic metal and metal oxide nanoparticles and their composites have been the subject of many studies. However, because there are so few commercialized goods based on it, it is urgently necessary for researchers working in this field to concentrate on creating commercial solutions based on biogenic metal and metal oxide nanoparticles and their composites for treating diabetes. However, several plants have been shown to have diabetic-lowering effects. To that end, they deserve serious consideration as a raw material for fabricating metal oxide nanoparticles to investigate their potential for use in fighting diabetes.

8.5.4 Antioxidant Activity

Inside the cell, oxidative stress is caused when there is a disproportion between the creation of reactive oxygen species and the amount of nitrogen available. Because of their impressive antioxidant capabilities, nanoparticles play a crucial role in the biomedical field by scavenging and neutralizing free radicals. Nanoparticles with two oxidation states, like cerium oxide, have been the subject of much study. Exposure of rat brain tissues to cerium oxide nanoparticles resulted in a rise in thiol content and the initiation of caspase-3 activity, demonstrating the remarkable antioxidant characteristics of cerium oxide nanoparticles [90]. Levan polysaccharide was also used to coat cerium oxide nanoparticles. When tested against H_2O_2 in NIH3T3 cells, both compounds proved to be effective antioxidants. The water solubility and stability of cerium oxide nanoparticles are paired with the antioxidant, anti-tumor, and anti-inflammatory actions of its derivatives. Cerium oxide nanoparticles synthesized in a green manner have features that make them useful in a number of therapeutic contexts [91]. Nanoparticles are ideal for a variety of medical uses because of their tiny size and high surface-to-volume ratio, which allows them to function as synthetic antioxidants in the body. Several of the illnesses that ROS (reactive oxygen species) cause by upsetting the body's normal redox balance are shown here, and nanoparticles' potential as antioxidants in the body is discussed. Zinc oxide nanoparticles have been shown to have potent antioxidant activity by scavenging 2,2-diphenyl-1-picrylhydrazyl (DPPH) radicals in a number of tests [92, 93]. These nanoparticles have been synthesized from plant extracts such Citrus paradise and Cassia fistula. It follows that plants high in citric acid are more effective antioxidants. Scientists may study the plants that produce citric acid to learn how to synthesize nanoparticles of metals and metal oxides for use as antioxidants. Zinc oxide nanoparticles synthesized from Lavandula vera leaf extract showed similar results in a comparison investigation with zinc sulphate ($ZnSO_4$). Enhanced DPPH radical scavenging activity demonstrated that the zinc oxide nanoparticles synthesized using a green process were more effective antioxidants than zinc sulphate [94].

8.5.5 Drug Delivery

Nanoparticle encapsulation of pharmaceuticals is a promising and effective medication delivery technology. In addition, various polymer-based nanoparticles have been included to further boost bioavailability. Treatment of cancer cells with metallic nanoparticle treatments was more effective since the therapeutic dose may be reduced. Injectable chemotherapy is harmful and poorly tolerated by patients. This makes it difficult to get the therapeutic substance to the tumor cells where needed. New research shows that doxorubicin (DOX)-loaded FeO nanoparticles actively suppress cancer cells by associating with the urokinase plasminogen activator receptor. When compared to non-targeted nanoparticles, targeted FeO-DOX particles significantly reduced tumor growth. This verifies the effectiveness of the dual-targeted FeO- DOX nanoparticles in a highly regarded model for enhanced drug delivery strategies. Nanogels based on a novel hybrid of quantum dots based on bismuth oxide and polyvinyl alcohol have been reported in the context of novel applications in heat sensing, bioimaging, and drug delivery. In the presence of Zn metal ions, recently published research demonstrates the MOF's potential as a drug carrier. The resulting MOF was used to load ibuprofen (IBU), and its chemical formula is

$[NH_2(CH_3)_2]$-$[Zn_2(HL)L0.5]$n n(8DMF $5H_2O$). The results demonstrate that Zn ions may store up to 50% more IBU than other ions and have longer discharge duration of 96 h in simulated bodily fluid.

Metallic nanoparticles, or clusters, offer various important and useful properties for medical use. According to the findings, metal nanoparticles have been classified as ahuman immunodeficiency virus (HIV) preventative-therapeutic agent. In previous research, silver nanoparticles have been shown to directly combat viral infection through their ability to bind to glycoproteins. This network of connections blocks the various viral bindings, reducing HIV infection. Metallic nanoparticles are well known to be effective antiviral agents against respiratory syncytial virus, herpes simplex virus, and influenza A virus. Angiogenesis is the formation of new blood vessels, which may occur usually and in response to illness. There are several disorders in which it plays a crucial role, including arthritis and cancer. Nanoparticles may be manufactured using synthetic techniques and employed in multiple nanomedical and biological applications. Under stable settings, angiogenesis is tightly controlled by a few proangiogenic growth factors and a few antiangiogenic factors. However, preparing these nanoparticles commercially is necessary to bring down their price. Sustainable, low-cost, environmentally friendly, and chemical-free natural resources are preferred for producing these nanoparticles. For the sake of science, it is crucial to create nanoparticles with a single-size distribution. However, the precise technique for creating nanoparticles remains unclear. Therefore, the process by which nanoparticle form and size may be modified should be investigated shortly. The desire to increase nanoparticle utilization in therapeutic applications while decreasing toxicity is another significant obstacle.

Rheumatoid arthritis is a joint condition caused by a malfunctioning immune system. The disease targets the connective tissue between bones. Researchers at Australia's University of Wollongong have developed a novel treatment for arthritis. There are a few small downsides to their new antidrug that works by binding to Au nanoparticles. According to new evidence, Au nanoparticles may both target macrophages for destruction and dampen inflammation. For example, a particle with a size of 50 nm or less was more successful in boosting cell immunity without any adverse effects, suggesting that the size of Au nanoparticles is an essential element in mitigating their toxicity. To add to this, titanium oxide nanoparticles (TiO_2) were synthesized utilizing the bulb extract of Ledebouria revolute, which showed potent anticancer action against A549 cells. Inhibiting the larvicidal activity of Aedes aegypti mosquitoes led researchers to choose Ledebouria revolute as one of the therapeutic plants they would study [46].

8.5.6 Tissue Engineering

Recently, nanoparticles have garnered a lot of interest in the field of tissue engineering. Practical replacements for many sorts of damaged tissues are developed using engineering and biological concepts. Nanoparticles with desirable characteristics for growing and repairing tissues include low toxicity, adaptability, contrasting agent actions, targeted delivery possibilities, and precise behavioural control. The effectiveness and function of the nanoparticles in tissue restoration have been improved by the use of composites in their fabrication. Ti, Au, Ag, and many more were among the metal nanoparticles synthesized, and all of them shown promising applications. Due to the proliferation of tumor cells, one potential use of Au nanoparticles is in regenerative medicine, namely in the field of tissue transplantation. Nanoparticles made from bioactive glass ceramics (BGC nanoparticles) also show great promise and are widely employed for bone tissue regeneration. According to recent studies, the alkaline phosphate activities were boosted when dense BGC nanoparticles were encapsulated in gelatin made from chitosan polymer. Results indicated that bone growth was the most rapid (80%) in the dense chitosan gelatin hydrogel containing bioactive nanoparticles. Recent studies have investigated the potential of BGC nanoparticles, are composed of a multifunctional poly-citrate-siloxane elastomer, in regenerating bone tissue. Bone healing is another use for titanium nanoparticles. It has widespread use in orthopedics and dentistry. It has qualities including higher fracture resistance, enhanced flexibility, and a high strength-to-weight ratio. They had weaker bioactivities because they did not promote cell adhesion and proliferation. As a result, apatite coatings are favoured as bioactive materials [46].

8.5.7 Gene Delivery System

Gene delivery systems provide great promise for treating a wide range of genetic problems via gene therapy (GT). In order to safely and effectively transmit genetic material, a reliable and effective carrier is needed. Organic and inorganic nanohybrids (ONHs and IONHs) are promising possibilities to overcome this challenge. The organic portion of NHs is primarily responsible for gene transfer [95]. In the presence of a magnetic field, iron oxide (Fe_3O_4), for instance, may transfect and internalize a gene because of its superparamagnetic characteristic. It has been reported that polyethyleneimine (PEI)-functionalized gold-nickel-silicon (Au-Ni-Si) nanospheres may transmit targeted gene transfection by magnetically guided delivery [96]. The nanoclusters of the gene delivery method develop a nanocomposite, which binds to the positively charged pDNA. Bimetallic nanoclusters were utilized to monitor the uptake of pDNA by cancer cells; the luminescence was created by the composite NPs, which were internalized by the cells. It has been shown, for instance, that Au-Ni nanorods may be used for the targeted delivery of genes [97].

8.5.8 Environmental Pollution Control

Metallic and metal oxide NPs have established themselves as important players in gas sensing due to the rapid development of nanotechnological applications. Toxic and flammable gases, such as zinc oxide (ZnO), tin oxide (SnO), and titanium dioxide (TiO_2), have been identified in ambient quantities due to industrial emissions [35, 98, 99]. Antagonic sensor response for detecting ammonia ($-NH_3$) was shown to be enhanced in V-ZnO: Ca powder synthesized through the sol–gel process, as compared to V, ZnO, and Ca individually. This is significant for detecting petrol contamination in the environment, medicinal applications, and the automotive chemical industry. In both biological and environmental contexts, even only 1 ppm of ammonia is crucial. Short-term ammonia exposure limits are set at 35 ppm, whereas long-term limits are set at 25 ppm [100]. Heavy metals, including arsenic (As), lead (Pb), mercury (Hg), and cadmium (Cd) may be found in the environment at deficient concentrations. Yet, they are toxic to both people and ecosystems. It will cause widespread water contamination and the spread of several illnesses. Electrochemical sensors based on metal oxide (MOX) are now being utilized to detect trace heavy metals in the environment [101].

8.5.9 Antibacterial Activity

Antibiotics are being developed to treat a wide variety of bacterial infections. However, the low dependability is related to the developing antibiotic resistance and toxicity. Metal and metal oxide nanostructures, on the other hand, have more promise than medicines for treating bacterial infections. Antibacterial therapy is one potential use for nanoceria-doped composite nanofibres, which exhibit toxicity against both Gram-positive and Gram-negative bacterial strains [102, 103]. More recently, researchers have detailed the several ways cerium oxide nanoparticles may be biosynthesized and their impacts on living tissues, characteristics, and potentials in diagnostic and therapeutic sectors. Combining, capping, stabilizing, or doping metal and metal oxide nanoparticles with other elements, such as polysaccharides and metals, has recently been described as a way to improve their biological capabilities. According to the literature, iron-doped copper oxide nanoparticles displayed improved antibacterial activities by suppressing bacterial colony growth and biofilm formation [104]. A similar potent antibacterial action was shown by copper nanoparticles based on chitosan against Vibrio parahaemolyticus [105]. Further, as illustrated in Fig. 8.17, biogenic metal and metal oxide nanoparticles demonstrated antibacterial efficiency via a variety of mechanisms, including damaging the plasma membrane (which leads to bursting of the bacterial cells), the generation of excess ROS in the bacterial cell (which will increase oxidative stress inside the cell), and the slow release of the drug (which will target the organelle).

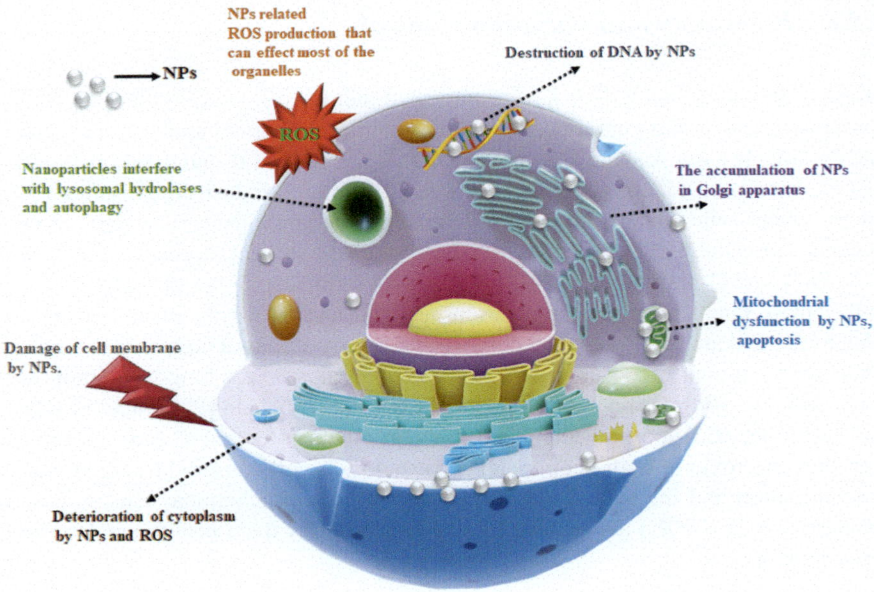

Fig. 8.17 Metals and metal oxides are shown to generate reactive oxygen species (ROS), which in turn increases oxidative stress inside the cell and damages the cell wall, all of which contribute to their antibacterial action

8.6 Conclusion and Future Prospective

This chapter provides an in-depth exploration of metallic and metal oxide nanohybrids, emphasizing their significant emergent properties and versatile applications in comparison with individual metal nanoparticles. Zn, Fe, Gd, Mg, and Cu metal and ZnO, TiO_2, and FeO nanostructures and their nanohybrids offer greater advantages compared to single-metal nanoparticles. Their unique attributes stem from the synergistic effects arising from combining the individual properties of two or three metals or their metal oxide counterparts, resulting in the creation of entirely new hybrid properties. This chapter stands out for its specific focus on biologically derived metallic nanoparticles and their versatile uses in both diagnostic and therapeutic contexts. It serves as a valuable resource for understanding various types of biogenic metal and metal oxide nanoparticles. In the context of the ongoing pandemic, there is a growing need for progress in public health research, biotechnology, nanotechnology, and environmental technology. Consequently, this review contributes to the advancement of knowledge in the field of biomedicine and the application of nanotechnology in its development and enhancement. Furthermore, it sheds light on the diverse applications of bioinspired metal and metal oxide nanoparticles in the context of pandemic situations, highlighting their biomedical utility.

Acknowledgements The authors are grateful to the Council of Scientific and Industrial Research (CSIR), New Delhi, India, for providing financial support (Project grant no. 22(0847)/20/EMR-II, dated 10.12.2020). The authors also gratefully acknowledge the National Institute of Technology (NIT), Arunachal Pradesh, India, for assistance and support.

Compliance with ethical standards

Conflict of Interest The authors declare that the work was conducted in the absence of any commercial or financial relationships that could be construed as potential conflicts of interest.

References

1. D. Kim, J. Kim, Y.I. Park, N. Lee, T. Hyeon, Recent development of inorganic nanoparticles for biomedical imaging. ACS Cent. Sci. **4**, 324–336 (2018)
2. X. Wu, H. Liu, J. Liu, K.N. Haley, J.A. Treadway, J.P. Larson, M.P. Bruchez, Immunofluorescent labeling of cancer marker Her2 and other cellular targets with semiconductor quantum dots. Nat. Biotechnol. **21**, 41–46 (2003)
3. C.G. Hadjipanayis, M.J. Bonder, S. Balakrishnan, X. Wang, H. Mao, G.C. Hadjipanayis, Metallic iron nanoparticles for MRI contrast enhancement and local hyperthermia. Small **4**, 1925–1929 (2008)
4. D.A. Giljohann, D.S. Seferos, W.L. Daniel, M.D. Massich, P.C. Patel, C.A. Mirkin, Gold nanoparticles for biology and medicine, Sph. Nucleic Acids, 55–90 (2020)
5. J. Zhou, Y. Yang, C.Y. Zhang, Towards biocompatible semiconductor quantum dots: from biosynthesis and bioconjugation to biomedical application. Chem. Rev. **115**, 11669–11717 (2015)
6. K.R. Singh, V. Nayak, J. Singh, A.K. Singh, R.P. Singh, Potentialities of bioinspired metal and metal oxide nanoparticles in biomedical sciences. RSC Adv. **11**, 24722–24746 (2021)
7. J. Das, J.W. Han, Y.J. Choi, H. Song, S.G. Cho, C. Park, J.H. Kim, Cationic lipid-nanoceria hybrids, a novel nonviral vector-mediated gene delivery into mammalian cells: investigation of the cellular uptake mechanism. Sci. Rep. **6**, 29197 (2016)
8. P. Dong, K.P. Rakesh, H.M. Manukumar, Y.H.E. Mohammed, C.S. Karthik, S. Sumathi, H.L. Qin, Innovative nano-carriers in anticancer drug delivery-a comprehensive review. Bioorg. Chem. **85**, 325–336 (2019)
9. T. Klaus, R. Joerger, E. Olsson, C.G. Granqvist, Silver-based crystalline nanoparticles, microbially fabricated. Proc. Natl. Acad. Sci. U.S.A. **96**, 13611–13614 (1999)
10. Y. Konishi, K. Ohno, N. Saitoh, T. Nomura, S. Nagamine, H. Hishida, T. Uruga, Bioreductive deposition of platinum nanoparticles on the bacterium Shewanella algae. J. Biotechnol. **128**, 648–653 (2007)
11. I. Willner, R. Baron, B. Willner, Growing metal nanoparticles by enzymes. Adv. Mater. **18**, 1109–1120 (2006)
12. Y. Yoon, P.L. Truong, D. Lee, S.H. Ko, Metal-oxide nanomaterials synthesis and applications in flexible and wearable sensors. ACS Nanosci. Au **2**, 64–92 (2021)
13. K.K.W. Lo (ed), *Inorganic and Organometallic Transition Metal Complexes with Biological Molecules and Living Cells* (Academic Press, 2016)
14. Q. Zhao, C. Huang, F. Li, Phosphorescent heavy-metal complexes for bioimaging. Chem. Soc. Rev. **40**, 2508–2524 (2011)
15. D. Tessore, R. Roberto, M. Ugo, S. Pizzotti, M. Quici, F.D. Cavazzini, Angelis, Terpyridine Zn(II), Ru(III), and Ir(III) complexes: the relevant role of the nature of the metal ion and of the ancillary ligands on the second-order nonlinear response of terpyridines carrying electron donor or electron acceptor groups. Inorg. Chem. **44**, 8967–8978 (2005)

16. A. Das, T. Ringu, S. Ghosh, N. Pramanik, Processing and biomedical applications of novel eco-sustainable fluconazole-loaded zinc oxide (ZnO) encapsulated chitosan (CS) biopolymer nanocomposite by inhibiting microbe species against candidiasis. Mater. Today Commun. **37**, 107071 (2023)
17. T. Ringu, S. Ghosh, A. Das, N. Pramanik, Zinc oxide nanoparticles: an excellent biomaterial for bioengineering applications. Emerg. Mater. **5**, 1629–1648 (2022)
18. M. Stefanidou, C. Maravelias, A. Dona, C. Spiliopoulou, Zinc: a multipurpose trace element. Arch. Toxicol. **80**, 1–9 (2006)
19. H. Wang, V.C. Jordan, I.A. Ramsay, M. Sojoodi, B.C. Fuchs, K.K. Tanabe et al., Molecular magnetic resonance imaging using a redox-active iron complex. J. Am. Chem. Soc. **141**, 5916–5925 (2019)
20. J. Gao, Y.G. Liu, Y. Zhou, L.M. Boxer, F.R. Woolley, R.A. Zingaro, Artificial Zinc(II) complexes regulate cell cycle and apoptosis-related genes in tumor cell lines. ChemBioChem **8**, 332–340 (2007)
21. X. Tian, S. Hussain, C. de Pace, L. Ruiz-Pérez, G. Battaglia, Zn(II) complexes for bioimaging and correlated applications. Chem. Asian J. **14**, 509–526 (2019)
22. Y. Zhang, X. Guo, W. Si, L. Jia, X. Qian, Ratiometric and water-soluble fluorescent zinc sensor of carboxamidoquinoline with an alkoxyethylamino chain as receptor. Org. Lett. **10**, 473–476 (2008)
23. E. Tomat, S.J. Lippard, Imaging mobile zinc in biology. Curr. Opin. Chem. Biol. **14**, 225–230 (2010)
24. J.H. McCullough, Iron restriction. Evol. Med. Public Health **2015**, 149–149 (2015)
25. J. Wade, D.J. Byrne, C.J. Ballentine, H. Drakesmith, Temporal variation of planetary iron as a driver of evolution. Proc. Natl. Acad. Sci. U.S.A. **118**, e2109865118 (2021)
26. T. Ganz, E. Nemeth, Iron homeostasis in host defence and inflammation. Nat. Rev. Immunol. **15**, 500–510 (2015)
27. K.D. Dubey, S. Shaik, Cytochrome P450—the wonderful nanomachine revealed through dynamic simulations of the catalytic cycle. Acc. Chem. Res. **52**, 389–399 (2019)
28. P. Papan, J. Kantapan, P. Sangthong, P. Meepowpan, N. Dechsupa, Iron (III)-quercetin complex: synthesis, physicochemical characterization, and MRI cell tracking toward potential applications in regenerative medicine. Contrast Media Mol. Imaging **2020**, 1–22 (2020)
29. E.M. Snyder, D. Asik, S.M. Abozeid, A. Burgio, G. Bateman, S.G. Turowski et al., A class of Fe(III) macrocyclic complexes with alcohol donor groups as effective T1 MRI contrast agents. Angew. Chem. **132**, 2435–2440 (2020)
30. P.B. Tsitovich, F. Gendron, A.Y. Nazarenko, B.N. Livesay, A.P. Lopez, M.P. Shores et al., Low-spin Fe(III) macrocyclic complexes of imidazole-appended 1,4,7-Triazacyclononane as paramagnetic probes. Inorg. Chem. **57**, 8364–8374 (2018)
31. S.M. Abozeid, M.S.I. Chowdhury, D. Asik, J.A. Spernyak, J.R. Morrow, Liposomal Fe (III) macrocyclic complexes with hydroxypropyl pendants as MRI Probes. ACS Appl. Bio Mater. **4**, 7951–7960 (2021)
32. D. Asik, R. Smolinski, S.M. Abozeid, T.B. Mitchell, S.G. Turowski, J.A. Spernyak et al., Modulating the properties of Fe (III) macrocyclic MRI contrast agents by appending sulfonate or hydroxyl groups. Molecules **25**, 2291 (2020)
33. L. An, Y. Cai, Q. Tian, J. Lin, S. Yang, Ultrasensitive iron-based magnetic resonance contrast agent constructed with natural polyphenol tannic acid for tumor theranostics. Sci. China Mater. **64**, 498–509 (2021)
34. H. Wang, A. Wong, L.C. Lewis, G.R. Nemeth, V.C. Jordan, J.W. Bacon et al., Rational ligand design enables pH control over aqueous iron magnetostructural dynamics and relaxometric properties. Inorg. Chem. **59**, 17712–17721 (2020)
35. C. Wang, L. Yin, L. Zhang, D. Xiang, R. Gao, Metal oxide gas sensors: sensitivity and influencing factors. Sensors **10**, 2088–2106 (2010)
36. M. Mauro, A. Aliprandi, D. Septiadi, N.S. Kehr, L. De Cola, When self-assembly meets biology: luminescent platinum complexes for imaging applications. Chem. Soc. Rev. **43**, 4144–4166 (2014)

37. S. Nasiri Sovari, F. Zobi, Recent studies on the antimicrobial activity of transition metal complexes of groups 6–12. Chemistry **2**, 418–452 (2020)
38. M.K. Tiwari, P.M. Hägglund, I.M. Møller, M.J. Davies, M.J. Bjerrum, Copper ion/H_2O_2 oxidation of Cu/Zn-Superoxide dismutase: Implications for enzymatic activity and antioxidant action. Redox Biol. **26**, 101262 (2019)
39. R. Memary, D. Giurco, G. Mudd, L. Mason, Life cycle assessment: a time-series analysis of copper. J. Clean. Prod. **33**, 97–108 (2012)
40. L. Lei, W. Xie, Z. Chen, Y. Jiang, Y. Liu, Metal ion chelation-based color generation for alkaline phosphatase-linked high-performance visual immunoassays. Sens. Actuators B Chem. **273**, 35–40 (2018)
41. A. Loas, S.J. Lippard, Direct ratiometric detection of nitric oxide with Cu(II)-based fluorescent probes. J. Mater. Chem. B **5**, 8929–8933 (2017)
42. A. Fatima, M.W. Ahmad, A.K.A. Al Saidi, A. Choudhury, Y. Chang, G.H. Lee, Recent advances in gadolinium based contrast agents for bioimaging applications. Nanomaterials **11**, 2449 (2021)
43. A.S. Thakor, J. Jokerst, C. Zavaleta, T.F. Massoud, S.S. Gambhir, Gold nanoparticles: a revival in precious metal administration to patients. Nano Lett. **11**, 4029–4036 (2011)
44. N. Li, P. Zhao, D. Astruc, Anisotropic gold nanoparticles: synthesis, properties, applications, and toxicity. Angew. Chem. Int. Ed. **53**, 1756–1789 (2014)
45. L.Y. Chen, C.W. Wang, Z. Yuan, H.T. Chang, Fluorescent gold nanoclusters: recent advances in sensing and imaging. Anal. Chem. **87**, 216–229 (2015)
46. A.A. Yaqoob, H. Ahmad, T. Parveen, A. Ahmad, M. Oves, I.M. Ismail et al., Recent advances in metal decorated nanomaterials and their various biological applications: a review. Front. Chem. **8**, 341 (2020)
47. P. Tan, H. Li, J. Wang, S.C. Gopinath, Silver nanoparticle in biosensor and bioimaging: clinical perspectives. Biotechnol. Appl. Biochem. **68**, 1236–1242 (2021)
48. D. Ndiaye, M. Sy, A. Pallier, S. Même, I. de Silva, S. Lacerda et al., Unprecedented kinetic inertness for a Mn^{2+}-bispidine chelate: a novel structural entry for Mn^{2+}-based imaging agents. Angew. Chem. Int. Ed. **59**, 11958–11963 (2020)
49. K. Chen, P. Li, C. Zhu, Z. Xia, Q. Xia, L. Zhong et al., Mn (II) complex of lipophilic group-modified ethylenediaminetetraacetic acid (EDTA) as a new hepatobiliary MRI contrast agent. J. Med. Chem. **64**, 9182–9192 (2021)
50. G.S. Loving, S. Mukherjee, P. Caravan, Redox-activated manganese-based MR contrast agent. J. Am. Chem. Soc. **135**, 4620–4623 (2013)
51. T.E. Hutchinson, A. Bashir, M. Yu, R.J. Beyers, C.R. Goldsmith, An overly anionic metal coordination environment eliminates the T1-weighted response of quinol-containing MRI contrast agent sensors to H_2O_2. Inorg. Chim. Acta **496**, 119045 (2019)
52. M. Yu, M.B. Ward, A. Franke, S.L. Ambrose, Z.L. Whaley, T.M. Bradford et al., Adding a second quinol to a redox-responsive MRI contrast agent improves its relaxivity response to H_2O_2. Inorg. Chem. **56**, 2812–2826 (2017)
53. M. Carofiglio, S. Barui, V. Cauda, M. Laurenti, Doped zinc oxide nanoparticles: synthesis, characterization and potential use in nanomedicine. Appl. Sci. **10**, 5194 (2020)
54. L.C. Mimun, Rare earths based NIR luminomagnetic nanoparticles and their multimodal applications in biomedical imaging. University of Texas at San Antonio (2015)
55. H. Dong, S.R. Du, X.Y. Zheng, G.M. Lyu, L.D. Sun, L.D. Li et al., Lanthanide nanoparticles: from design toward bioimaging and therapy. Chem. Rev. **115**, 10725–10815 (2015)
56. H.M. Xiong, ZnO nanoparticles applied to bioimaging and drug delivery. Adv. Mater. **25**, 5329–5335 (2013)
57. A. Das, T. Ringu, S. Ghosh, N. Pramanik, A comprehensive review on recent advances in preparation, physicochemical characterization, and bioengineering applications of biopolymers. Polym. Bull. **80**, 7247–7312 (2022)
58. H.M. Xiong, Photoluminescent ZnO nanoparticles modified by polymers. J. Mater. Chem. **20**, 4251–4262 (2010)

59. Y. Liu, K. Ai, Q. Yuan, L. Lu, Fluorescence-enhanced gadolinium-doped zinc oxide quantum dots for magnetic resonance and fluorescence imaging. Biomaterials **32**, 1185–1192 (2011)
60. J. Jiang, J. Pi, J. Cai, The advancing of zinc oxide nanoparticles for biomedical applications. Bioinorg. Chem. Appl. (2018)
61. H.M. Xiong, Y. Xu, Q.G. Ren, Y.Y. Xia, Stable aqueous ZnO@ polymer core-shell nanoparticles with tunable photoluminescence and their application in cell imaging. J. Am. Chem. Soc. **130**, 7522–7523 (2008)
62. X. Tang, E.S.G. Choo, L. Li, J. Ding, J. Xue, Synthesis of ZnO nanoparticles with tunable emission colors and their cell labeling applications. Chem. Mater. **22**, 3383–3388 (2010)
63. T. Mao, M. Liu, L. Lin, Y. Cheng, C. Fang, A study on doping and compound of zinc oxide photocatalysts. Polymers **14**, 4484 (2022)
64. M. Geppert, M. Himly, Iron oxide nanoparticles in bioimaging–an immune perspective. Front. Immunol. **12**, 688927 (2021)
65. M.G. Montiel Schneider, M.J. Martín, J. Otarola, E. Vakarelska, V. Simeonov, V. Lassalle et al., Biomedical applications of iron oxide nanoparticles: current insights, progress and perspectives. Pharmaceutics **14**, 204 (2022)
66. Z.F. Yin, L. Wu, H.G. Yang, Y.H. Su, Recent progress in biomedical applications of titanium dioxide. Phys. Chem. Chem. Phys. **15**, 4844–4858 (2013)
67. D. Ziental, B. Czarczynska-Goslinska, D.T. Mlynarczyk, A. Glowacka-Sobotta, B. Stanisz, T. Goslinski et al., Titanium dioxide nanoparticles: prospects and applications in medicine. Nanomaterials **10**, 387 (2020)
68. M.A. Jhonsi, Carbon quantum dots for bioimaging, in *State of the Art in Nano-bioimaging* (2018), pp. 35–53
69. S. Rita Kakkar, ZnO quantum dots for biomedical applications. Adv. Mater. Lett. **4**, 876–887 (2013)
70. P.G. Luo, S. Sahu, S.T. Yang, S.K. Sonkar, J. Wang, H. Wang et al., Carbon "quantum" dots for optical bioimaging. J. Mater. Chem. B **1**, 2116–2127 (2013)
71. S.C. Gopinath, T. Lakshmipriya, M.M. Arshad, M.N. Uda, Y. Al-Douri, Nanoelectronics in biosensing applications, in *Nanobiosensors for Biomolecular Targeting* (Elsevier, 2019), pp. 211–224
72. B. Nikhil, J. Pawan, F. Nello, E. Pedro, Introduction to biosensors. Essays Biochem. **60**, 1–8 (2016)
73. W. Kim, J.C. Lee, G.J. Lee, H.K. Park, A. Lee, S. Choi, Low-cost label-free biosensing bimetallic cellulose strip with SILAR-synthesized silver core–gold shell nanoparticle structures. Anal. Chem. **89**, 6448–6454 (2017)
74. I. Şerban, A. Enesca, Metal oxides-based semiconductors for biosensors applications. Front. Chem. **8**, 354 (2020)
75. S. Nadzirah, S.C. Gopinath, N.A. Parmin, A.A. Hamzah, M.A. Mohamed, E.Y. Chang, C.F. Dee, State-of-the-art on functional titanium dioxide-integrated nano-hybrids in electrical biosensors. Crit. Rev. Anal. Chem. **52**, 637–648 (2022)
76. I. Ocsoy, B. Gulbakan, M.I. Shukoor, X. Xiong, T. Chen, D.H. Powell, W. Tan, Aptamer-conjugated multifunctional nanoflowers as a platform for targeting, capture, and detection in laser desorption ionization mass spectrometry. ACS Nano **7**, 417–427 (2013)
77. S.K. Nune, P. Gunda, P.K. Thallapally, Y.Y. Lin, M. Laird Forrest, C.J. Berkland, Nanoparticles for biomedical imaging. Expert Opin. Drug Deliv. **6**, 1175–1194 (2009)
78. Y. Zhang, J. Li, H. Jiang, C. Zhao, X. Wang, Rapid tumor bioimaging and photothermal treatment based on GSH-capped red fluorescent gold nanoclusters. RSC Adv. **6**, 63331–63337 (2016)
79. S. Kumar, V.K. Meena, P.P. Hazari, S.K. Sharma, R.K. Sharma, Rose Bengal attached and dextran coated gadolinium oxide nanoparticles for potential diagnostic imaging applications. Eur. J. Pharm. Sci. **117**, 362–370 (2018)
80. E.S. Lee, B. Shuter, J. Chan, M.S. Chong, J. Ding, S.H. Teoh et al., The use of microgel iron oxide nanoparticles in studies of magnetic resonance relaxation and endothelial progenitor cell labeling. Biomaterials **31**, 3296–3306 (2010)

81. W. Xu, P. Cui, E. Happonen, J. Leppanen, L. Liu, J. Rantanen et al., Tailored synthesis of PEGylated bismuth nanoparticles for x-ray computed tomography and photothermal therapy: one-pot, targeted pyrolysis, and self-promotion. ACS Appl. Mater. Interfaces **12**, 47233–47244 (2020)
82. S. Farahani, E. Gorji, R. Rahnamafar, S. Fazli, H. Khosravi, M. Pakravan et al., EP-1750: enhanced radiotherapy by novel class of radiosensitizers based bismuth and gadolinium nanoparticles. Radiother. Oncol. **123**, S963–S964 (2017)
83. S. Prabhu, S. Vinodhini, C. Elanchezhiyan, D. Rajeswari, Retracted: Evaluation of antidiabetic activity of biologically synthesized silver nanoparticles using Pouteria sapota in streptozotocin-induced diabetic rats. J. Diabetes **10**, 28–42 (2018)
84. S. Al-Quraishy, M.A. Dkhil, A.E. Abdel Moneim, Anti-hyperglycemic activity of selenium nanoparticles in streptozotocin-induced diabetic rats. Int. J. Nanomed. **10**, 6741–6756 (2015)
85. H.H. Ahmed, M.D. Abd El-Maksoud, A.E. Abdel Moneim, H.A. Aglan, Pre-clinical study for the antidiabetic potential of selenium nanoparticles. Biol. Trace Elem. Res. **177**, 267–280 (2017)
86. A.C. Maritim, A. Sanders, J.B. Watkins III., Diabetes, oxidative stress, and antioxidants: a review. J. Biochem. Mol. Toxicol. **17**, 24–38 (2003)
87. N. Pourkhalili, A. Hosseini, A. Nili-Ahmadabadi, S. Hassani, M. Pakzad, M. Baeeri et al., Biochemical and cellular evidence of the benefit of a combination of cerium oxide nanoparticles and selenium to diabetic rats. World J. Diabetes **2**, 204 (2011)
88. N. Pourkhalili, A. Hosseini, A. Nili-Ahmadabadi, M. Rahimifard, M. Navaei-Nigjeh, S. Hassani et al., Improvement of isolated rat pancreatic islets function by combination of cerium oxide nanoparticles/sodium selenite through reduction of oxidative stress. Toxicol. Mech. Methods **22**, 476–482 (2012)
89. A. Bayrami, S. Parvinroo, A. Habibi-Yangjeh, S. Rahim Pouran, Bio-extract-mediated ZnO nanoparticles: Microwave-assisted synthesis, characterization and antidiabetic activity evaluation. Artif. Cells Nanomed. Biotechnol. **46**, 730–739 (2018)
90. A. Ranjbar, S. Soleimani Asl, F. Firozian, H. HeidaryDartoti, S. Seyedabadi, M. Taheri Azandariani, M. Ganji, Role of cerium oxide nanoparticles in a paraquat-induced model of oxidative stress: emergence of neuroprotective results in the brain. J. Mol. Neurosci. **66**, 420–427 (2018)
91. S.J. Kim, B.H. Chung, Antioxidant activity of levan coated cerium oxide nanoparticles. Carbohydr. Polym. **150**, 400–407 (2016)
92. B. Kumar, K. Smita, L. Cumbal, A. Debut, R.N. Pathak, Sonochemical synthesis of silver nanoparticles using starch: a comparison. Bioinorg. Chem. Appl. **2014**, 1–7 (2014)
93. D. Suresh, P.C. Nethravathi, H. Rajanaika, H. Nagabhushana, S.C. Sharma, Green synthesis of multifunctional zinc oxide (ZnO) nanoparticles using Cassia fistula plant extract and their photodegradative, antioxidant and antibacterial activities. Mater. Sci. Semicond. Process. **31**, 446–454 (2015)
94. Z. Salari, A. Ameri, H. Forootanfar, M. Adeli-Sardou, M. Jafari, M. Mehrabani, M. Shakibaie, Microwave-assisted biosynthesis of zinc nanoparticles and their cytotoxic and antioxidant activity. J. Trace Elem. Med. Biol. **39**, 116–123 (2017)
95. N. Zhao, L. Yan, X. Zhao, X. Chen, A. Li, D. Zheng et al., Versatile types of organic/inorganic nanohybrids: from strategic design to biomedical applications. Chem. Rev. **119**, 1666–1762 (2018)
96. X. Xu, S. Hou, N. Wattanatorn, F. Wang, Q. Yang, C. Zhao et al., Precision-guided nanospears for targeted and high-throughput intracellular gene delivery. ACS Nano **12**, 4503–4511 (2018)
97. A.K. Salem, P.C. Searson, K.W. Leong, Multifunctional nanorods for gene delivery. Nat. Mater. **2**, 668–671 (2003)
98. W. Guan, N. Tang, K. He, X. Hu, M. Li, K. Li, Gas-sensing performances of metal oxide nanostructures for detecting dissolved gases: a mini review. Front. Chem. **8**, 76 (2020)
99. A. Mirzaei, J.H. Lee, S.M. Majhi, M. Weber, M. Bechelany, H.W. Kim, S.S. Kim, Resistive gas sensors based on metal-oxide nanowires. J. Appl. Phys. **126** (2019)
100. E. Fazio, M. Hjiri, R. Dhahri, L. El Mir, G. Sabatino, F. Barreca et al., Ammonia sensing properties of V-doped ZnO: Ca nanopowders prepared by sol–gel synthesis. J. Solid State Chem. **226**, 192–200 (2015)

101. R. Ding, Y.H. Cheong, A. Ahamed, G. Lisak, Heavy metals detection with paper-based electrochemical sensors. Anal. Chem. **93**, 1880–1888 (2021)
102. A.R. Unnithan, A.R.K. Sasikala, Y. Sathishkumar, Y.S. Lee, C.H. Park, C.S. Kim, Nanoceria doped electrospun antibacterial composite mats for potential biomedical applications. Ceram. Int. **40**, 12003–12012 (2014)
103. K. Krishnamoorthy, M. Veerapandian, L.H. Zhang, K. Yun, S.J. Kim, Surface chemistry of cerium oxide nanocubes: toxicity against pathogenic bacteria and their mechanistic study. J. Ind. Eng. Chem. **20**, 3513–3517 (2014)
104. A. Pugazhendhi, S.S. Kumar, M. Manikandan, M. Saravanan, Photocatalytic properties and antimicrobial efficacy of Fe doped CuO nanoparticles against the pathogenic bacteria and fungi. Microb. Pathog. **122**, 84–89 (2018)
105. N.Y. Nguyen, B.N. An, M.V. Le, H.A. Hoang, Antibacterial activity of copper nanoparticles-chitosan composite against Vibrio parahaemolyticus. Biocontrol Sci. **25**, 159–165 (2020)

Chapter 9
Advances in Graphene-Based Nanostructures for Multimodal Bioimaging

Wessam Wanas, Moataz Soliman, and Shaker Ebrahim

Abstract Graphene-based nanostructures including graphene quantum dots (GQDs), reduced graphene oxide (RGO), and graphene oxide (GO) have captured the attention of the whole scientific research field due to their unique optical, structural, mechanical, and electrical properties. Graphene and its derivatives are able to be simply converted by small molecules such as drugs, dyes, nanoparticles (NPs), polymers, or biomolecules to produce graphene functionalized materials for various modes of bioimaging purposes due to their adaptable surface modification and ultra-high surface area characteristics. This chapter covers a comprehensive overview of using graphene and its derivatives for various molecular imaging modalities such as optical imaging (fluorescence, up-conversion imaging), magnetic resonance imaging (MRI), Raman imaging, photoacoustic imaging (PAI), and multimodal imaging. The following sections will go through each imaging modality in detail. Finally, we will examine briefly the potential and problems challenges for regarding their future bioimaging applications.

9.1 Introduction

Medical imaging methods are fast and powerful tools for giving valuable details about regarding in vivo systems, the disease of the internal body organs, its progress, and the treatment response that has shown significant progress to reduce mortality in patients that result from an insufficient understanding of in vivo systems. Among these are methods like direct visualization through dissection, which was one of the main processes of imaging for about 600 years up to the end of the nineteenth century. Radiography entered a new era when X-rays were discovered by Wilhelm Conrad Roentgen in 1895, ushering in a new type of medical imaging. By making organs and blood arteries more visible between 1906 and 1912, the use of pharmacological

W. Wanas (✉) · M. Soliman · S. Ebrahim
Department of Materials Science, Institute of Graduate Studies and Research, Alexandria University, Alexandria, Egypt
e-mail: Wessam.Wanas@alexu.edu.eg

N. Kalarikkal et al. (eds.), *Multimodal Biomedical Imaging Techniques*,
Biological and Medical Physics, Biomedical Engineering,
https://doi.org/10.1007/978-981-96-1124-9_9

contrast agents revolutionized the area of medical radiography. In the 1960s, it gave rise to an independent branch of bioimaging known as angiography, which was used to view the interior or lumen of blood vessels and organs, with a focus on the heart chambers, arteries, and veins. Typically, an X-ray-based technique called fluoroscopy is used to visualize the blood vessel after injecting a radio-opaque contrast agent there. Nuclear biomedical imaging became a popular approach around the same time, owing to the development of specialized gamma cameras. The concepts of sonar, or ultrasound, which were widely explored through the Second World War, were used in diagnostic imaging [1].

For developing computerized tomography (CT), G. N. Hounsfield and A. M. Cormack won the Nobel Prize in medicine in the 1970s. Another turning point in the field of bioimaging was in 1971. When Raymond Damadian demonstrated that the nuclear magnetic relaxation periods of tissues and tumors varied, it inspired researchers to employ *Magnetic Resonance Imaging* (MRI) to examine disorders. Subsequently, a standard for storing and sending data in medical imaging is called Digital imaging and communications in medicine (DICOM). Beginning in the 1970s and carrying on till the present day. Visible light imaging (or near-visible), positron emission tomography (PET), photoacoustic imaging (PAI), microwave imaging, single-photon emission computerized tomography (SPECT), and thermal imaging are a number of additional techniques that have been used in medical imaging [2, 3].

Today's biomedical imaging techniques provide views of the biological processes and mechanisms inside the body, including variation in cellular signaling, receptor kinetics, and interactions of molecules through membranes [4, 5].

Methods of medical imaging may often be divided into two categories: functional or structural information: Functional imaging is associated with the physiology of diverse and dynamic bioprocesses at the cellular or whole-organ level, which is often related to molecular imaging. Tissue metabolism, sensing molecular targets, blood flow, and the processes of diffusion, perfusion, and biodistribution can be visualized using this technology. Structural imaging is originally associated with the morphology, shape, and size of body items. This imaging technique can focus on vessels, bones, and organs to examine anatomical or physical abnormalities in the body that might help direct medical interventions [6].

Images clearly delineate morphology, physiological activities, and internal structure (anatomy) at the various described scales because of differences in contrast across regions. External pharmaceutical contrast enhancement elements, also known as contrast agents (CAs) or identifiable probes, can be administered to the patients in order to enhance a modality's sensitivity and/or specificity to a targeted region of interest (ROI), which can improve the visualization of the internal organs structure by altering the contrast between healthy and diseased areas. This demonstrates the ongoing requirement for the creation of fresh diagnostic contrast agents [7–11]. Nanotechnology has recently prompted attempts to improve biological detection and imaging due to their targeting properties. Nanoparticles have increased the effects of permeability and retention (EPR) in tumors due to their small size, with proportionate increases in local tumor contrast agent (CA) concentrations [12]. Carbon nanomaterials have emerged as a dominant material in a variety of bioimaging applications [13–15].

Carbon (C), the sixth chemical element in the periodic table and the essential element of life, forms the fundamental chemistry for life on Earth due to being able to establish permanent bonds with other atoms of carbon, sulfur, oxygen, nitrogen, and a number of other elements in the periodic tables. Carbon-based substances with a variety of forms and properties are categorized depend on their C–C bonding, which is based on sp, sp^2, or sp^3 hybrid orbitals.

Carbon nanostructures of different allotropes as graphene, fullerenes, and nanotubes have advantages in biomedical studies due to their numerous surface functionalization options and exceptionally high surface area properties. Among its allotropes, graphene and its derivatives are easily modified by biomolecules, dyes, polymers, nanoparticles and drugs to produce graphene-based nanomaterials for various bioimaging applications [16, 17]. Graphene was discovered in 1859 by British scientist Benjamin Collins Brodie. However, graphene gained notice in the scientific world when it was developed as a single sheet of material by Novoselov et al. using the scotch tape method. The state-of-the-art in the most common imaging modalities using graphene nanostructures will be discussed in this chapter in order to determine the best strategy of diagnosis and possible strategy of therapy. We demonstrate how nanostructures contrast agents are already lighting new paths in the field of bioimaging and nanomedicine.

9.2 Properties of Graphene

Graphene represents the unique allotrope of carbon that shows each atom of carbon is securely bound to its classmates by a special electronic cloud, which offers numerous remarkable challenges for quantum physics. Graphene itself can be found in a variety of shapes, each one of them has outstanding applications in addition to the peculiar quantum hall phenomenon [18, 19]. The electrical and quantum characteristics of graphene are still a subject of vital research, as was already indicated. Every carbon atom in graphene is sp^2 hybridized, meaning it has three bonds with various neighbors. S, Px, and Py orbitals are combined to form the sp^2 hybridization. In the hexagonal phase, three different carbon atoms bond covalently to each other and are basically sp^2 hybridized, leaving one free electron for each carbon atom. This free electron is held in the Pz orbital, which creates the pi bond and is located above the plane Fig. 9.1. It is interesting to note that the graphene Pz orbital is essential to the physical and chemical behavior. A characteristic property of graphene is the presence of a zero bandgap, which creates a number of new prospects for the Production of innovative artificial, man-made materials with tunable bandgaps that can be useful for the development of the next generation of computing [18, 20, 21]. Numerous studies have revealed that graphene

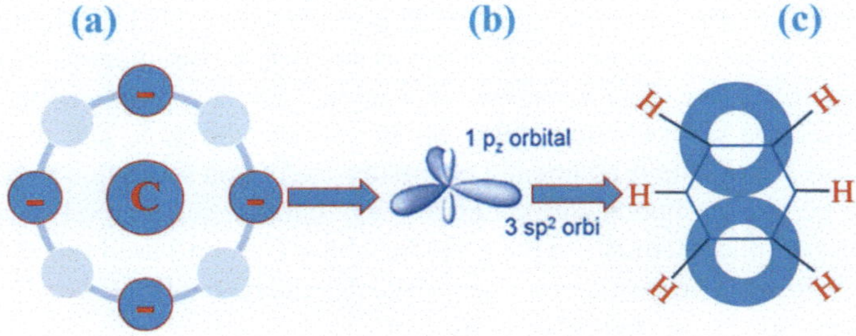

Fig. 9.1 **a–c** Graphene's fundamental bonding characteristics [22]

has exceptional electrical, mechanical, and optical properties, which could make it a very useful material.

9.2.1 Electronic Properties

Both practically and theoretically, it has been demonstrated that graphene is a nearly ideal electronic conductor: Since graphene has a zero band gap semiconductor (the valance and conduction bands are touching, as shown in Fig. 9.2, as well as the electron momentum k is linearly associated to energy E near the edges of the 2D first Brillouin zone. In this sector, electrons act like massless Dirac fermions, and finally, graphene sheets show evidence of ballistic transport. Low energy charge carriers are best classified as massless Dirac fermions (electrons without mass or neutrinos carrying the charge of an electron), which makes them quasiparticles that may be subject to the phenomena of quantum electrodynamics (QED). Dirac equations can be used to explain that. The following equation describes the energy of the carriers:

$$E = \hbar \cdot v_F \cdot \sigma \cdot k \qquad (9.1)$$

where v_F is the Fermi velocity, σ is the Pauli matrix, and k is the quasiparticle momentum.

In common conductor, charge carriers (either electrons or holes) scatter as they pass through the material, losing momentum in the process. Ballistic transport describes the ability of charge carriers in graphene to travel up to 0.3 μm^2 without being scattered. The Dirac points are six locations in k-space where the graphene conduction and valence bands intersect. The band structure disperses linearly as cones with slope of the energy previously described, where Fermi velocity is 10^6 m/s, close to the Dirac points [22, 23].

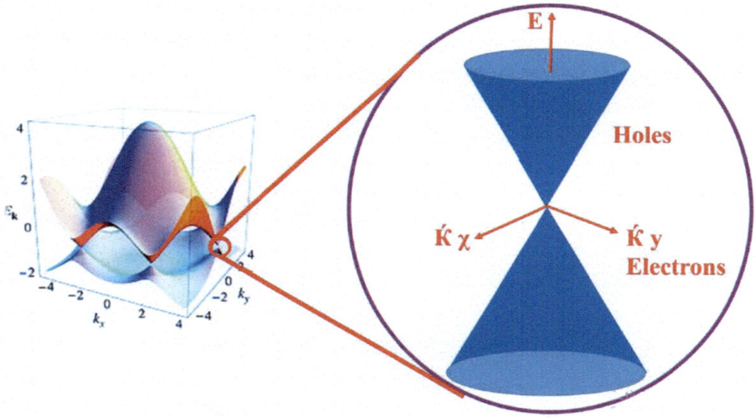

Fig. 9.2 Graphene electronic band structure: (right) concentrate on a linear structure near a K point; (left) 3D representation of a unit cell in reciprocal lattice space

9.2.2 Thermal Properties

If the temperatures of the two kinds of solid materials are different, heat will spontaneously flow from the side of high temperature to the side of low temperature. This is known as heat conduction [24]. Thermal conductivity (λ) is an index of physical properties that defines a material's capacity for conducting heat and is stated by the formula below [25]:

$$\lambda = 1/3\, C\, v\, L \tag{9.2}$$

where C is the capacity per unit volume, v is the phonon's propagation velocity, and L is the mean free path of the phonon, which is the distance the phonon travels from the initial scattering stage to the second scattering stage.

For graphene family, the thermal conductivity at room temperature mainly depends on the size of the mean free path of phonons, which is related to the collision and scattering of phonons and the plane size of the graphite mesh $L\alpha$. The higher $L\alpha$, the higher L and the higher λ. In the case of polycrystalline structure of graphene, the smaller the grain size, the lower the material's performance. This is because the reducing grain size raises the grain boundaries number in the structure; it consequently generates more defects that leads to scatter phonons, resulting in reduced the mean free path. In the synthesis processes of graphene, to prevent thermal conductivity reduction, the concentration of defects should be decreased.

There are two types of heat flow directions in graphene: in-plane and out-of-plane. Because of covalent sp^2 bonding between carbon molecules, in-plane heat flow is more prevalent than out-of-plane heat flow, whereas out-of-plane heat flow is produced by weak van der Waals coupling. In-plane heat flow is directed to graphene-based materials through a particular network. Interfaces, edges, and phonon scattering can all be used to control heat flow. Graphene's specific thermal capabilities result from its two-dimensional structure, opening up a significant space for new developments in heat-flow physics and maybe taking the lead in thermal devices [26, 27].

9.2.3 Mechanical Properties

One of graphene's most notable characteristics is its fundamental strength. Graphene is the most powerful material ever discovered owing to the strength of its long carbon bonds is 0.142 Nm, with an ultimate tensile strength of 130 gigapascals, compared to 375,7 megapascals for Aramid (Kevlar) or 400 million pascals for A36 structural steel. Not only is graphene is highly durable, but it is also extremely lightweight which only 0.77 mg per square meter (for guidance, one square meter of paper weighs within a thousand times more). Graphene single sheet enough in size to fill a full football field has been estimated to weigh less than one gram. The unique mechanical capabilities of this amazing material in graphene sp^2 are mostly due to C–C bonding. Different factors may be used to define the static mechanical features of graphene such as the breaking stress (σ_{int}) and strain (ε_{int}), the Young's modulus (Y), the bending rigidity (B) and the Poisson ratio (ν). According to these criteria, the response of a graphene to deflections and strain is ideal [26].

9.2.4 Optical Properties

Physical characteristics of graphene are unique. The charge carriers that exist in graphene obey the equation of relativistic massless Dirac; consequently, they have been identified as the Dirac fermions. At the Dirac point in the first Brillouin zone, these Dirac fermions in graphene show a distinctive linear band structure. In the presence of a strong electromagnetic field, these carriers display unique linear and nonlinear optical features. The remarkable optical characteristics, both linear and nonlinear, result from graphene's unconventional low-energy electronic band structure. By applying a magnetic field, graphene's linear and nonlinear optical response may be adjusted in the terahertz zone [28].

9.2.4.1 Linear Optical (LO) Response of Graphene

The exceptional optical features of graphene can be attributed to its unique two-dimensional band structure. According to graphene's linear optical characterization, it absorbs around $\pi\alpha \approx 2.3\%$ of the incidence of red light and $\pi\alpha \approx 2.6\%$ of the incidence of green light, where α is the constant of fine structure [29, 30].

Linear Optical Absorption

The process of changing light energy into another form of energy is called absorption. The following physical mechanisms are involved in the linear absorption of light. In insulating materials, strong linear absorption happens when light energy exceeds the band gap. A single photon's absorption excites every electric charge carrier across the band. The electrical excitation energy is either partially transformed to fluorescent light or to heat (i.e., numerous phonons) throughout this process. At long wavelengths, multi-phonon absorption also produces strong linear absorption. Semiconductors operating in the visible portion of the electromagnetic spectrum may undergo similar phenomena. The absorption coefficient in linear optical absorption is independent of light intensity.

Additionally, in the presence of a magnetic field perpendicular to the graphene sheet, graphene shows the well-known magneto-optic effect. The Faraday effect is brought about by light being transmitted through a magneto-optic substance. When a magnetic field is applied perpendicularly to monolayer graphene, the magneto-optical absorption coefficient displays several absorption peaks. Nevertheless, in the traditional 2D platforms [31–33], the absorption coefficient has only one peak. This is due to the fact that the Landau level index affects the energy differential between two Landau levels (LLs). For both intra- and interband transitions in graphene between distinct LLs, the selection rule is Ln to Ln ± 1. The absorption coefficient of graphene has a number of distinct peaks that can only be attributed to these permitted transitions; in the case of phosphorene [34], this absorption coefficient is much different.

Linear Optical Transparency

Among the unique optical features of graphene and other two-dimensional materials is a simple transparency of $T \approx 1 - \pi\alpha$ for visible light. The majority of theoretical explanations for this "universal" opacity focus on the graphene bands' linear dispersion and use a model that couples light to the electron's crystal momentum. A formulation of interband absorption like this is not permitted in band structure theory, though, as it confuses the canonical momentum operator with the crystal momentum label [35].

9.2.4.2 Nonlinear Optical (NLO) Response of Graphene

Nonlinear optics (NLO) is an advanced form of linear optics that has been made possible by laser technology. This field is primarily concerned with new optical effects and phenomena that result from the interaction of strong radiation that is coherent with matter [36].

Graphene-based material exhibits exceptional optoelectronic and optical nonlinearities, as well as rapid response rates and a wide spectrum range. The electronic structure of graphene can be altered by all of these many chemical changes, including oxidation and functionalization. As a result, they can be used to fine-tune the nonlinear response and facilitate the development of more common applications. Several NLO outcomes, which include saturable absorption (SA), four-wave mixing (FWM), two-photon absorption (TPA), reverse saturable absorption (RSA), and optical limiting, can be detected in graphene structures [37].

Saturable Absorption

SA is a material feature in which light absorption reduces with a rise in light intensity. At a high sufficient incident intensity, atoms in a saturable absorber material's ground state are excited into an upper energy state, but the rates during which they decay back into the ground state is not sufficient to enable them to decay earlier than the ground state is depleted, and the absorption eventually saturates. When the excitation intensity is low, after photo-excitation, the linear optical transition causes electron–hole recombination to occur rapidly which allows the electron and hole distribution to equilibrium. As the intensity of the excitation increases, the concentration of photogenerated carriers increases (far more than the intrinsic hole carrier and electron densities in graphene at ambient temperature) [38]. Due to the Pauli blocking principle, when the incident light intensity is high enough, the produced carriers occupy the valence bands. This prevents additional electron excitation at the valence band and enables photon transmission without loss. Subsequently, SA can be achieved due to this Pauli blocking process.

Graphene is a perfect saturable absorber that is a vital mode-locking element in an ultrafast laser system due to broad wavelength range, significant nonlinearity, and its fast response times.

Four-Wave Mixing

Monolayer graphene has a susceptibility value that is eight orders of magnitude greater than the comparable nonlinearity in common dielectric materials. As a result, the massive nonlinearity of graphene can be utilized as agent in FWM imaging, considerably improving the visibility of the graphene flakes [39].

In the region of near-infrared frequencies, the susceptibility is slightly wavelength dependent. When compared to semiconducting nanostructures, which also have interband transitions, the dispersion properties are very different. In the light of this property, graphene is a potential material for broadband nonlinear applications. Stronger nonlinearities are believed to be possible at lower frequencies such as terahertz and far-infrared ones.

Two-Photon Absorption

TPA is the simultaneous absorption of two photons of different or identical frequencies. Graphene is considered as a candidate substance for optoelectronic devices and next-generation photonic. Because of the distinct band structure of many layers of graphene, TPA would occur in graphene and would vary with the graphene layer numbers. The SA is produced by the Pauli blocking of single-photon absorption, and the TPA results from interband electron transitions [40].

Reverse Saturable Absorption

Materials that have RSA will be less transmissive when excited. This produces the opposite result compared to SA and is thus termed RSA. In principle, long-lived excited state absorption levels are not exhibited in graphene due to the fast recombination of the electron–hole. When graphene is functionalized and distributed as single sheets in heavy-atom solvents or certain film matrices, it shows a new excited-state absorption mechanism [41].

Optical Limiting

Low transmittance for high incident light intensity and high transmittance for low incident light intensity are characteristics of optical limiters. These devices have importance needs for optical sensors and for protecting human eyes because retinal damage can happen when intensities rise over certain limits.

Wang et al. realized a significant broadband optical limiting response in the graphene dispersions using liquid-phase exfoliated method. Nonlinear scattering (NLS), originating from the thermally induced solvent bubbles and microplasmas, is responsible for this nonlinear behavior. The surface tension of the solvent has a strong influence on the optical limiting performance of each of the graphene dispersions.

9.3 Preparation Processes of Graphene Based Nanostructures

Due to the promise of graphene, numerous organizations, including the National Graphene Association and the National Graphene Institute, are developing and promoting the commercialized products that use graphene. Additionally, the European Commission founded the Graphene Flagship, a 150-partner organization with a 1billion EUR overall budget, for a long time commercialization of graphene [42]. Graphene research and development is constantly expanding, with research topics ranging from scaled production and new uses to the amazing properties of its derivatives. Graphene needs to be produced at a larger scale to lower the cost per unit while retaining acceptable product quality if it is to be widely used in commercial products.

Scientists have been developing many graphene production processes; as a result, many strategies for exfoliating the graphite to graphene have been presented. Many challenges are associated with the preparation processes of graphene to find a viable and well-fitting manufacturing technology for mass-scale graphene fabrication and to produce defect-free graphene. In general, graphene production may be divided into two strategies as top-down and bottom-up processes. Furthermore, according to the requirements of the final use, each strategy can be applied. The processes are categorized as being described in Fig. 9.3.

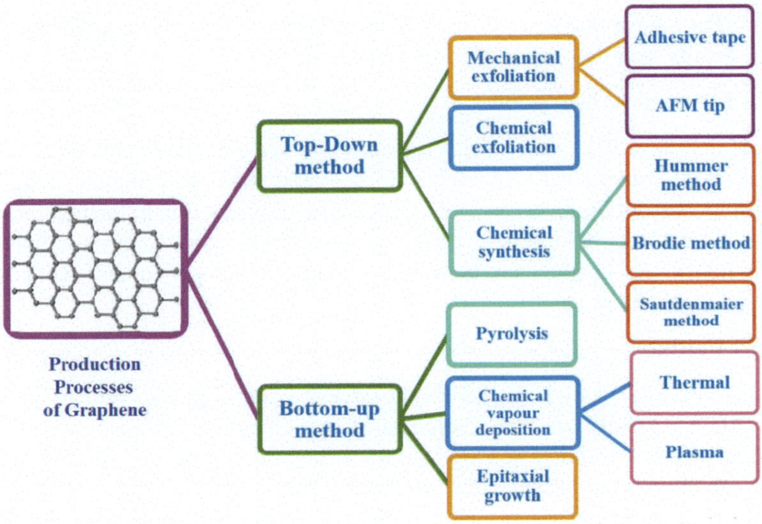

Fig. 9.3 Categorizes of graphene synthesis processes

9.3.1 Top-Down Methods

The top-down approach's fundamental base is to exfoliate the graphite that is utilized as a starting point. Top-down approaches include mechanical exfoliation, chemical exfoliation, and chemical synthesis. The main goal of these techniques is to manufacture monolayer flakes or few-layer of graphene by overcoming the van der Waals forces between the layers. It is relatively straightforward to use the top-down method for industrial-scale graphene manufacturing. Although traditional top-down techniques, such as Hummer graphite oxidation, need for a regulated reaction and invariably, these techniques result in an abundance of structural imperfections that reduce electrical conductivity [43].

9.3.2 Bottom-Up Methods

Using a bottom-up approach, a beginning material is deposited using a controlled process that regulates factors like pressure, temperature and flow rate. Bottom-up approaches include pyrolsis, chemical vapour deposition and epitaxial growth. A high-quality graphene with certain defects in structure and excellent electrical properties may be prepared using the bottom-up approach. However, since a limited amount graphene is produced, it can only be used in only a few of applications [44].

9.4 Functionalization of Graphene Based Nanostructures

In 2004, graphene and its derivatives attracted attention in materials, biological, chemistry, physics communities. Following that, the new members on the family of graphene and their derivatives such as graphene dots and graphene nanoribbon were extensively explored. Figure 9.4 shows different derivatives of graphene.

The two faces of graphene; the edges, and the defect sides can be used for functionalization through a stable covalent or weak non-covalent bonding like π–π interactions, hydrogen bonds and van der Waals interactions [45] Fig. 9.5. The electrical conductivity of graphene is partially lost during covalent functionalization due to the introduction of defects, but the sp^2 structure is maintained during non-covalent functionalization.

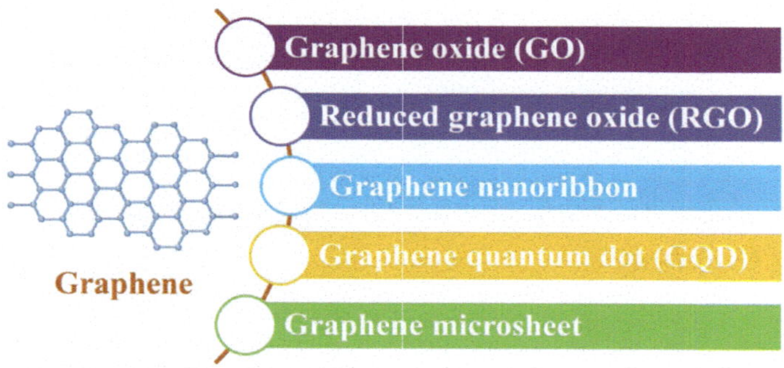

Fig. 9.4 Different derivatives of graphene

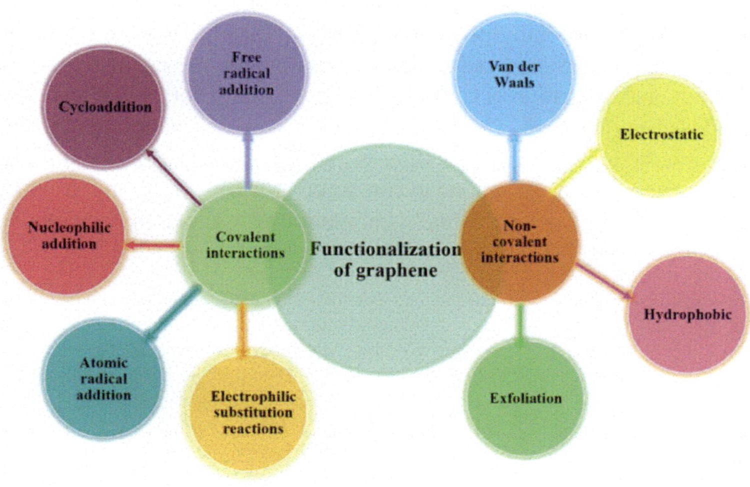

Fig. 9.5 Functionalization of graphene nanostructures

9.4.1 Covalent Interactions for Functionalization of Graphene

One or more sp^2 carbon atoms of the aromatic structure undergo rehybridization as a result of covalent surface modification of graphene, which results in a strain in the plane and electronic conjugation loss. Furthermore, due to the variation of hybridization, the severely constrained sp^2 carbon atoms should extend out of the plane in a tetrahedral sp^3 geometry. Consequently, the edge of carbon atoms react better in covalent reactions [46]. The organic covalent functionalization is helpful for enhancing the material's dispersibility in both water and organic solvents as well as for combing the properties of graphene with those of other functional materials like polymers or chromophores [47, 48]. Functionalization can occur at the defect sites, edges and on the surface. The defects include carbon lattice damage, adatoms and structural imperfections [49, 50]. Covalent modifications may be categorized in free radical addition, cycloaddition, nucleophilic addition, atomic radical addition, and electrophilic substitution reactions.

9.4.2 Non-covalent Interactions for Functionalization of Graphene

Graphene's large specific surface area promotes the physical adsorption of molecules through non-covalent interactions including van der Waals, electrostatic, and hydrophobic [51]. This approach of functionalization is a vital way to efficiently and easily immobilize molecules without affecting the electrical network [52]. The non-covalent functionalization is essential for the immobilization of DNA, enzymes, proteins, and drugs, especially in the context of devices due to even minor changes to the electronic properties of the π system that can completely alter the system's characteristics and structure [53, 54]. There has been a lot of research done on graphene-based π complexes. Several categories of reactions have been divided such as anion–π interactions, π–π, H–π, cation–π and non-polar gas–π interactions [55]. Non-covalent interactions of graphene can be created by exfoliating graphite with various surfactants, such as cetyltrimethylammonium bromide, triphenylene, sodium cholate, polyvinylpyrrolidone, or pyrene derivatives [56].

9.5 Applications of Graphene in Multimodal Bioimaging

9.5.1 Optical Imaging

Optical imaging is a nonsurgical technology that produces precise images of tissues and organs as well as smaller components as cells and also molecules using visible light and specific characteristics of photons [57]. Its advantages over other imaging modalities include being low in cost, highly sensitive, non-ionizing radiation, capable of real-time imaging, and having the capacity to multiplex [58]. As a result of tissue autofluorescence, light absorption by proteins, heme groups, and even water (above 900 nm), as well as poor tissue penetration (0–2 cm) and strong tissue scattering of photons in the visible light region (395–600 nm), this modality suffers. Near-infrared (NIR) window (650–900 nm) and second NIR window (1000–1700 nm) imaging techniques have been investigated in order to solve these issues due to their advantages of decreased autofluorescence, decreased tissue scattering, and higher depth of penetration for in vivo imaging [59–61]. Graphene-based nanostructures were actively promoted for optical imaging, such as fluorescence imaging (FL) and up-conversion imaging.

9.5.2 Fluorescence Bioimaging

The fluorescence phenomenon is now essential for producing imaging technologies for biomedical diagnosis and biological research. The term "fluorescence" describes the ability of some materials to emit light at a specific wavelength after absorbing light of a shorter wavelength for a brief period of time known as the "fluorescence lifetime." Fluorescent probes have important functions for identifying the target molecules and improving the fluorescence signal in fluorescence bioimaging. The qualities of fluorescent probes have a significant impact on the accuracy of fluorescence imaging.

The ideal graphene has a zero optical band gap and is made entirely of sp^2 hybridized carbon atoms, which shows no fluorescence. GQDs, RGO, and GO have a mixture of sp^2 and sp^3 carbons. The addition of functional groups expands the optical band gap and causes fluorescence. Functional groups containing oxygen are prevalent in GO to create RGO, partial functional groups in GO must undergo chemical reduction. Inherent from synthesis procedures of GQDs that include both sp^2 and sp^3 carbon atoms, GQDs have an optical band gap that varies with size due to quantum confinement. Cheap, photo-stable, non-toxic, soluble in water, biocompatible, and ecologically safe are all attributes of GO and GQDs. The fluorescence they produce is excitation wavelength dependant. Furthermore, its 2D surface exhibits powerful non-covalent interactions with adsorbed biomolecules through the π-π interactions, hydrogen, or electrostatic bonding, offering a chemically adaptable platform for bio-conjugation. Thus, in recent years, both GO and GQDs have found a wide

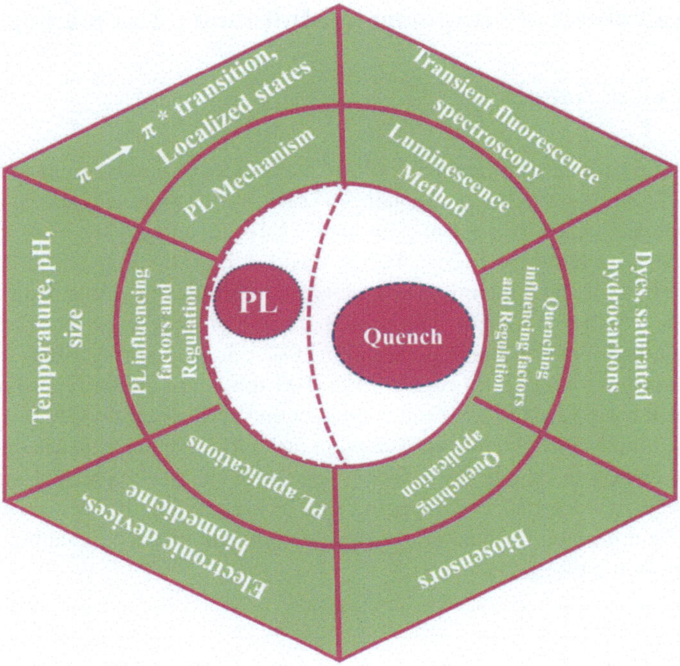

Fig. 9.6 Graphene nanostructures mechanisms [63]

range of applications in fluorescence sensors. They may be used as both an efficient fluorescence quencher and a tuneable fluorophore [62, 63].

Although various potential processes have been hypothesized based on their experimental phenomena and density functional theory (DFT) calculations, the photoluminescence (PL) and fluorescence quenching mechanisms of graphene nanostructures in the ultraviolet and visible light range remain unclear at this time. Clarifying the PL and fluorescence quenching mechanisms is also crucial for developing the understanding of carbon materials and their potential for imaging application as Fig. 9.6.

9.5.3 Up-Conversion Bioimaging

Up-conversion nanoparticles (UCNPs) can convert lower-energy excitation near-infrared (NIR) light into higher-energy emission visible light using process of multiphoton absorption. Additionally, the ability to use both NIR excitation and NIR emission (NIR-to-NIR up-conversion) is very pertinent for in vivo imaging of complicated species since it allows for deep tissue penetration and results in less absorption and scattering of bio-tissues and organs. These characteristics make UCNPs one-of-a-kind optical tools for biological investigations, as do their great stability, minimal cytotoxicity, good photostability, and non-photoblinking or photobleaching. UCNPs can be employed as probes to detect and target certain cells with high sensitivity and selectivity by causing surface conjugation with particular targeting ligands (such as peptides, antibodies, and small molecule medicines) [64]. A small animal up-conversion luminescence imaging platform using UCNPs had been developed by Bogdan et al. It was decided to use two external continuous wave (CW), power-adjustable 980 nm lasers as the excitation sources [65, 66] Fig. 9.7. Beam expanders in this system turned the radiation into beams with higher diameters that subsequently uniformly illuminated the small animal's whole body. Weak up-conversion emission was captured using an electron-multiplying charge-coupled device (EMCCD). The up-conversion materials' various excitation and emission wavelengths were taken into account when tuning the short-pass filter. By altering the optical filters between the sample and the EMCCD, up-conversion emission signals might be obtained in various channels. By substituting CW lasers at 543 or 633 nm for the 980 nm laser as the excitation source, this setup for the small-animal up-conversion luminescence imaging system might also be adjusted for triplet–triplet annihilation (TTA) based up-conversion systems. According to Zarghami et al., the design and synthesis of colloidal graphene nanoparticles doped with neodymium cations and cobalt were used in biological imaging. Carboxylic groups that can be bioconjugated with biomolecules adorn the surface of nanoparticles. Here, cobalt and neodymium ions were needed to be doped into the structure of the graphene quantum dots to enhance the quantum yield of the up conversion property. The intensity of the up-conversion phenomena increased by 41% for cobalt and 100% for neodymium-doped graphene quantum dots. DFT computations were used to simulate the aforementioned structures and to determine their characteristics [67].

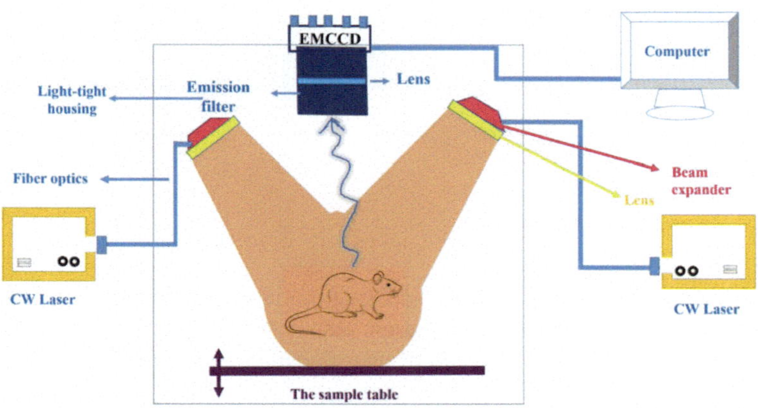

Fig. 9.7 Diagram showing the small-animal up-conversion luminescence imaging system's experimental setup [68], copyright, American Chemical Society, 2009

9.5.4 Magnetic Resonance Imaging

Magnetic resonance imaging is an effective imaging tool for medical diagnosis, Magnetic contrast agents are used to improve accuracy and sensitivity. Graphene has superior mechanical and physicochemical qualities, as well as high biocompatibility. Researchers are interested in graphene-based magnetic nanocarriers because they have reduced toxicity, increased chemical stability, and amazing functionalization potential. Magnetic nanocarriers based on graphene have caught the attention of researchers who are looking for superior contrast agents with minimal toxicity, extraordinary functionalization potential, and enhanced chemical stability. Their large specific surface area inhibits uncontrolled nanoparticle aggregation and hence dramatically boosts MRI efficiency. MRI uses strong magnetic fields and radio waves to generate detailed computerized images of the internal body structures [69]. Contrast agents (CAs) are employed in MRI to reduce the relaxation duration of nuclei within human tissues, enhancing visibility of internal body structures. Gadolinium (Gd), a common MRI contrast agent, is toxic and has been shown to have a generally depressive effect on many biological systems. Moreover, there is a possibility that free Gd^{3+} ions could be generated from the chelated Gd CAs, which could be neurotoxic and harm cardiovascular health by inhibiting channels of calcium in the body [70]. Graphene nanomaterials have numerous diagnostic applications, one of which is contrast agents in MRI. Among them, graphene oxide that has been extensively employed as a nanoplatform for creating MRI contrast agents.

GO has numerous advantages. The presence of hydrophilic oxygen-containing groups in its structure leads to increase dispersibility of water which is similar to superparamagnetic iron oxide nanoparticles (SPIONs) that are good contrast agents. Interestingly, combining them with reduced graphene oxide minimizes clumping, enhancing efficiency and circulation time [71]. Also, as previously indicated, the poisonous character of Gd^{3+} ions can be reduced by combining Gadolinium oxide(Gd_2O_3) with GO, while increasing solubility in aqueous solutions. Wang et al. designed graphene oxide/iron oxide nanoparticles (GO/IONPs) using a solvothermal procedure to diagnose regional lymph nodes (RLN) of pancreatic cancer metastases [72].

9.5.5 Raman Imaging

Raman spectroscopy investigates the inelastic scattering of phonons generated by modes of molecular vibrational excitation. Raman imaging is a useful analytical instrument that includes the potential of vibrational spectroscopy and is capable of solving biochemical and chemical problems in a nondestructive manner due to its extremely high signal-to-noise ratio, multiplexing capabilities, and minimal photobleaching [73]. Surface-enhanced Raman scattering (SERS) materials can improve the Raman signals of analytical substances nearby the surfaces of metal nanomaterials by upwards of ten orders of magnitude [74]. This enables high-sensitivity probing and imaging of cells using SERS labels. GOs have instinctively powerful D and G peaks, which can be enhanced by the combination of with metal NPs [75–77]. Liu et al. demonstrated that folic acid conjugated silver/graphene oxide nanocomposite FA-Ag/GO nanocomposite can be designed for SERS imaging targeting on folate receptor (FR). Transmission electron microscopy (TEM) images of Ag/GO nanocomposite are shown in Fig. 9.8a, b at different magnifications. As shown in Fig. 9.8c–h, FA-Ag/GO nanocomposites were used as SERS labels for selective labelling and imaging of cancer cells. Significant signals of Raman are seen on the HeLa cells plasma membrane of Fig. 9.8f–h, but not seen on the surface of FR-negative A549 cells (Fig. 9.8c–e). As shown in Fig. 9.8i and j, Raman signals can be seen in the cytoplasm of the cells surface. These findings point to FA- Ag/GO nanocomposite is potential SERS labels [76].

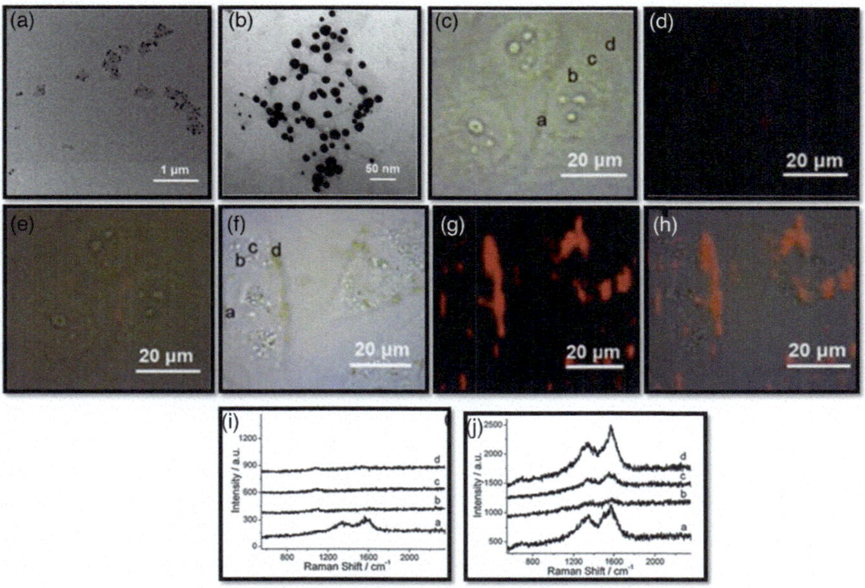

Fig. 9.8 **a** and **b** TEM images of GO/Ag nanocomposite at different magnifications. **c–h** FA-Ag/GO nanocomposites were used for imaging of cancer cells. The Raman spectra of the various places indicated in (**c**) and (**f**), respectively, are shown in (**i**) and (**j**) [76], copyright, Royal Society of Chemistry, 2013

9.5.6 Radionuclide-Based Imaging

For the diagnosis of cancer, nuclear imaging techniques such as planar imaging (gamma camera), single photon emission computed tomography (SPECT/CT), and positron emission tomography/computed tomography (PET/CT) have been frequently employed in clinics. Biocompatibility plays an essential role in the nuclear medicine field due to the decreased clearance of radioactive elements, which can enhance the harmful effects of produced radio compositions. Different nanomaterials have been produced as nanocarriers for the transport of radionuclides in order to reduce needle side effects, improve the therapeutic effect of radiotherapy, and achieve the targeted delivery of radionuclides [77]. Among these, graphene nanostructures serve as promising platforms in PET/SPECT imaging. Radionuclides have been effectively delivered in vitro and in vivo using materials based on graphene. In fact, to lessen the toxicity of the radioactive Iodine ^{131}I, Chen and colleagues developed graphene-based radioisotopes carrier. When compared to free ^{131}I isotopes, synthetic delivery systems were more effective against cancer cells [78]. In addition to minimizing the toxicity of graphene materials, surface functionalization can also improve radionuclide labelling affinity.

9.5.7 Photoacoustic Imaging

Low-energy electromagnetic waves can penetrate deeper than short wavelengths, and radio frequency waves scatter far less in biological samples, which may be advantageous for deep-tissue/organ imaging such as photoacoustic imaging (PAI). Using the phatoacoustic (PA) effect, the imaging system turns the absorbed nonionizing pulses of laser (short electromagnetic energy pulses) into heat, resulting in an acoustic emission due to thermoelastic expansion. Actually, within pulsed laser exposure, contrast agents or tissues absorb light and result in a pressure grow by localized rapid thermoelastic expansion, which generates acoustic waves with wide-band that can be recorded by ultrasound transducers and analyzed using equivalent reconstruction algorithms.

In the family of graphene nanostructures, reduced graphene oxide acts an indispensable role as contrast agent of PA that can absorb NIR light better than GOs owing to its wider sp^2 domains [79]. On the other hand, RGOs have lower water solubility due to the elimination of graphene oxide hydrophilicity after reduction. Patel et al. (2013) used microwave heating to create direct production of less-oxygenated nanosize graphene sheets that is called microwave-enabled low-oxygen graphene (ME-LOGr) [80].

Figure 9.9a shows the nitronium oxidation mechanism with microwave assistance. This method leads the graphite oxidation processes toward direct formation of graphene nanosheets rather than GO nanosheets. According to atomic force microscopy (AFM), the ME-LOGr nanosheets have a lateral diameter of (10 $\pm$ 4)

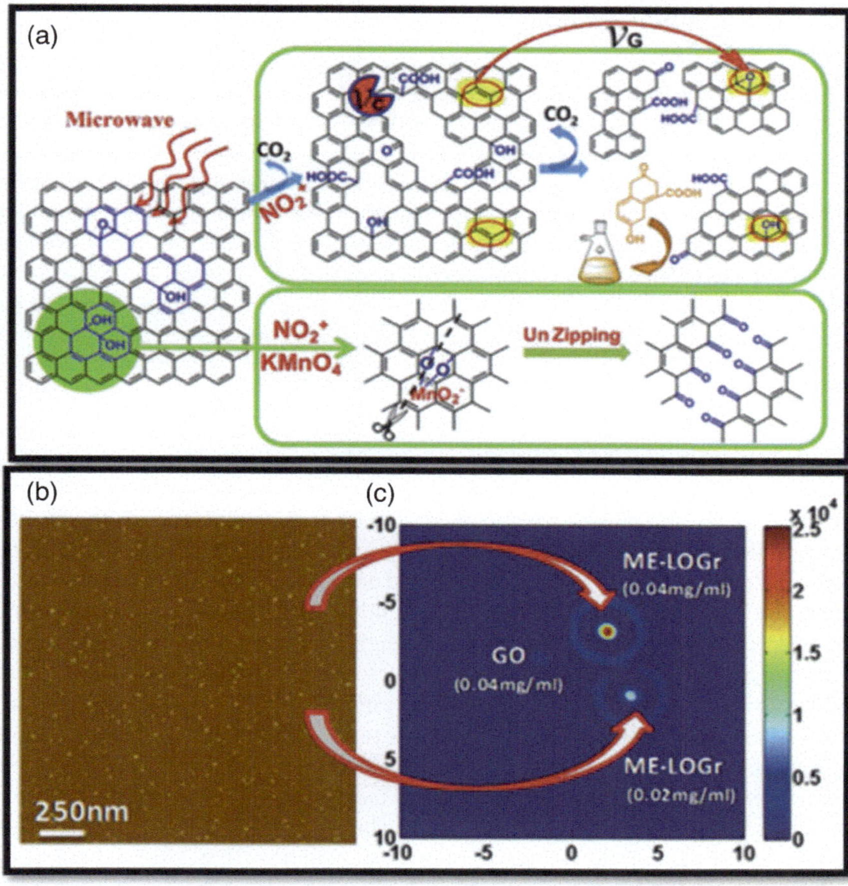

Fig. 9.9 **a** Schematic of nitronium oxidation mechanisms by microwave-assisted, **b** AFM images of ME-LOGr nanosheets, **c** PA signal of different concentrations of graphene nanosheets and GO that illuminated with 800 nm [80], copyright, American Chemical Society, 2013

nm as shown in Fig. 9.9b, which can produce photoacoustic signals with NIR excitation as shown in Fig. 9.9c that are easily diffused in water [80]. To improve the GO absorption cross section, indocyanine green (ICG) was loaded onto it via π–π stacking reaction that has a high absorbance in the NIR region, and the indocyanine green graphene oxide nanocomposite (ICG-GO) was shown to be an ultrahigh sensitivity PA contrast agent [81].

9.5.8 Multimodal Imaging

The concept of combining different imaging modalities for biomedical purposes has recently gained attraction. Multimodality imaging refers to promoting the benefits of every imaging modality and gathering all information from several imaging modalities to improve diagnostic efficiency and accuracy [61, 82, 83]. In the meantime, including multimodal detectability into a single drug would eliminate the added load on the body's blood clearance that comes with administering many dosages of medicines [84]. Because of their specific high surface area and multifunctional chemistry, graphene and its derivatives have been used as fundamental components for the development of multimodal imaging contrast agents.

9.6 Cytotoxicity of Graphene

Graphene and its derivatives' toxicity were investigated using several cell lines and animal models. Reactive oxygen species (ROS)-mediated necrosis causes cell death, which results in plasma membrane damage and apoptosis. The toxicity of graphene is determined by in vitro features such as layer number surface functionalization, lateral size, hydrophobicity, dosage, and stiffness. Significant data were collected on the generation of biocompatible graphene and their toxicity from different investigations conducted on different cells and animal models in a review [85]. Furthermore, Syama et al. presented strategies for reducing graphene toxicity by developing biocompatible graphene nanostructures using plant extracts, microorganisms, and biocompatible polymers [86–88]. Surface modifications may tackle the issue of toxicity in different tissues by selectively targeting infected areas. The impact and severity of toxicity are also affected by the duration of exposure in the human body, the dose, and the mode of administration [88].

9.7 Recent Challenges and Future Prospects

The development of graphene nanostructures which tries to prevent their negative effects without compromising the intended nanomedical functionalities is vital to maximize the performance of bioimaging. Comprehending the potential biophysicochemical interactions at bio-nano interfaces of graphene and their outcomes may yield vital data for the development of a nano- model, which in turn may direct the safe design of nanomaterials for medical use and innovate new avenues for enhancing their efficacy.

Efforts to reduce the toxicities of biomedical graphene nanomaterials could be focused on various stages of their life cycles. Personal protection equipment and remote control procedures are necessary and beneficial during the synthesis or application of graphene, but they cannot minimize or remove the intrinsic toxicity of these materials completely. As a result, efforts to reduce the potential risk of nanomaterials during their design are more appealing and significantly facilitate the clinical implementation of nanomedicines.

9.8 Conclusion

We discussed the most recent developments in graphene-based nanomaterials as image contrast agents for various bioimaging applications in this chapter. For FL imaging, functionalization of graphene-based nanostructures has been frequently used, due to their inherent photoluminescence. GOs have been used for Raman imaging because of their inherent strong Raman signals with D and G peaks. This technique can be improved by incorporating silver metallic nanoparticles. Radionuclide-labeled graphene nanostructures have been investigated for PET/SPECT imaging. IONP-functionalized graphene materials have been developed for MRI. PAI can be used with RGOs or GOs modified with molecules that have strong optical absorbance in the NIR region. Multimodal imaging can be achieved by combining GOs with other materials that have specific features. Graphene-based nanomaterials will definitely have promising impact in bioimaging in the near future.

References

1. S.K. Ghosh, Human cadaveric dissection: a historical account from ancient Greece to the modern era. Anat. Cell Biol. **48**, 153–169 (2015)
2. P. Raghavendra, T. Pullaiah, Biomedical imaging role in cellular and molecular diagnostics in *Advances in Cell and Molecular Diagnostics*, ed. by P. Raghavendra, T. Pullaiah (Academic Press, 2018), pp. 85–111
3. R. Robb, Biomedical imaging: past, present and predictions. Med. Imaging Technol. **24**, 25 (2006)
4. Y. Bagheri, A.A. Ali, M. You, Current methods for detecting cell membrane transient interactions. Front. Chem. **8**, 603259 (2020)
5. B. Föllmer et al., Roadmap on the use of artificial intelligence for imaging of vulnerable atherosclerotic plaque in coronary arteries. Nat. Rev. Cardiol. **21**, 51–64 (2024)
6. M.A. Saleh, A.A. Ali, K. Ahmed, A.M. Sarhan, A brief analysis of multimodal medical image fusion techniques. Electronics **12**, 97 (2023)
7. J. Wallyn, N. Anton, S. Akram, T.F. Vandamme, Biomedical imaging: principles, technologies, clinical aspects, contrast agents, limitations and future trends in nanomedicines. Pharm. Res. **36**, 78 (2019)
8. M.A. Hahn, A.K. Singh, P. Sharma, S.C. Brown, B.M. Moudgil, Nanoparticles as contrast agents for in-vivo bioimaging: current status and future perspectives. Anal. Bioanal. Chem. **399**, 3–27 (2011)
9. J. Key, J.F. Leary, Nanoparticles for multimodal in vivo imaging in nanomedicine. Int. J. Nanomed. **9**, 711–726 (2014)

10. X. Li, N. Anton, G. Zuber, T. Vandamme, Contrast agents for preclinical targeted X-ray imaging. Adv. Drug Deliv. Rev. **76**, 116–133 (2014)
11. M. Elsabahy, G.S. Heo, S.M. Lim, G. Sun, K.L. Wooley, Polymeric nanostructures for imaging and therapy. Chem. Rev. **115**, 10967–11011 (2015)
12. A. Girigoswami, K. Girigoswami, Potential applications of nanoparticles in improving the outcome of lung cancer treatment. Genes (Basel) **14**, 1370 (2023)
13. H. Liu et al., A review of carbon dots in synthesis strategy. Coord. Chem. Rev. **498**, 215468 (2024)
14. Y. Nagai et al., Bright NIR-II fluorescence from biocompatible gel-coated carbon nanotubes for in vivo imaging. Carbon **218**, 118728 (2024)
15. A. Clermont-Paquette, K. Larocque, A. Piekny, R. Naccache, Shining a light on cells: amine-passivated fluorescent carbon dots as bioimaging nanoprobes. Mater. Adv. **3** (2024)
16. W. Wanas, S.A. Abd El-Kaream, S. Ebrahim, M. Soliman, M. Karim, Cancer bioimaging using dual mode luminescence of graphene/FA-ZnO nanocomposite based on novel green technique. Sci. Rep. **13**, 27 (2023)
17. A. Valimukhametova et al., Five near-infrared-emissive graphene quantum dots for multiplex bioimaging. 2D Materials **11**, 025009 (2024)
18. S.K. Tiwari, R.K. Mishra, S.K. Ha, A. Huczko, Evolution of graphene oxide and graphene: from imagination to industrialization. ChemNanoMat **4**, 598–620 (2018)
19. Z. Hughes, T. Walsh, Computational chemistry for graphene-based energy applications: progress and challenges. Nanoscale **7**, 6883–6908 (2015)
20. S.K. Tiwari et al., Magical allotropes of carbon: prospects and applications. Crit. Rev. Solid State Mater. Sci. **41**, 257–317 (2016)
21. Y. Yang et al., Bottom-up fabrication of graphene on silicon/silica substrate via a facile soft-hard template approach. Sci. Rep. **5**, 13480 (2015)
22. S.K. Tiwari, S. Sahoo, N. Wang, A. Huczko, Graphene research and their outputs: status and prospect. J. Sci. Adv. Mater. Devices **5**, 10–29 (2020)
23. Y. Zhang, Y.W. Tan, H.L. Stormer, P. Kim, Experimental observation of the quantum Hall effect and Berry's phase in graphene. Nature **438**, 201–204 (2005)
24. P. Wang et al., A superhydrophobic/electrothermal synergistically anti-icing strategy based on graphene composite. Compos. Sci. Technol. **198**, 108307 (2020)
25. A.E. da Costa Júnior et al., A self-assembly of graphene oxide@Fe3O4/metallo-phthalocyanine nanohybrid materials: synthesis, characterization, dielectric and thermal properties. J. Mater. Sci. **52**, 9546–9557 (2017)
26. K. Bilisik, M. Akter, Graphene nanocomposites: a review on processes, properties, and applications. J. Ind. Text. **51**, 3718S-3766S (2021)
27. F. Ebrahimi, E. Heidari, A review on modeling, synthesis, and properties of graphene, in *Graphene—new trends and developments* (Intech Open, London, 2015)
28. V. Kumar, Linear and nonlinear optical properties of graphene: a review. J. Electron. Mater. **50**, 3773–3799 (2021)
29. R.R. Nair et al., Fine structure constant defines visual transparency of graphene. Science **320**, 1308 (2008)
30. S.-E. Zhu, S. Yuan, G. Janssen, Optical transmittance of multilayer graphene. Europhys. Lett. **108**, 17007 (2014)
31. H. Dakhlaoui, Linear and nonlinear optical absorption coefficients and refractive index changes in GaN/AlxGa(1−x)N double quantum wells operating at 1.55 μm. J. Appl. Phys. **117**, 135705 (2015)
32. E.B. Al et al., Effects of applied electric and magnetic fields on the nonlinear optical properties of asymmetric GaAs/Ga1-xAlxAs double inverse parabolic quantum well. Opt. Mater. **47**, 1–6 (2015)
33. U. Yesilgul et al., Linear and nonlinear optical properties in an asymmetric double quantum well under intense laser field: effects of applied electric and magnetic fields. Opt. Mater. **58**, 107–112 (2016)

34. C.V. Nguyen et al., Linear and nonlinear magneto-optical properties of monolayer phosphorene. J. Appl. Phys. **121**, 045107 (2017)
35. D.J. Merthe, V.V. Kresin, Transparency of graphene and other direct-gap two-dimensional materials. Phys. Rev. B **94**, 205439 (2016)
36. L. Kang, Z. Lin, Deep-ultraviolet nonlinear optical crystals: concept development and materials discovery. Light Sci. Appl. **11**, 201 (2022)
37. M. Sedaghat Nejad, A. Ghasempour Ardakani, Significant enhancement of third harmonic generation in graphene layers placed on a random grating supporting plasmonic Anderson localized states. Opt. Laser Technol. **161**, 109123 (2023)
38. J.M. Dawlaty, S. Shivaraman, M. Chandrashekhar, F. Rana, M.G. Spencer, Measurement of ultrafast carrier dynamics in epitaxial graphene. Appl. Phys. Lett. **92**, 042116 (2008)
39. E. Hendry, P.J. Hale, J. Moger, A.K. Savchenko, S.A. Mikhailov, Coherent nonlinear optical response of graphene. Phys. Rev. Lett. **105**, 097401 (2010)
40. H. Yang et al., Giant two-photon absorption in bilayer graphene. Nano Lett. **11**, 2622–2627 (2011)
41. G.-K. Lim et al., Giant broadband nonlinear optical absorption response in dispersed graphene single sheets. Nat. Photonics **5**, 554–560 (2011)
42. N. Kumar et al., Top-down synthesis of graphene: a comprehensive review. FlatChem **27**, 100224 (2021)
43. M.S.A. Bhuyan, M.N. Uddin, M.M. Islam, F.A. Bipasha, S.S. Hossain, Synthesis of graphene. Int. Nano Lett. **6**, 65–83 (2016)
44. J.Y. Lim et al., Recent trends in the synthesis of graphene and graphene oxide based nanomaterials for removal of heavy metals—a review. J. Ind. Eng. Chem. **66**, 29–44 (2018)
45. D.W. Boukhvalov, M.I. Katsnelson, Chemical functionalization of graphene. J. Phys. Condens. Matter **21**, 344205 (2009)
46. L. Yan et al., Chemistry and physics of a single atomic layer: strategies and challenges for functionalization of graphene and graphene-based materials. Chem. Soc. Rev. **41**, 97–114 (2012)
47. V. Georgakilas et al., Functionalization of graphene: covalent and non-covalent approaches, derivatives and applications. Chem. Rev. **112**, 6156–6214 (2012)
48. L. Rodríguez-Pérez, M. Herranz, N. Martín, The chemistry of pristine graphene. Chem. Commun. (Camb.) **49**, 3721–3735 (2013)
49. M. Quintana, J.I. Tapia, M. Prato, Liquid-phase exfoliated graphene: functionalization, characterization, and applications. Beilstein J. Nanotechnol. **5**, 2328–2338 (2014)
50. F. Banhart, J. Kotakoski, A.V. Krasheninnikov, Structural defects in graphene. ACS Nano **5**, 26–41 (2011)
51. T. Kuila et al., Chemical functionalization of graphene and its applications. Prog. Mater. Sci. **57**, 1061–1105 (2012)
52. D. Tasis, N. Tagmatarchis, A. Bianco, M. Prato, Chemistry of carbon nanotubes. Chem. Rev. **106**, 1105–1136 (2006)
53. N.J. Singh, H.M. Lee, S.B. Suh, K.S. Kim, De novo design approach based on nanorecognition toward development of functional molecules/materials and nanosensors/nanodevices. Pure Appl. Chem. **79**, 1057–1075 (2007)
54. K.S. Kim, P. Tarakeshwar, J.Y. Lee, Molecular clusters of π-systems: theoretical studies of structures, spectra, and origin of interaction energies. Chem. Rev. **100**, 4145–4186 (2000)
55. K.E. Riley, M. Pitonák, P. Jurecka, P. Hobza, Stabilization and structure calculations for noncovalent interactions in extended molecular systems based on wave function and density functional theories. Chem. Rev. **110**, 5023–5063 (2010)
56. I.A. Vacchi, C. Ménard-Moyon, A. Bianco, Chemical functionalization of graphene family members. Phys. Sci. Rev. **2**, 20160103 (2017)
57. S.M. Janib, A.S. Moses, J.A. MacKay, Imaging and drug delivery using theranostic nanoparticles. Adv. Drug Deliv. Rev. **62**, 1052–1063 (2010)
58. J. Wang et al., Imaging-guided delivery of RNAi for anticancer treatment. Adv. Drug Deliv. Rev. **104**, 44–60 (2016)

59. K. Park et al., New generation of multifunctional nanoparticles for cancer imaging and therapy. Adv. Func. Mater. **19**, 1553–1566 (2009)
60. A.M. Smith, M.C. Mancini, S. Nie, Bioimaging: second window for in vivo imaging. Nat. Nanotechnol. **4**, 710–711 (2009)
61. P. Huang et al., Dye-loaded ferritin nanocages for multimodal imaging and photothermal therapy. Adv. Mater. **26**, 6401–6408 (2014)
62. P. Zheng, N. Wu, Fluorescence and sensing applications of graphene oxide and graphene quantum dots: a review. Chem. Asian J. **12**, 2343–2353 (2017)
63. X. Xiao, Y. Zhang, L. Zhou, B. Li, L. Gu, Photoluminescence and fluorescence quenching of graphene oxide: a review. Nanomaterials (Basel) **12**, 2444 (2022)
64. M. González-Béjar, L. Francés-Soriano, J. Pérez-Prieto, Upconversion nanoparticles for bioimaging and regenerative medicine. Front. Bioeng. Biotechnol. **4**, 47 (2016)
65. N. Bogdan, F. Vetrone, R. Roy, J.A. Capobianco, Carbohydrate-coated lanthanide-doped upconverting nanoparticles for lectin recognition. J. Mater. Chem. **20**, 7543–7550 (2010)
66. M. Yu et al., Laser scanning up-conversion luminescence microscopy for imaging cells labeled with rare-earth nanophosphors. Anal. Chem. **81**, 930–935 (2009)
67. A. Zarghami, M. Dolatyari, H. Mirtagioglu, A. Rostami, High-efficiency upconversion process in cobalt and neodymium doped graphene QDs for biomedical applications. Sci. Rep. **13**, 10277 (2023)
68. L. Xiong et al., High contrast upconversion luminescence targeted imaging in vivo using peptide-labeled nanophosphors. Anal. Chem. **81**, 8687–8694 (2009)
69. S. Gulati et al., Magnetic nanocarriers adorned on graphene: promising contrast-enhancing agents with state-of-the-art performance in magnetic resonance imaging (MRI) and theranostics. Mater. Adv. **3**, 2971–2989 (2022)
70. F. Wang, E. Peng, B. Zheng, S.F.Y. Li, J.M. Xue, Synthesis of water-dispersible Gd2O3/GO nanocomposites with enhanced MRI T1 relaxivity. J. Phys. Chem. C **119**, 23735–23742 (2015)
71. M. Llenas et al., Microwave-assisted synthesis of SPION-reduced graphene oxide hybrids for magnetic resonance imaging (MRI). Nanomaterials **9**, 1364 (2019)
72. S. Wang et al., Magnetic graphene-based nanotheranostic agent for dual-modality mapping guided photothermal therapy in regional lymph nodal metastasis of pancreatic cancer. Biomaterials **35**, 9473–9483 (2014)
73. J.M. Yoo, J.H. Kang, B.H. Hong, Graphene-based nanomaterials for versatile imaging studies. Chem. Soc. Rev. **44**, 4835–4852 (2015)
74. E.C. Le Ru, E. Blackie, M. Meyer, P.G. Etchegoin, Surface enhanced raman scattering enhancement factors: a comprehensive study. J. Phys. Chem. C **111**, 13794–13803 (2007)
75. Q. Liu et al., Cell imaging by graphene oxide based on surface enhanced Raman scattering. Nanoscale **4**, 7084–7089 (2012)
76. Z. Liu et al., Graphene oxide based surface-enhanced Raman scattering probes for cancer cell imaging. Phys. Chem. Chem. Phys. **15**, 2961–2966 (2013)
77. Y.K. Kim, S.W. Han, D.H. Min, Graphene oxide sheath on Ag nanoparticle/graphene hybrid films as an antioxidative coating and enhancer of surface-enhanced Raman scattering. ACS Appl. Mater. Interfaces **4**, 6545–6551 (2012)
78. L. Chen et al., Radionuclide (131)I labeled reduced graphene oxide for nuclear imaging guided combined radio- and photothermal therapy of cancer. Biomaterials **66**, 21–28 (2015)
79. G. Lalwani, X. Cai, L. Nie, L.V. Wang, B. Sitharaman, Graphene-based contrast agents for photoacoustic and thermoacoustic tomography. Photoacoustics **1**, 62–67 (2013)
80. M.A. Patel et al., Direct production of graphene nanosheets for near infrared photoacoustic imaging. ACS Nano **7**, 8147–8157 (2013)
81. Y.W. Wang et al., Dye-enhanced graphene oxide for photothermal therapy and photoacoustic imaging. J. Mater. Chem. B **1**, 5762–5767 (2013)
82. J. Ma et al., Folic acid-conjugated LaF3:Yb, Tm@SiO2 nanoprobes for targeting dual-modality imaging of upconversion luminescence and X-ray computed tomography. J. Phys. Chem. B **116**, 14062–14070 (2012)

83. G.B. Shevalkar, M.K. Prajapati, K. Mali, A brief overview on theranostic applications of graphene and graphene-based nanomaterials in *Nanomaterial-Based Drug Delivery Systems*, ed. by C.V. Pardeshi (Springer, 2023), pp. 295–325
84. Z. Cheng, A. Al Zaki, J.Z. Hui, V.R. Muzykantov, A. Tsourkas, Multifunctional nanoparticles: cost versus benefit of adding targeting and imaging capabilities. Science **338**, 903–910 (2012)
85. L. Ou et al., Toxicity of graphene-family nanoparticles: a general review of the origins and mechanisms. Part. Fibre Toxicol. **13**, 57 (2016)
86. S. Syama, P.V. Mohanan, Safety and biocompatibility of graphene: a new generation nanomaterial for biomedical application. Int. J. Biol. Macromol. **86**, 546–555 (2016)
87. C. McCallion, J. Burthem, K. Rees-Unwin, A. Golovanov, A. Pluen, Graphene in therapeutics delivery: problems, solutions and future opportunities. Eur. J. Pharm. Biopharm. **104**, 235–250 (2016)
88. H. Zhang, G. Grüner, Y. Zhao, Recent advancements of graphene in biomedicine. J Mater Chem B **1**, 2542–2567 (2013)

Chapter 10
Tailoring Nanoscale Metal–Organic Frameworks (MOFs) for Medical Imaging Applications

Omotayo Aderinoye and Sreejith Raveendran

Abstract Nanotechnology has significantly advanced, applying the core principles of biological, physical, and chemical sciences at the nanoscale to revolutionize various industries, including medicine. A key focus within this domain is the development and application of nanoscale Metal–Organic Frameworks (MOFs) for medical imaging. MOFs, with their unique structural and functional properties, offer promising enhancements in diagnostic accuracy and treatment outcomes. By manipulating objects at atomic and molecular levels, these nanoparticles, which measure less than 100 nm, exhibit exceptional physiochemical characteristics ideal for bioimaging applications. This chapter delves into the customization of MOFs for medical imaging, examining their potential to provide high-resolution images, improve contrast, and target specific tissues or diseases with precision. Through this exploration, we aim to underscore the transformative impact of MOFs in medical imaging, highlighting their role in advancing diagnostic techniques, enhancing therapeutic interventions, and contributing to economic growth and social welfare by improving patient care and medical research.

10.1 Introduction

Nanotechnology has found its way into many industries as a result of rapid growth in nanoscience. Nanotechnology is a process that combines the basic attributes of biological, physical, and chemical sciences and these occur at the minute scale of nanometers. It involves the reduction in the size of particles used, nanoparticles (NPs) and to target a particular site. Nanotechnology has found its application in different industries especially agriculture, medicine, healthcare, cosmetics, etc. [1]

O. Aderinoye · S. Raveendran (✉)
School of Health and Life Sciences, Teesside University, Middlesbrough TS1 3BX, UK
e-mail: Sreejith.R@tees.ac.uk

S. Raveendran
National Horizon Centre, Teesside University, Darlington DL1 1HG, UK

N. Kalarikkal et al. (eds.), *Multimodal Biomedical Imaging Techniques*,
Biological and Medical Physics, Biomedical Engineering,
https://doi.org/10.1007/978-981-96-1124-9_10

Over the years, Man has looked for remedies that could miraculously relieve sickness and injury pain. Many academics, however, feel that applications of nanotechnology in medicine can be essential in achieving this. Nanotechnology can radically transform medical research and result in creation of a new improvement to human. There is a good line for nanotechnology's diagnostic, therapeutic, and preventive applications. This technology can intentionally modify the body, among many problems. The nano drugs employed have demonstrated that bioavailability is enhanced, side effects are eliminated, and therapeutic medicine is absorbed more effectively. Nanotechnology, therefore, can contribute to economic growth while improving the social welfare of people [2]. In 1959, physicist Richard Feynman gave a talk on creating objects at the atomic and molecular levels, which initiated the concept of nanotechnology [2]. Nanotechnology employs the use of NPs in all its applications. NPs are a class of materials which includes particulate substances and have a dimension that is less than 100 nm as its name indicates. The use of NPs was birthed when researchers found that the size of a particle affects the physiochemical property of a substance. These characteristics, such as colour, size, and shape, make it possible for utilization in bioimaging [3].

10.2 Application of Nanotechnology

Nanotechnology has found its use in many fields of study. Below are some of the fields that have incorporated the use of nanotechnology with lots of positive outcomes.

10.2.1 Agriculture and Food Industry

Nanotechnology application started in 2003 in the United States of America and since then more research has gone into every aspect of food and agriculture, starting from the very beginning up to packaging processes. It has also found its use in food additives and food preservatives. Examples of NPs used are titanium dioxide (TiO2), iron oxide, zinc oxide, and silver-silica [4].

10.2.2 Medicine

Researchers have looked into nanotechnology as a unique tool in medical research, and it is currently considered the most promising technology of the twenty-first century. Such extensive research is ongoing in the fields of nephrology, cardiovascular disease, cancer therapy, and gene therapy. NPs are utilized in targeting cancer cells, as well as in cancer molecular imaging, which allows genetic alterations to be input into cancer cells via these devices [2].

Due in large part to their nanoscale size, NPs have a large surface-area-to-volume ratio that enables them to swiftly circulate throughout the bloodstream and absorb large volumes of medications. Because of their enhanced surface area, they have unique properties that enable them to be utilized in more medicinal applications. These properties include enhanced mechanical, magnetic, optical, and catalytic properties [2].

10.2.3 Diagnosis/Theranostic

Nanotechnology has also found great use in diagnosis and treatment combined. This has allowed the development of highly effective and precise treatments in cancer therapy. A significant example is the use of MRI dual mode T1/T2 contrast agents together with iron-based NPs to get high-resolution images [5].

Carbon-based hybrid nanogels are quickly becoming innovative nanoplatforms that can monitor biochemical and biophysical parameters both in vivo and in vitro, as well as intelligently deliver drugs in response to environmental stimuli for imaging-guided and combinatorial therapies that offer multimodal diagnosis, enhanced therapeutic efficacy, and reduced side effects in various disease treatments [6] represented in Fig. 10.1.

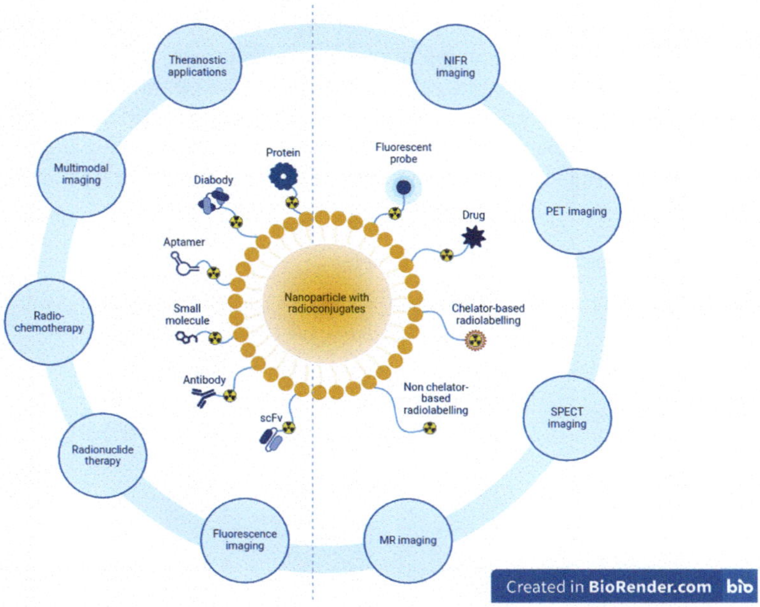

Fig. 10.1 Schematic representation of application of nanoplatform networks created by using biorender

Gold-based NPs have been used in many studies for therapeutic applications in the treatment of cancer. It has a high atomic number and, a great photoelectric absorption coefficient, its renal clearance is good and very biocompatible. All these characteristics make them a good fit for use in radiotherapy. The radiation-absorbing properties of AuNPs can be influenced by their size, shape, concentration, surface functionalization, and intracellular distribution [7].

10.3 Types of Nanoparticles

NPs are classified based on their composition into three: Organic NPs, carbon-based NPs, and inorganic NPs [8]. Table 10.1 shows the examples of the different NPs.

10.3.1 Organic Nanoparticles

Proteins, carbohydrates, lipids, polymers, and other organic substances can all be found in this class of NPs. Dendrimers, liposomes, micelles, and protein complexes like ferritin as shown in Fig. 10.2 are the most well-known examples of this family. These NPs are generally non-toxic, and biodegradable, and in certain instances—like with liposomes—may have a hollow center. Heat and light, as well as other forms of electromagnetic radiation, can affect organic NPs. Furthermore, they are frequently created by non-covalent intermolecular interactions, which increase their labile character and provide an exit path for the body. The possible range of use of organic NPs is determined by several factors, including composition, surface shape, stability, and carrying capacity. Organic NPs are now used in biomedicine in the targeted delivery of drugs and cancer therapy [8].

Table 10.1 Table showing types of NPs and examples

Nanoparticles	Examples
Organic nanoparticles	Dendrimers, liposomes, micelles, and ferritin
Carbon-based nanoparticles	Fullerene, carbon black, carbon quantum dots, and nanodiamonds
Inorganic nanoparticles	Gold NPs-as nanospheres, nanoshells, nanorods, nanoclusters, silver NPs, iron oxide, and quantum dots

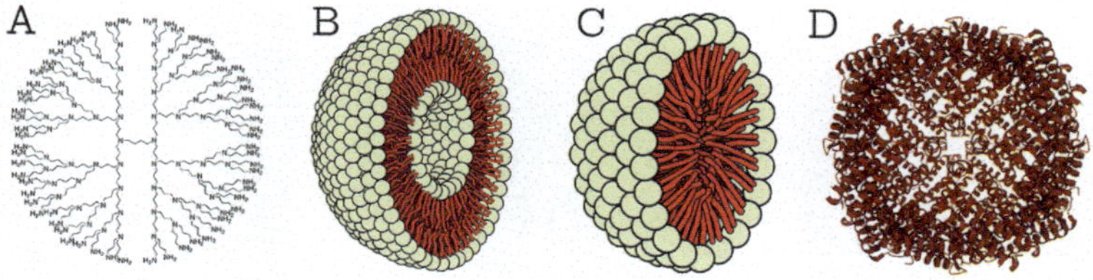

Fig. 10.2 Types of organic NPs: Dendrimers (**a**), Liposomes (**b**), Micelles (**c**), and Ferritin (**d**) [8]

10.3.2 Carbon-Based Nanoparticles

NPs which are entirely composed of carbon atoms belong to this class. Carbon quantum dots, carbon black NPs, and fullerenes (Fig. 10.3) are well-known representatives of this class [8]. Ellipsoids, hollow spheres, and tubes are some of the forms in which carbon is present in these nanomaterials [9]. Fullerenes are carbon molecules with a closed-cage structure that is symmetrical. The strongly fused spherical particles that make up carbon black NPs resemble grapes. Carbon NPs that are discrete and quasi-spherical, with diameters less than 10 nm, make up carbon quantum dots. At the nanoscale, carbon-based NPs combine the unique physicochemical properties of carbon bonds with the characteristic features of sp2-hybridized carbon bonds. Carbon-based NPs have a wide range of applications, including drug delivery, energy storage, and bioimaging, because of their distinctive electrical conductivity, high strength, electron affinity, optical, thermal, and sorption properties [8].

A new type of carbon-based nanoparticle is a nanodiamond with a size between 4 and 10 nm. It is known by another name, detonation diamonds; these diamond NPs are a particular class of carbon-based NPs with a spherical shape, distinct magnetic and optical behaviour, a greater surface area, and a unique lattice structure (Fig. 10.4). They are chemically stable and have a high absorption rate. This absorption rate makes it useful in the production of skincare products [9]. Nanodiamonds can be produced by explosives in a chamber filled with inert gases or water coolant resulting in a detonation soot (diamonds mixed with carbon allotropes and some impurities) of 4–5 nm diameter [10].

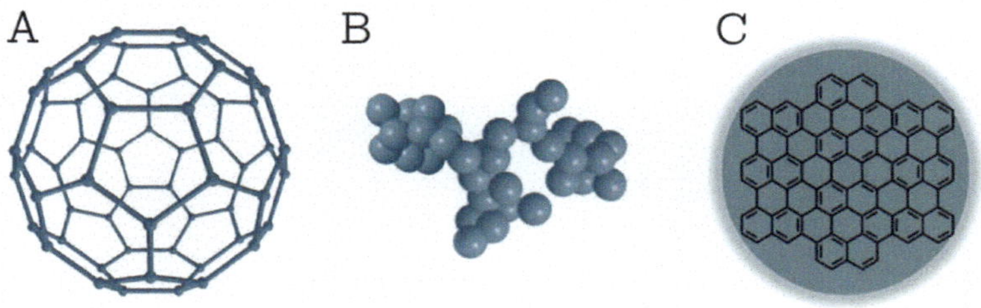

Fig. 10.3 Types of carbon-based NPs; Fullerene (**a**), Carbon black (**b**), and Carbon quantum dots (**c**) [8]

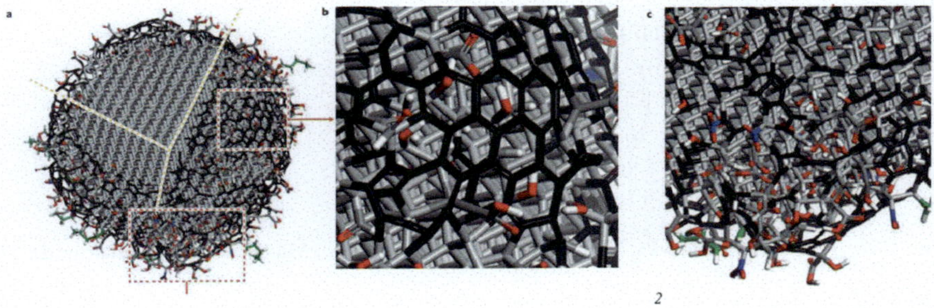

Fig. 10.4 **a** Purified nanodiamond, **b**, **c** Closeup view showing sp^2 carbon chains (black) and oxygen groups in blue colour (**d**) [10]

10.3.3 Inorganic Nanoparticles

Inorganic NPs are not made of carbon or organic materials. Examples are metal (gold, silver), ceramics, and semiconductors. Metal NPs can be monometallic, bimetallic, or polymetallic which are entirely composed of metal precursors. These NPs have distinct optical and electrical properties because of their localized surface plasmon resonance features. Certain metal NPs also have special biological, magnetic, and thermal characteristics. Because of this, they are becoming more and more crucial building blocks for the creation of nanodevices with a wide range of pharmacological, chemical, biological, and physical uses [8]. Figures 10.5 and 10.6 show examples of inorganic NPs.

10.4 Nanonetworks

Nanonetwork as the name implies is the interconnection of several independent nanoplatforms to form a single nanomachine. These nanonetworks can be useful in targeted and effective drug delivery, developing nanosensors that are deployed in the monitoring of glucose, cholesterol, and other electrolytes in the human body [13]. Nanonetworks allow for the combination of diagnosis, therapy, and monitoring under

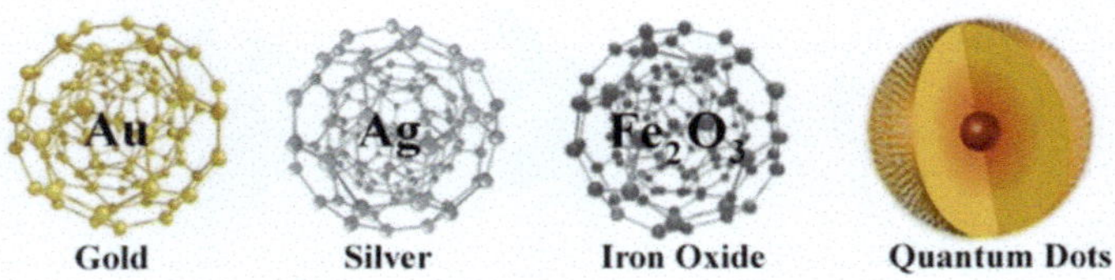

Fig. 10.5 Inorganic NPs [11]

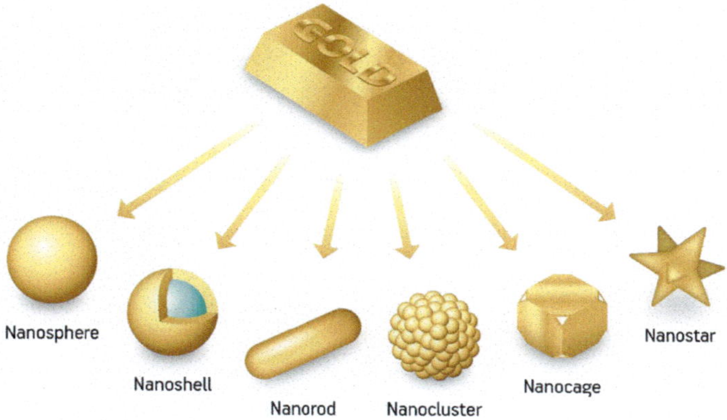

Fig. 10.6 Examples of metal NPs made from gold [12]

a single system or platform [14]. A widely used nanoplatform network is the metal–organic nanoplatform network. Metals and metal oxides have gained wide use in the medical field due to physical and chemical properties that can be modified.

Scientists and surgeons can now correctly and completely comprehend the pathophysiology and pathological processes of complex diseases, including cancer, thanks to the rapid advancement of imaging technologies. Theranostic platforms have integrated these imaging technologies into diagnosis and treatment [15].

Metal–organic frameworks (MOFs) are newly developed porous organic–inorganic hybrid materials that are divided into nodes made of metal ions or ion clusters and linkers made of organic ligands. Thanks to their intriguing properties, which include easy synthesis, varied compositions, high surface areas and porosities, customizable sizes, multiple physicochemical properties, simple surface functionalization, and good biocompatibility, have been utilized as intelligent carriers for cancer theranostics on a large scale in recent years [15].

10.5 Metal–Organic Frameworks

Various approaches have been designed to enhance therapeutic efficacy while reducing undesirable side effects in treatment. Various nanocarriers such as dendrimers, organic micelles, metal oxide NPs, quantum dots, crystalline mesoporous silica, and others were developed to tackle these issues. Metal–organic frameworks (MOFs) have also been utilized for these purposes, with various organic compounds and metals being used. By mixing metal and ligand precursor solutions or adding solvent to a solid metal salt and ligand combination, MOFs are created in the liquid phase [16]. MOFs have properties that make them very useful. MOFs are renowned for having pore sizes that can be adjusted, incredibly large surface areas,

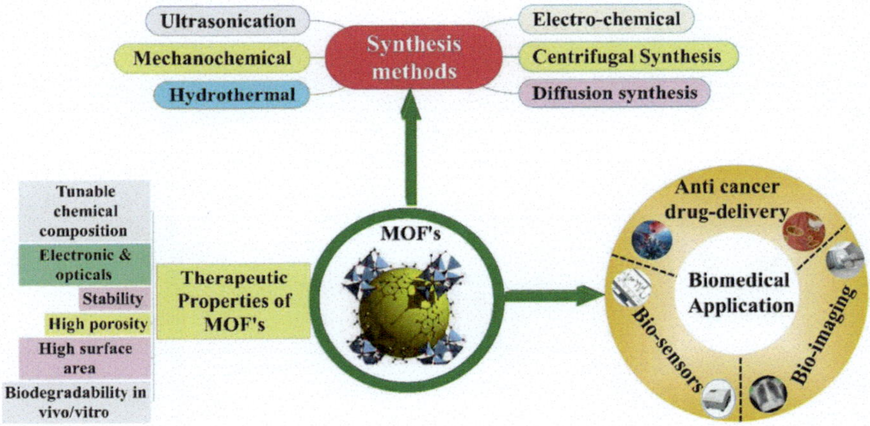

Fig. 10.7 Schematic representation of synthesis, properties and applications of MOFs [16] (Munawar et al. 2023)

and customizable interior surface qualities [17]. Figure 10.7 illustrates the different methods of producing MOFs and their valuable properties and also its application in medicine.

10.5.1 Structural Property of MOFs

MOFs have diverse crystalline structures and this makes them very useful for various functions. In comparison with other porous materials like zeolites and carbon, MOFs are tunable. Its tunability can be achieved by carefully selecting the secondary building units and the linkers, as shown in Fig. 10.8, with the desired pore size, structure, and desired functionality to help achieve the specific application [18, 19].

10.5.2 Physical Property of MOFs

MOFs have highly ordered pores which are permanent. The majority of MOFs created to date are microporous (pore diameter < 2 nm) and have demonstrated excellent adsorption abilities for a variety of gases, such as carbon dioxide and hydrogen. However, their applications in storage, separation, catalysis, and medication delivery are limited by the fact that such tiny pores cannot support big items, anchor molecular functions, or facilitate rapid mass diffusion and transfer [18]. MOFs also have a large surface area, are biocompatible and biodegradable, and have high loading efficiency. The water solubility of MOFs, degradability and biocompatibility allow the body

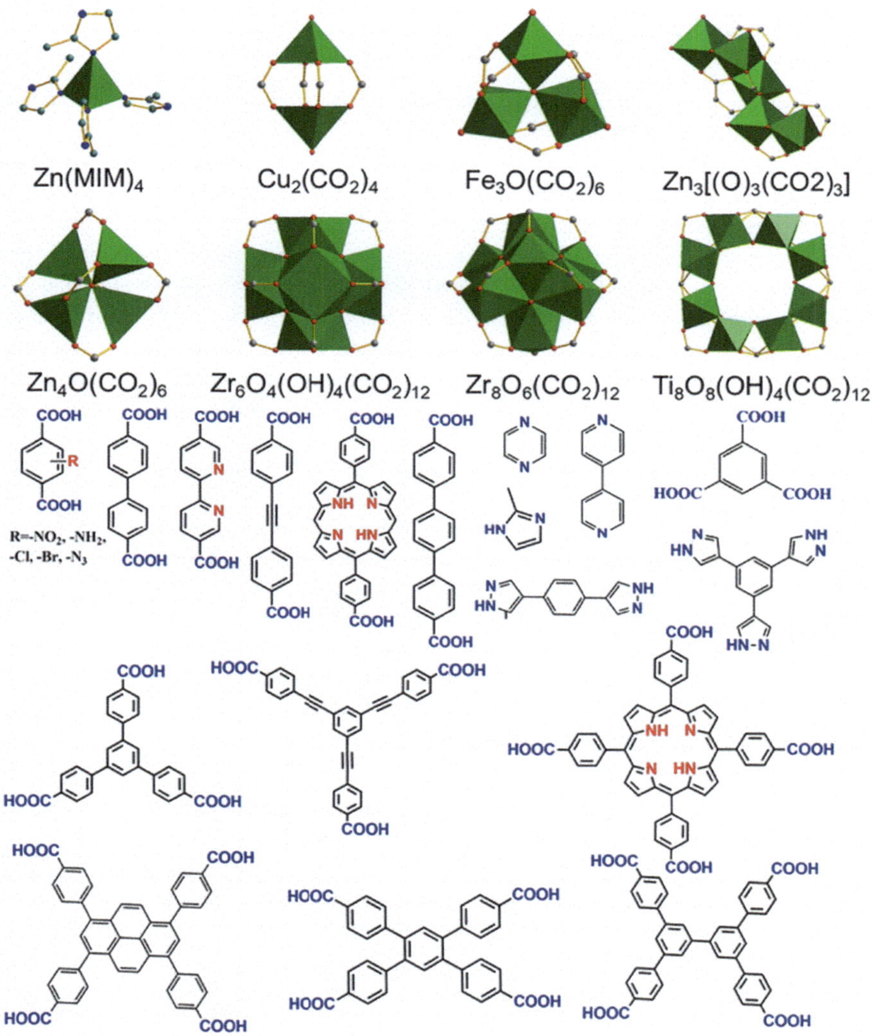

Fig. 10.8 Illustration of secondary building units and organic linkers [18] (Jiao et al. 2019)

to better utilize and recycle a medication [20]. In recent years, MOFs have found significant use in bioimaging (Medical Imaging).

10.6 Medical Imaging

Medical imaging is the method and procedure of using images of the inside of the body to visualize the function of certain organs or tissues and for clinical analysis and medical intervention. The fields of medical diagnosis, analysis, and historical documentation are seeing an increase in interest in and use of imaging technologies. Medical professionals are more effective in reaching a fair and objective decision quickly when they use computer-aided imaging tools, which allow for a quantitative assessment of the images being evaluated [21]. Before 1985, physicians had no way of knowing what was going on in the human body. However, Wilhelm Conrad Roentgen, on November 8, 1985, discovered what he called "X-rays" while working with a Crookes cathode ray tube. This set in motion what is called a revolution in medical diagnosis, and this has led to further research and more discoveries of other medical imaging technologies [22].

Aside from the use of medical imaging, it has also found its application in theranostics (therapeutics and diagnostics), thereby allowing for finding what the problem is and fixing it immediately. This approach has been used successfully in therapies for neuroendocrine tumours with peptide receptor radionuclide therapy (PRRT) and prostate cancer using prostate-specific membrane antigen (PSMA)-directed radioligand therapy (PRLT) [22].

10.6.1 MOFs In Medical Imaging

To construct MOFs with imaging agents, there are four main methods: (i) encapsulating the agents within the pores of MOFs during synthesis; (ii) incorporating them as intrinsic structural components of MOFs; (iii) immersing previously synthesized MOFs in a solution of imaging agents and loading them inside the pores by size effect, ion-exchange, or post-synthetic modification, which is collectively known as the post-insertion method; and (iv) attaching imaging agents to the surface of MOFs through covalent conjugation, or electrostatic or hydrophobic or hydrophilic interactions [23].

The image modes can be classified into two groups: (1) images that provide structural information, like computed tomography (CT), magnetic resonance imaging (MRI), and ultrasound; and (2) images that provide functional and molecular information, like PET, SPECT, optical imaging, and magnetic resonance spectroscopy [20].

10.6.2 Magnetic Resonance Imaging (MRI) and MOFs

MRI is one of the most common clinical examination tools that provide both anatomical and biochemical information and it has proved to be one of the most adaptable, noninvasive, and nonionizing imaging methods [24]. For the identification and diagnosis of a wide range of clinical diseases, MRI is very sensitive in evaluating anatomical structures, organs, and soft tissues with a high resolution. MRIs can be used to stage tumours and assess how effectively a treatment is working by providing contrast between healthy and diseased tissues [25]. MRI as an imaging technique can penetrate living tissue without depth restrictions; however, the overlapping signals of background signals from endogenous glucose, creatine, glutamate, nucleic acids, peptides, and other macromolecules make it challenging to extract target information [26]. The signal intensity and contrast in MR images are determined by proton density, longitudinal relaxation (T_1) and spin–spin (T_2) relaxation, and other factors. According to the main impact (weighting), such as proton density or T_1 and T_2 (or T_2*) relaxation, distinct MR contrasts are formed for the same object from the same location. The sequence of radiofrequency (RF) pulses and gradients—small linear magnetic fields superposed to a static magnetic field—that produce the MR imaging sequences determine the degree of weighting. These combinations and their order are utilized for slice selection, spatial encoding, and other purposes [27].

While MRI has found successful application in the field of diagnosis, more can still be done to improve it further. As a result, more studies have been done and different contrast agents developed. One such contrast agent is MOFs. Contrast agents are utilized in clinical practice to enhance image quality, improve pathological process detectability, and help identify diseases from healthy tissue. Contrast chemicals are also employed in experimental medicine to monitor drug delivery systems or to visualize transplanted cells with imaging labelling. The main goal is to provide a contrast in the images. The majority of MRI contrast agents used are gadolinium ion complexes or superparamagnetic magnetite NPs. They work by shortening the relaxation time T_1 and T_2. Due to the use of contrast agents in vivo, an ideal contrast agent should have the following qualities: (i) sufficient solubility or dispersibility in the body or water fluids; (ii) high relaxivities (r1, r2); (iii) chemical stability; (iv) nontoxicity and dependable pharmacokinetic and pharmacodynamic properties, such as circulation time and biodegradability; (v) a surface that is easily modifiable (e.g., for a targeting ligand attachment); and perhaps even (vi) response to stimuli [27]. MOFs are chemically stable, soluble in water and body fluids, degradable and have tunable surfaces. These properties make them suitable for use as contrast agents. MOFs are comprised of metal ions or clusters, which is a desirable property for MRI to be carried out. High metal content leads to the high efficiency of MOFs as a contrast agent. MOF pores also offer the advantage of including other contrast agents into it. Another approach to using MOFs is the integration of MRI-active metal oxide NPs such as Fe_2O_3, and Fe_3O_4 NPs [27]. Gd(III), Mn(II), and Fe(III)-based MOFs are the most appropriate because of their magnetic properties, which are caused by the quantity of unpaired electrons. The first Gd(III)-MOFs were reported on by Rieter et al. in 2006 [28] as MRI contrast agents. Afterwards, several additional Gd(III)-MOFs with primarily carboxylate ligands have been documented. These ligands include benzene-1,4-dicarboxylate and benzene-1,3,5-tricarboxylate [29], as well as N-(4-carboxybenzyl)- (3,5-dicarboxyl)pyridinium bromide or 5-boronobenzene-1,3-dicarboxylate [30]. Over the past few years, the field of Gd-MOFs for MRI has seen a modest shift in research focus, in line with a broader trend toward multifunctional materials. To prepare agents for multimodal imaging or theranostic agents, greater focus has been placed on mixing existing Gd-MOFs with other materials rather than creating new ones. Over the past few years, the field of Gd-MOFs for MRI has seen a modest shift in research focus, in line with a broader trend toward multifunctional materials. To prepare agents for multimodal imaging or theranostic agents, greater focus has been placed on mixing existing Gd-MOFs with other materials rather than creating new ones. Fe(iii)-MOFs and Mn(ii)-MOFs have also been studied and developed as contrasting agents for MRI [27].

10.6.3 Computed Tomography (CT) and MOFs

CT is a combination of X-ray and information technology, i.e., X-ray whose readings are recorded, displayed, and stored on a computer [31]. CT is used for visualization of internal organs such as the liver, lungs, cardiovascular system, gastrointestinal system, and even bones. For CT to be used successfully in medical applications, it has to use a contrasting agent. Such CT contrast has to be minimally toxic and be used in the lowest dose possible to provide the maximum contrast between the target organ and its background. Examples of contrast agents that have been used are iodine and barium; however, these have more than one undesirable effect in patients [32]. Recent studies have shown the use of nanomaterials in CT as a contrast agent. Different NPs have been experimented with; however, MOFs have proven to be very successful due to its crystalline structure, tunability, porosity, and chemical stability. An example of MOFs is NU-901, which is made up of tetratopic 1,3,6,8-tetrakis(p-benzoate)-pyrene (TBAPy) linkers and Zr6(μ3-OH)4(μ3- O)4(OH2)4 nodes. Its thermal and chemical stability motivated research into the potential of an isostructural bismuth-based counterpart as a potent contender for a novel CT contrast agent which was studied by Robison et al. 2019. It was found that a combination of bismuth with NU-901 MOFs (Bi-NU-901) (Fig. 10.9) was more effective compared to other contrast agents and less volume of it needed and this study will lead to more research into designing more MOFs for future use as contrast agents.

10.6.4 Ultrasound and MOFs

Ultrasound has been used for over 40 years as a diagnostic tool, especially in obstetrics in the medical field. A prototype was created by engineer, John Reid and physician,

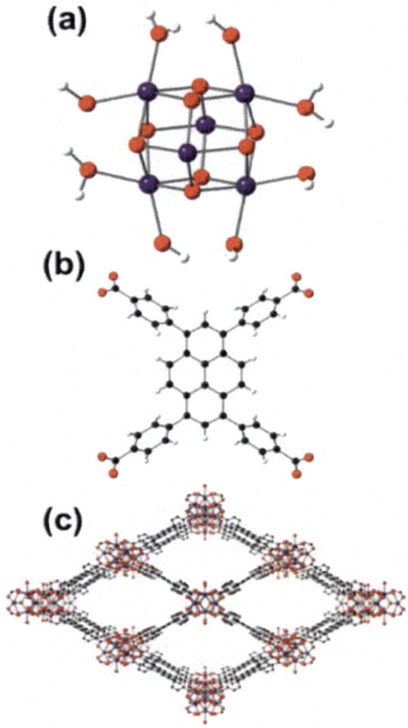

Fig. 10.9 Bi-NU-901 MOF cluster with **a** bismuth metal nodes; **b** 1,3,6,8-tetrakis(p-benzoate)pyrene linkers; and **c** cluster formed exhibiting topology [32] (Robison et al. 2019)

John Wild of the University of Minnesota Medical School in the United States. Components taken from a neighbouring navy laboratory are used in this 15-MHz B-mode ultrasonic mechanical scanner. They were able to show off ultrasound's ability to image and characterize malignant tissues [33]. Ultrasound is used in obstetrics and gynaecology due to its non-invasiveness, cost-effectiveness and its capability to give real-time imaging. Its use has also been found in mammography for breast cancer imaging [33]. Two-dimensional (2D) images are created by ultrasound scanners, and a needle-steering mechanism installed on the transducer allows a needle to be inserted into the image plane in a predictable path. The transducer is first adjusted until the scanning plane crosses the puncture target, making it visible on the image. Once the target is crossed by the puncture line on the monitor, the transducer is angulated. A needle is then continually watched as it approaches and strikes the target after being placed through the guiding device [34]. Some factors determine the type of imaging produced by an ultrasound scan and they include (i) echogenicity of tissue being scanned, (ii) angle of incidence- if the angle of ultrasound waves hitting the surface being scanned, a clearer image is obtained. This is important for when carrying out an ultrasound scan guided needle insertion, (iii) ultrasound wave frequency—Midrange frequency probes (5–10 MHz) and high-frequency probes (10–15 MHz) have less penetration but offer superior resolution. It is therefore recommended to use high-frequency probes ultrasound imaging is best for superficial structures (2–4 cm). Although the picture quality will be significantly worse, a low-frequency probe (2–4 MHz) is recommended for ultrasound imaging of deep structures [35]. MOFs can be used in conjunction with ultrasound to deliver a sonodynamic therapy (SDT) for cancer treatment such as pancreatic cancer. In SDT, the sonosensitizers that have accumulated in tumours are released and tissue oxygen is transformed into very reactive oxygen species (ROS) like singlet oxygen ($^1O_2$), which can kill tumour cells, by using low-intensity ultrasound stimulation. MOFs can be used as sonosensitizers to generate ROS in the SDT approach of treatment [36]. One example of such MOFs is the zirconium-based MOFs (Zr-MOFs) which have shown great advantage both as a sonosensitizer in itself and also as a carrier of 2,2-azobis[2-(2-imidazolin-2-yl)propane] dihydrochloride (AIPH), which is a water-soluble azo compound that has the ability generate alkyl radicals under ultrasound stimulation. Zr-MOFs have photophysical properties, low toxicity, stable structure, tunable surface, and high drug-loading efficiency [36]. Aside from delivering gases to tumour cells, MOFs can also be used to deliver enzymes or enzyme mimics into target sites, which are then stimulated using an ultrasound [37].

10.6.5 Photoacoustic Imaging

Optoacoustic imaging, also known as photoacoustic imaging (PAI) or optoacoustic imaging, is a hybrid imaging technique that combines optical light and ultrasound detection [38, 39]. PAI produces high-resolution pictures and is a good method for medical imaging. The biological sample in PAI is often illuminated with a nanosecond pulsed laser (pulse duration < 10 ns). A rise in temperature results from the molecules' conversion of the optical energy into heat. Utilizing ultrasonic transducers, acoustic waves are detected as a result of the thermoelastic expansion caused by temperature rise. The auditory signal travels through biological tissue for a significantly longer time without experiencing appreciable attenuation since sound scatters 1000 times less than light [40]. Mainly, PAI makes use of the chromophores—such as haemoglobin, melanin, lipid, and water—that naturally absorb light in human tissue. The distinct absorption spectra of individual chromophores enable their relative quantification by PAI at various wavelengths, which also facilitates the investigation of physiological alterations in illnesses to comprehend their underlying mechanisms and devise efficient management strategies [40]. For additional uses, PAI may distinguish between deoxy- and oxy-hemoglobin (HbO_2) and measure the saturation in the blood arteries, which is an important feature of ischemia, hypoxia, or hypoxemia physiological conditions [41]. PAI has been used along with photothermal therapy (PTT) in the treatment of tumour cells. Due to the desirable properties of MOFs, it is used in image-guided treatment using the PAI/PTT method in Fig. 10.10. It, therefore, has been functionalized with polydopamine, which has high photothermal conversion efficiency, thus providing a good contrast agent in PAI image-guided therapy [42].

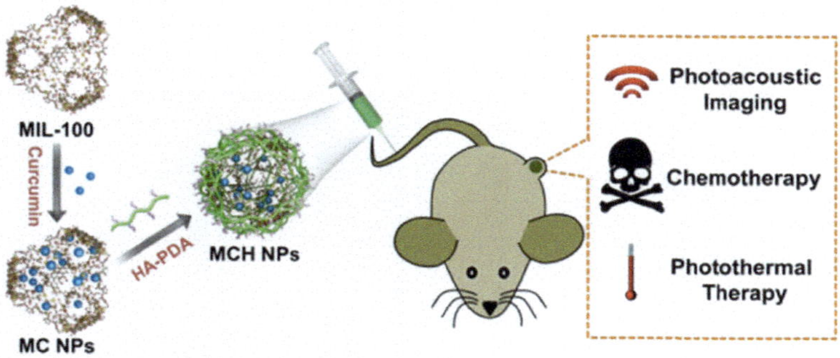

Fig. 10.10 Schematic representation of a PAI guided PTT using MOFs functionalized with Polydopamine [42] (Zhang et al. 2018)

10.6.6 Single-Photon Emission Computed Tomography

Single-photon emission computed tomography (SPECT) is used for brain imaging [43]. The living brain and different neurotransmitter systems in health and disease have been the subject of intensive investigation over the past few decades using PET and SPECT as neuroreceptor mapping technology [44]. Using a radioactive tracer that is injected into the bloodstream and then absorbed by specific organs, SPECT creates a three-dimensional image of the tracer's distribution achieved by using specialized nuclear medicine cameras. It allows the assessing the perfusion and functionality of specific tissues which sets it apart from other imaging techniques [45]. Commonly used radioactive tracers or probes are technetium-99 m (Tc), iodine-123, and thallium-201. The choice is based on the goal of the scan, what type of risk it may pose to the patient and the cost [45]. SPECT also allows for in-vivo studies in patients over a long period and its combination with CT (SPECT-CT system) allows for better localization of tracer activity determination. [46]

As with other imaging techniques, MOFs are also used in SPECT imaging. Radioisotopes can be chelated with MOFs to monitor tumour accumulation. [47] Recent studies by Chao et al. 2018 [48] showed the chelation Hafnium nanoscale coordinated polymers (NCPs) linked with tetrakis (4-carboxyphenyl) porphyrin (TCPP) to $^{99\,m}$ Tc utilizing the porphyrin ring of the MOFs to form ^{99m}Tc-Hf-TCPP NCPs (Fig. 10.11) which is used for SPECT imaging guided radioisotope treatment.

10.6.7 Positron Emission Tomography (PET)

PET is an imaging technique which measures physiological function using a combination of blood flow, metabolism, neurotransmitters, and radiolabeled medicines. As a disease process progresses or in reaction to a particular stimulus, PET's quantitative studies enable the monitoring of relative changes over time. [49] Detection

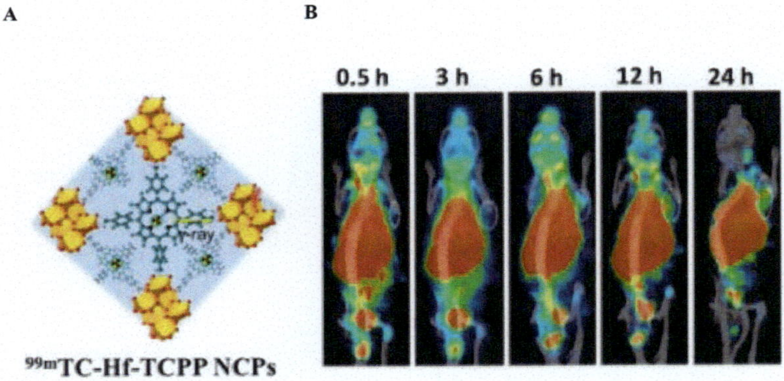

Fig. 10.11 **a** Chelate radioisotope formed with ^{99m}Tc with Hf-TCPP NCPs which is injected into mice. **b** SPECT images of tumour-bearing mice post intravenous injection at different times with the highest tumour accumulation at the 24 h time (Zhu et al. 2019). [47]

of radioactivity following the injection of a tiny amount of radioactive tracer is the principle behind its use. Tracers are often labelled with oxygen-15, fluorine-18, carbon-11, or nitrogen-13 [49]. Radioactive tracer isotope which is incorporated with a biomolecule which has an affinity for the target organ is injected into the bloodstream and as the radioisotopes transition into stable elements, they experience positron emission decay. These positrons each move a brief distance within tissue before they contact an adjacent electron and lose its kinetic energy. The annihilation of electrons and positrons results in the production of two gamma rays, or photons, travelling in opposing directions. The annihilation of electrons and positrons results in the production of two gamma rays, or photons, travelling in opposing directions, detectors outside the patient's body, as shown in Fig. 10.12, then pick up these anti-parallel rays which then reconstruct the image. [50]

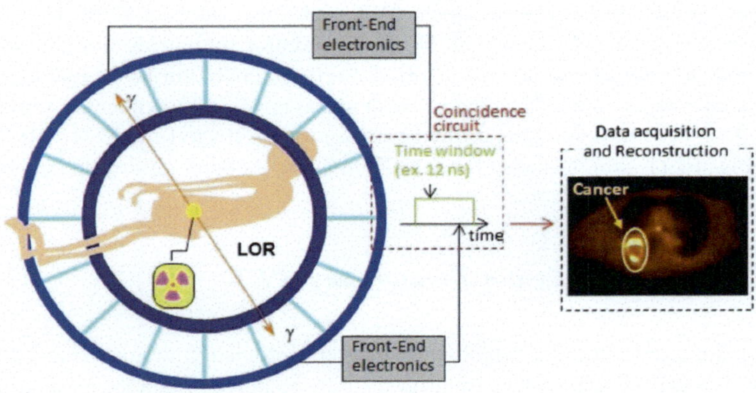

Fig. 10.12 Schematic illustration of PET Imaging (Ullah et al. 2016) [50]

Table 10.2 Various imaging techniques and the MOFs used in them

Metal organic frameworks	Imaging technique
Gd(iii)-MOFs	MRI
Fe(iii)-MOFs	MRI
Mn(ii)-MOFs	MRI
Bi-NOU-901	CT
Zr-MOFs	Ultrasound, PET, CT
^{99m}Tc-Hf-TCPP	SPECT

PET imaging offers stronger quantitative capacity, deeper signal penetration, and improved detection sensitivity (down to the picomolar range) when compared to other imaging methods. [51] Due to MOFs having high cargo delivery or drug loading efficacy, they can be used in enhanced PET-guided drug delivery into tumour cells. Studies by Chen et al. [51] designed a radioactive nanomaterial containing UiO-66 and zirconium with a benzoic acid linker (^{89}Zr-UiO-66) with good radiochemical stability, easy functionalization. This radioactive cargo is then loaded with drugs like doxorubicin which also fluoresce when it comes in contact with tumour cells.

Many PET detectors use scintillators or scintillating counters for detection; therefore, MOFs can be created by connecting clusters of high-atomic-number (Z) elements by the use of scintillating dyes as ligands which intensify the ionizing radiation interaction and increase the ligand luminescence [52].

MOFs have found successful applications in various medical imaging, which have been used for imaging-guided therapy, especially in the treatment of tumour cells. Below are examples (Table 10.2) of MOFs and the imaging techniques they have been used for. These successful applications, as previously stated, are due to the crystal structure of MOFs, its tunability, being able to be functionalized, degradable and also chemical stability and stability in body fluids.

10.7 Conclusion

Undoubtedly, this chapter has discussed the various uses of nanotechnology both historically and currently its application, particularly in the simultaneous delivery of medical therapies and bioimaging. Since they are the foundation of MOFs, which are being used to enhance patient health and change treatment to be more customized, particularly in different cancer diagnoses, the many forms of nanoparticles have also been extensively discussed. In addition to its various potential physicochemical qualities, MOFs have intriguing properties such a large surface area, porosity, adaptability, and configurable sizes that make them ideal for usage as linkers. These qualities should not be overlooked. But as technology advances and more medical diseases are discovered, more sophisticated techniques for theranostics—a field that uses medical imaging to diagnose patients—as well as for medical imaging diagnosis itself must be created in order to support treatment. Even if it might seem like there are already enough MOFs produced, more can be done. In order to meet the growing demand for better imaging, more MOFs should be investigated and developed for use in medical imaging. These include increased compatibility with other linkers, metals, and drugs they may carry, less interaction with drugs they may carry, and fewer side effects when introduced into the human body. Further research is also required to determine the potential applications of MOFs in current imaging and diagnostic systems which have little or no information of using MOFs, thereby making available different imaging systems that can be adapted for different use in diagnosis and guided treatment.

References

1. S. Malik, K. Muhammad, Y. Waheed, Nanotechnology: a revolution in modern industry. Molecules **28**, 661 (2023)
2. A. Haleem, M. Javaid, R.P. Singh, S. Rab, R. Suman, Applications of nanotechnology in medical field. Glob. Health J. **7** (2023)
3. I. Khan, K. Saeed, I. Khan, Nanoparticles: properties, applications and toxicities. Arab. J. Chem. **12**, 908 (2017)
4. X. He, H. Deng, H. Hwang, The current application of nanotechnology in food and agriculture. J. Food Drug Anal. **27**, 1 (2019)
5. J. Li, X. Li, S. Gong, C. Zhang, C. Qian, H. Qiao, M. Sun, Dual-mode avocado-like all-iron nanoplatform for enhanced T_1/T_2 MRI-guided cancer theranostic therapy. Nano Lett. **20**, 4842 (2020)
6. H. Wang, Q. Chen, S. Zhou, Carbon-based hybrid nanogels: a synergistic nanoplatform for combined biosensing. Bioimaging Responsive Drug Deliv. **47**, 4198 (2018)
7. A. Khorasani, D. Shahbazi-Gahrouei, A. Safari, Recent metal nanotheranostics for cancer diagnosis and therapy: a review. Diagnostics **13**, 833 (2023)
8. N. Joudeh, D. Linke, Nanoparticle classification, physicochemical properties, characterization, and applications: a comprehensive review for biologists. J. Nanobiotechnol. **20** (2022)
9. M. Rizwan, A. Shoukat, A. Ayub, B. Razzaq, M.B. Tahir, Types and classification of nanomaterials, in *Nanomaterials: Synthesis, Characterization, Hazards and Safety*, vol. 31 (2021)
10. V.N. Mochalin, O. Shenderova, D. Ho, Y. Gogotsi, The properties and applications of nanodiamonds. Nat. Nanotechnol. **7**, 11 (2011)
11. R. Itani, A. Al Faraj, SiRNA conjugated nanoparticles—a next generation strategy to treat lung cancer. Int. J. Mol. Sci. **20**, 6088 (2019)
12. L. Freitas de Freitas, G. Varca, J. dos Santos Batista, A. Benévolo Lugão, An overview of the synthesis of gold nanoparticles using radiation technologies. Nanomaterials **8**, 939 (2018)
13. D. Malak, O.B. Akan, Molecular communication nanonetworks inside human body. Nano Commun. Netw. **3**, 19 (2012)
14. U.A.K. Chude-Okonkwo, R. Malekian, B.T. Maharaj, A.V. Vasilakos, Molecular communication and nanonetwork for targeted drug delivery: a survey. IEEE Commun. Surv. Tutor. **19**, 3046 (2017)
15. J. Yang, Y. Yang, Metal-organic framework-based cancer theranostic nanoplatforms. View **1** (2020)
16. J. Munawar, M.S. Khan, S.E. Zehra Syeda, S. Nawaz, F.A. Janjhi, H. Ul Haq, E.U. Rashid, T. Jesionowski, M. Bilal, Metal-organic framework-based smart nanoplatforms for biosensing, drug delivery, and cancer theranostics. Inorg. Chem. Commun. **147**, 110145 (2023)

17. H.-C. Zhou, J.R. Long, O.M. Yaghi, Introduction to metal-organic frameworks. Chem. Rev. **112**, 673 (2012)
18. L. Jiao, J.Y.R. Seow, W.S. Skinner, Z.U. Wang, H.-L. Jiang, Metal-organic frameworks: structures and functional applications. Mater. Today **27**, 43 (2019)
19. M. Safaei, M.M. Foroughi, N. Ebrahimpoor, S. Jahani, A. Omidi, M. Khatami, A review on metal-organic frameworks: synthesis and applications. TrAC, Trends Anal. Chem. **118**, 401 (2019)
20. Y. Liu, T. Jiang, Z. Liu, Metal-organic frameworks for bioimaging: strategies and challenges. Nanotheranostics **6**, 143 (2022)
21. A.P. James, B.V. Dasarathy, Medical image fusion: a survey of the state of the art. Inf. Fusion **19**, 4 (2014)
22. E. Bercovich, M.C. Javitt, Medical imaging: from roentgen to the digital revolution, and beyond. Rambam Maimonides Med. J. **9** (2018)
23. F. Demir Duman, R.S. Forgan, Applications of nanoscale metal–organic frameworks as imaging agents in biology and medicine. J. Mater. Chem. B **9**, 3423 (2021)
24. N. Jon Shah, *Hybrid MR-PET Imaging* (Royal Society of Chemistry, 2018)
25. A.R. Padhani et al., Diffusion-weighted magnetic resonance imaging as a cancer biomarker: consensus and recommendations. Neoplasia **11**, 102 (2009)
26. Y. Yang et al., Coloring ultrasensitive MRI by tunable metal-organic frameworks. Chem. Sci. (2021)
27. H. Bunzen, D. Jirák, Recent advances in metal-organic frameworks for applications in magnetic resonance imaging. ACS Appl. Mater. Interfaces **14**, 50445 (2022)
28. W.J. Rieter, K.M.L. Taylor, H. An, W. Lin, W. Lin, Nanoscale metal−organic frameworks as potential multimodal contrast enhancing agents. J. Am. Chem. Soc. **128**, 9024 (2006)
29. Y. Liu, C. Zhang, C. Xu, C. Lin, K. Sun, J. Wang, X. Chen, L. Li, A.K. Whittaker, H.-B. Xu, Controlled synthesis of up-conversion luminescent Gd/Tm-MOFs for PH-responsive drug delivery and UCL/MRI dual-modal imaging. Dalton Trans. **47**, 11253 (2018)
30. H. Zhang, Y. Shang, Y.-H. Li, S.-K. Sun, X.-B. Yin, Smart metal-organic framework-based nanoplatforms for imaging-guided precise chemotherapy. ACS Appl. Mater. Interfaces **11**, 1886 (2018)
31. A. Hessenbruch, A brief history of x-rays. Endeavour **26**, 137 (2002)
32. L. Robison et al., A bismuth metal-organic framework as a contrast agent for x-ray computed tomography. ACS Appl. Bio Mater. **2**, 1197 (2019)
33. K.K. Shung, Diagnostic ultrasound: past, present, and future. J. Med. Biol. Eng. **31**, 371 (2011)
34. H.H. Holm, B. Skjoldbye, Interventional ultrasound. Ultrasound Med. Biol. **22**, 773 (1996)
35. A. Boezaart, B. Ihnatsenka, Ultrasound: basic understanding and learning the language. Int. J. Shoulder Surg. **4**, 55 (2010)
36. C. Zhang et al., Metal–organic framework (MOF)-based ultrasound-responsive dual-sonosensitizer nanoplatform for hypoxic cancer therapy. Adv. Healthc. Mater. **11** (2021)
37. Y. Zhou, T. Yang, K. Liang, R. Chandrawati, Metal-organic frameworks for therapeutic gas delivery. Adv. Drug Deliv. Rev. **171**, 199 (2021)
38. M. Xu, L.V. Wang, Photoacoustic imaging in biomedicine. Rev. Sci. Instrum. **77** (2006)
39. V. Ntziachristos, D. Razansky, Molecular imaging by means of multispectral optoacoustic tomography (MSOT). Chem. Rev. **110**, 2783 (2010)
40. A.B.E. Attia, G. Balasundaram, M. Moothanchery, U.S. Dinish, R. Bi, V. Ntziachristos, M. Olivo, A review of clinical photoacoustic imaging: current and future trends. Photoacoustics **16**, 100144 (2019)
41. M. Li, Y. Tang, J. Yao, Photoacoustic tomography of blood oxygenation: a mini review. Photoacoustics **10**, 65 (2018)
42. Y. Zhang, L. Wang, L. Liu, L. Lin, F. Liu, Z. Xie, H. Tian, X. Chen, Engineering metal-organic frameworks for photoacoustic imaging-guided chemo-/photothermal combinational tumor therapy. ACS Appl. Mater. Interfaces **10**, 41035 (2018)
43. J. Intrator et al., A brain imaging (single photon emission computerized tomography) study of semantic and affective processing in psychopaths. Biol. Psychiat. **42**, 96 (1997)

44. P.P. Hazari, A. Pandey, S. Chaturvedi, A.K. Mishra, New trends and current status of positron-emission tomography and single-photon-emission computerized tomography radioligands for neuronal serotonin receptors and serotonin transporter. Bioconjug. Chem. **28**, 2647 (2017)
45. S. Yandrapalli, Y. Puckett, SPECT Imaging. https://www.ncbi.nlm.nih.gov/books/NBK564426/
46. A.B. Newberg, A. Alavi, *Single Photon Emission Computed Tomography*☆ (Elsevier EBooks, 2017)
47. W. Zhu, J. Zhao, Q. Chen, Z. Liu, Nanoscale metal-organic frameworks and coordination polymers as theranostic platforms for cancer treatment. Coord. Chem. Rev. **398**, 113009 (2019)
48. Y. Chao et al., Highly effective radioisotope cancer therapy with a non-therapeutic isotope delivered and sensitized by nanoscale coordination polymers. ACS Nano **12**, 7519 (2018)
49. A. Berger, Positron emission tomography. BMJ **326**, 1449 (2003)
50. M.N. Ullah, E. Pratiwi, J. Cheon, H. Choi, J.Y. Yeom, Instrumentation for time-of-flight positron emission tomography. Nucl. Med. Mol. Imaging **50**, 112 (2016)
51. D. Chen, D. Yang, C.A. Dougherty, W. Lu, H. Wu, X. He, T. Cai, M.E. Van Dort, B.D. Ross, H. Hong, In vivo targeting and positron emission tomography imaging of tumor with intrinsically radioactive metal-organic frameworks nanomaterials. ACS Nano **11**, 4315 (2017)
52. J. Perego et al., Composite fast scintillators based on high-Z fluorescent metal-organic framework nanocrystals. Nat. Photonics **15**, 393 (2021)

Chapter 11
Bioimaging Applications of Stimuli-Responsive Nanomaterials

Robin Augustine, Shashank Chetty, Betsabe Acosta Ramos, Rosita Primavera, Fantahun Biadglegne, Alexander Michael Vezeridis, and Avnesh S. Thakor

Abstract Bioimaging, essential for visualizing biological structures and processes, encompasses diverse modalities such as MRI, CT, PET, and optical imaging. Recent progress integrates stimuli-responsive nanomaterials, offering targeted imaging with improved contrast and specificity. This chapter explores how nanomaterials can change in response to factors like pH, temperature, and light, improving imaging quality. It covers how these nanomaterials are made and how they work, focusing on their use in optical, ultrasound, X-ray, and MRI imaging. Furthermore, it discusses multimodal imaging strategies, showcasing the potential of stimuli-responsive nanomaterials to revolutionize diagnostic and therapeutic approaches in medicine. Ongoing research promises continued refinement and expansion of these technologies, marking a transformative era in bioimaging.

11.1 Introduction

Bioimaging refers to the techniques and processes used to visualize biological materials, either in a living organism (i.e., cells, tissues, and organs), or in vitro for the purpose of medical diagnostics and scientific research. It encompasses a wide range of technologies and modalities that can capture the structure, function, physiology, and pathology of biological systems at various scales. Typical modalities used for in vivo bioimaging include X-ray, magnetic resonance imaging (MRI), computed tomography (CT), positron emission tomography (PET), ultrasound (US), fluorescence imaging (FI), and optical coherence tomography (OCT) [1]. Multimodal imaging

R. Augustine (✉) · S. Chetty · B. A. Ramos · R. Primavera · F. Biadglegne · A. S. Thakor
Department of Radiology, School of Medicine, Stanford University, 3155 Porter Drive, Palo Alto, CA 94304, USA
e-mail: robinaugustine9@gmail.com

A. M. Vezeridis
Department of Interventional Radiology, School of Medicine, Stanford University, Stanford, CA 94305, USA

N. Kalarikkal et al. (eds.), *Multimodal Biomedical Imaging Techniques*,
Biological and Medical Physics, Biomedical Engineering,
https://doi.org/10.1007/978-981-96-1124-9_11

is a complex application of bioimaging involving the combination of two or more modalities [2].

A recent advancement in the field of bioimaging involves the use of stimuli-responsive materials, whereby stimuli can be internally or externally applied to generate changes in material properties that can then be detected for selective imaging. The integration of stimuli-responsive materials in bioimaging holds immense potential to revolutionize imaging techniques and improve healthcare outcomes, especially given their compatibility and integration potential with existing imaging instruments.

Nanomaterials are materials with dimensions at the nanoscale, typically one of the dimensions less than 100 nm, and have unique properties due to their size and surface effects. Compared to conventional contrast agents, the use of nanomaterials has demonstrated improved signal intensity, enhanced accumulation at target sites (e.g., within tumors), ability to penetrate cell membranes and shorter circulation times in animal models [3]. Stimuli-responsive nanomaterials include carbon-based, silica-based, gold nanoparticles, iron oxide, and polymer stimuli-responsive nanomaterials [4]. These nanomaterials respond to various stimuli, including pH change, temperature change, application of a magnetic or electric field, and irradiation using ultraviolet (UV) or infrared (IR) light. For example, photo-responsive nanomaterials, like carbon dots, can be activated by light to emit fluorescent signals for imaging [5].

11.2 Working Principles of Stimuli-Responsive Nanomaterials

Stimuli-responsive nanomaterials can exhibit unique properties and behaviors in response to specific internal (e.g., intratumoral) or external stimuli, providing higher contrast. One class of stimuli-responsive nanomaterials is pH-responsive nanomaterials. These materials undergo changes in their physicochemical properties, such as swelling or shrinking, in response to variations in the surrounding pH. Several methods can be employed to synthesize pH-responsive nanomaterials [6]. One approach is to functionalize the surface of pre-existing nanomaterials with pH-responsive moieties like bis-chromophoric ratiometric fluorescent probes [7]. For example, polymeric nanoparticles can be coated with pH-sensitive polymers or grafted with pH-responsive functional groups to impart pH responsiveness [8]. Another method involves the synthesis of pH-responsive polymers or hydrogels in nanoscale dimensions, either through emulsion polymerization, nanoprecipitation, or other polymerization techniques [9].

Temperature-responsive nanomaterials, on the other hand, exhibit changes in their properties or structure with variations in temperature. One approach has been to incorporate thermoresponsive polymers (e.g., PEGylated poly(amide-imide)) into the nanomaterial structure [10]. These polymers then undergo a reversible phase transition, either at a lower critical solution temperature (LCST) or an upper critical solution temperature (UCST), resulting in a change in the property of the nanomaterial [11].

Light-responsive nanomaterials can be developed by incorporating photoactive molecules or nanoparticles into the nanomaterial structure [12]. These materials can undergo photochemical reactions or exhibit changes in their properties upon exposure to specific wavelengths of light [13]. For instance, magnetic-responsive nanomaterials can be synthesized by incorporating magnetic nanoparticles into a polymeric matrix, allowing the material to respond to magnetic fields [14].

The responses of stimuli-responsive nanomaterials include diffusion, disassembly or degradation, assembly or aggregation, switches in morphology, and the induction of autonomous motion. These kinds of responses result from changes in the physical properties of the material, such as its amphiphilicity, electrostatics, or steric, which bring about shape, size, or phase changes. Specifically, size plays a particularly important role for imaging because it influences biodistribution, blood circulation half-life, cellular uptake, tissue penetration, and targeting [15, 16].

11.3 Stimuli Responsive Nanomaterials in Optical Imaging

Optical imaging techniques have gained significant attention for bioimaging due to their high sensitivity, non-invasiveness, and ability to provide real-time imaging information [17]. Optical imaging encompasses a range of modalities, including fluorescence imaging, bioluminescence imaging, and photoacoustic imaging, which utilize different principles to visualize and study biological structures and processes [18].

Fluorescent stimuli-responsive nanomaterials, such as quantum dots, fluorescent nanoparticles, and stimuli-responsive dye conjugated nanomaterials, have gained prominence in bioimaging applications [19]. These nanomaterials exhibit changes in their fluorescence properties in response to specific stimuli, enabling real-time monitoring of biological events. For example, pH-sensitive fluorescent nanomaterials can be used to visualize pH variations in cellular environments, such as tumor tissues or organelles. Ultra-pH sensitive (UPS) nanoprobes remain dormant ('OFF', Non-fluorescent state) in the bloodstream at pH 7.4 [20]. Upon reaching tumors, they activate ('ON', Fluorescent state) in response to the acidic tumor environment (pH 6.5–6.8) or within organelles of tumor cells (pH 5.0–6.0) following their uptake via receptor-mediated endocytosis (Fig. 11.1). This technology enables precise visualization of the tumor microenvironment, offering insights for improved diagnostics and targeted interventions.

Plasmonic nanoparticles, such as gold or silver nanoparticles, exhibit unique optical properties due to localized surface plasmon resonance (LSPR) [22]. These nanoparticles can enhance optical imaging contrast through light scattering or absorption, resulting in improved visualization of biological targets. Stimuli-responsive

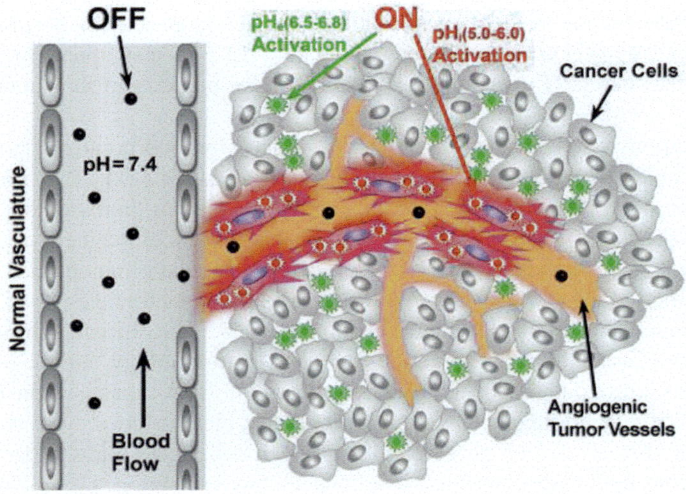

Fig. 11.1 Dynamic imaging of the tumor microenvironment using Ultra-pH sensitive (UPS) nanoprobes. UPS nanoprobes activate ('ON') in the acidic tumor microenvironment (pH 6.5–6.8) or within tumor endothelial cells (pH 5.0–6.0), transitioning from an 'OFF' state in the bloodstream (pH 7.4). Reproduced from Ref. [21] with the permission of Springer-Nature

plasmonic nanoparticles can undergo changes in their LSPR properties in response to specific stimuli, such as pH or temperature, enabling controlled modulation of their optical properties for enhanced imaging performance. For instance, in the tumor region, the fluorescence signal of a NIR dye, Cy5.5, was initially quenched by GO-Au nanoparticles conjugated with NIR dye-labeled matrix metalloproteinase-14 peptide substrate through surface plasmon resonance [23]. When the peptide was broken down by MMP-14, which is overexpressed in the tumor environment, a strong fluorescence signal showed up. This signal analysis facilitates the understanding of accumulation behaviors at the tumor site and guides subsequent photothermal therapy.

11.4 Stimuli-Responsive Nanomaterials in Ultrasound Imaging

Ultrasound imaging refers to a broad range of imaging techniques that use the complex interaction between soundwaves and tissues to produce images [24]. Here, an ultrasound transducer emits a brief soundwave (less than one microsecond in duration) and then captures the reflected wave that is then processed to generate an image. The advantage of ultrasound is its noninvasive and nonionizing nature, while also showing real-time physiological information such as blood flow direction and velocity [25]. Today, ultrasound equipment can even be miniaturized for use in blood vessels (i.e., intravascular probes) and endoluminal cavities (i.e., transesophageal and transrectal probes). Microbubbles (MBs) can be used as ultrasound contrast agents and are composed of 1–10 μm diameter gas-filled microbubbles stabilized by elastic lipid shells and filled with fluorocarbon gases [26, 27]. They resonate and oscillate nonlinearly under specific ultrasound pressures to produce harmonic frequencies to a much greater degree than tissue, thereby providing high contrast. Recently, a new scale of bubbles has been developed called Nanobubbles (NBs) with a diameter in the range of 10–100 nm [28]. These have also been shown to undergo transformation in response to stimuli such as temperature or pH to generate high contrast in ultrasound imaging for improved tissue imaging [29].

11.5 Stimuli-Responsive Nanomaterials in X-Ray Imaging

Ionizing radiation-based imaging including conventional X-rays (i.e., plain film radiography and mammography), fluoroscopy (i.e., conventional and digital subtraction angiography (DSA)) computerized tomography (CT) imaging, and dual-energy X-ray absorptiometry (DEXA), have revolutionized medical imaging by enabling non-invasive visualization and quantification of tissues of interest [30]. To increase tissue contrast using these techniques, stimuli-responsive polymers or liposomes can be used [31]. For instance, a smart nanoprobe was developed relying on a biocompatible click condensation reaction between 2-cyanobenzothiazole (CBT) and D-cysteine (Cys) (CBT-Cys click reaction). This nanoprobe demonstrated dual-stimuli responsiveness to acidic pH and glutathione (GSH) within the tumor microenvironment [32]. Upon reaching cancer cells, the probe releases a citraconic motif due to endosomal acidic pH, while the intracellular GSH simultaneously reduces the disulfide bond of cysteine (Cys). This process forms an active intermediate (1-Core), which undergoes spontaneous condensation via the CBT condensation reaction, resulting in an amphiphilic dimer (1-Dimer). The dimer self-assembles into nanoparticles (I-NPs) through π–π stacking, preventing them from being expelled by cells. This dual-mediated aggregation enhances the accumulation, and prolongs the retention time, of I-NPs in cancer cells, presenting a promising approach for improving CT contrast and precise tumor diagnosis in vivo (Fig. 11.2).

11.6 Stimuli-Responsive Nanomaterials in MRI Imaging

Magnetic resonance imaging (MRI), based on nuclear magnetic resonance (NMR), is a non-invasive imaging technique widely used in medicine and research. Here, certain atomic nuclei absorb radiofrequency waves when placed in an external magnetic field; any change in their polarization can then be detected, with these signals then processed to generate images. Paramagnetic and superparamagnetic nanomaterials,

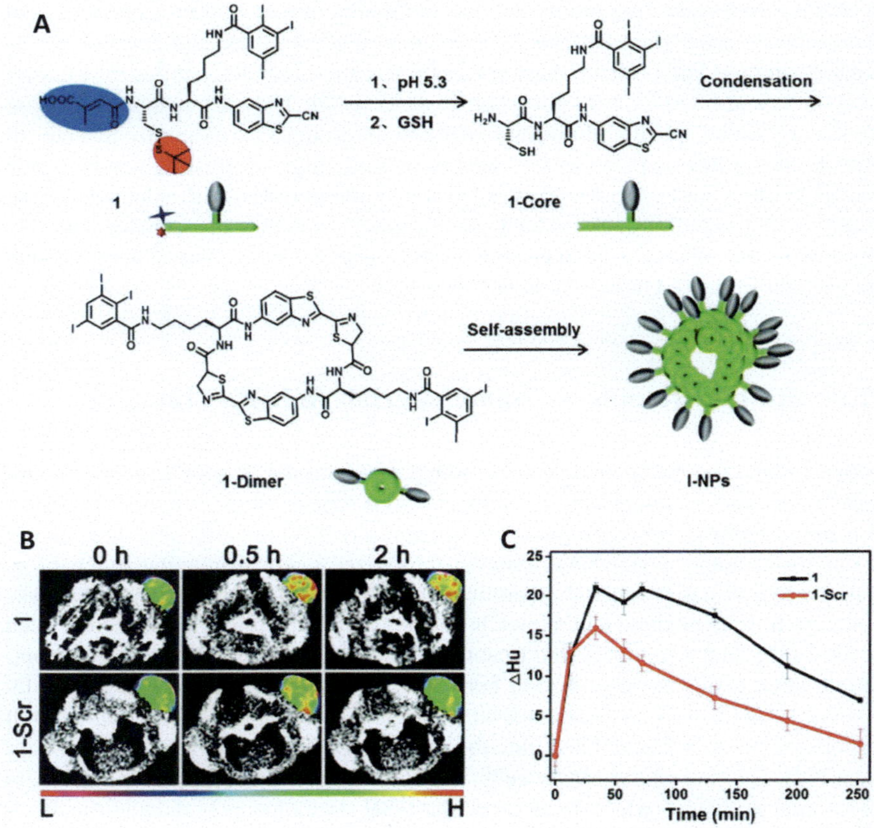

Fig. 11.2 **a** A schematic representation depicting the combined pH and GSH-controlled condensation and self-assembly process of I-NPs. **b** CT imaging experiments of living mice with a subcutaneous 4T1 cell xenograft at 0 h, 0.5 h and 2 h after intravenous injections of **1** (upper) or **1-Scr** (lower) via tail veins. **c** Dynamic contrast enhanced density (ΔHu) of tumor CT imaging at different time intervals. Reproduced from ref. [32] with the permission of Royal Society of Chemistry (RSC)

such as iron oxide nanoparticles, gadolinium-based nanoparticles, and manganese-based agents, have magnetic properties that alter the relaxation times of surrounding water molecules, resulting in enhanced MRI contrast [33]. The integration of stimuli-responsive nanomaterials in MRI imaging is now emerging to further enhance specificity and contrast [34]. In response to stimuli (i.e., variations in pH or enzymatic activity), these materials change their magnetic properties [35], such as their T1 [36] or T2 relaxation times [37].

11.7 Multimodal Imaging Applications of Stimuli-Responsive Nanomaterials

Multimodal imaging represents a robust and effective approach that integrates the utilization of two or more imaging modalities, thereby facilitating the acquisition of multiple complementary pieces of information pertaining to a disease [38, 39]. The integration of these conventional imaging techniques is performed to synergistically leverage their respective advantages while mitigating their individual limitations [40, 41].

Research has shown that the inclusion of acid-liable groups in hybrid organic–inorganic nanomaterials can enable stimuli activable multimodal imaging. Recently, Sun et al. synthesized honeycomb manganese oxide (MnO) nanospheres as a core material and then functionalized them with Chlorin e6 (Ce6, a photosensitizer)-conjugated core–shell upconversion nanoparticles (UCNPs) to develop an innovative multimodal contrast agent [42]. Under experimental conditions involving a mixture of hydrogen peroxide and a mildly acidic environment (stimuli), the nanospheres degraded, which subsequently released UCNPs-Ce6 from the nanosphere surface. Interestingly, the UCNPs-Ce6 conjugate resulted in an increased fluorescence and the increased concentration of Mn^{2+} (from MnO core) that increased MR contrast. This combination was further complemented by CT imaging due to the Gd/Yb elements in UCNPs, enabling comprehensive visualization of the anatomical structure.

In various studies, innovative nanosystems have been developed for activatable multimodal imaging. Tang et al. designed gold nanoparticles (AuNPs) conjugated with a ketal-linked NIR fluorophore (Cy5.5-FA), exhibiting enhanced CT contrast and quenching fluorescence at physiological pH [43]. Other investigations explored activatable imaging using mesoporous silica, gelatin nanolayer-protected MnO_2-coated UCNPs [44], and hollow porous manganese oxide nanoparticles [45] with diverse cargo. Dong et al. employed calcium carbonate ($CaCO_3$) to craft Ce6(Mn)@$CaCO_3$-PEG hybrid nanoparticles for MR and FL imaging in the acidic tumor microenvironment [46]. Notably, Bouché et al. utilized polyphosphazene-derived polymer nanogels releasing gold nanoparticles for enhanced CT contrast and reduced PA signal intensity upon degradation in the presence of reactive oxygen species (ROS) [47]. Yin et al. synthesized pH-responsive perylenediimide (PPDI) NPs, demonstrating augmented FL and PA signals in acidic environments [48]. Lastly, Wang et al. developed a caspase-3-responsive probe (1-RGD) for PET and FL imaging, where caspase-3 cleavage induced macrocyclization and nanoparticle self-assembly, enhancing PA signal intensity [49].

11.8 Conclusion

In conclusion, the field of nanomedicine has experienced a significant breakthrough through the advent of stimuli-responsive nanomaterials. Various imaging modalities, including X-rays, MRIs, and ultrasounds, play pivotal roles in this context. The achievement of this breakthrough relies on capitalizing on specific microenvironment conditions, such as pH, elevated levels of GSH and ROS, and the overexpression of specific enzymes to generate an imaging signal. The development of novel techniques and refining existing approaches will continue to improve the responsiveness, contrast and specificity of these new imaging agents can contribute to pushing the boundaries of stimuli-responsive nanomaterials for bioimaging applications. Unlike conventional imaging agents that remain continuously active, the strategic use of specific activation mechanisms in response to one or more stimuli enhances the signal-to-background of these imaging agents. The ongoing exploration of new synthesis protocols, multimodal capabilities, and diverse applications underscores the transformative impact that stimuli-responsive nanomaterials may have in the future.

Acknowledgements This work was supported by the National Institute of Diabetes and Digestive and Kidney Diseases (NIDDK) of the National Institutes of Health under Award Number DK129343. The content is solely the responsibility of the authors and does not necessarily represent the official views of the National Institutes of Health.

References

1. H.S. Lahoti, S.D. Jogdand, H. Lahoti, S.D. Jogdand, Bioimaging: evolution, significance, and deficit. Cureus **14** (2022)
2. S. Shaikh, F.U. Rehman, T. Du, H. Jiang, L. Yin, X. Wang, R. Chai, Real-time multimodal bioimaging of cancer cells and exosomes through biosynthesized iridium and iron nanoclusters. ACS Appl. Mater. Interfaces **10**, 26056 (2018)
3. S.H. Pham, Y. Choi, J. Choi, Stimuli-responsive nanomaterials for application in antitumor therapy and drug delivery. Pharmaceutics **12**, 1 (2020)
4. N. Lafuente-Gómez, A. Latorre, P. Milán-Rois, C. Rodriguez Diaz, Á. Somoza, Stimuli-responsive nanomaterials for cancer treatment: boundaries, opportunities and applications. Chem. Commun. **57**, 13662 (2021)
5. C. Zhao, J. Kang, Y. Li, Y. Wang, X. Tang, Z. Jiang, Carbon-based stimuli-responsive nanomaterials: classification and application. Cyborg Bionic Syst. **4** (2023)
6. Y. Lee, D.H. Thompson, Stimuli-responsive liposomes for drug delivery. Wiley Interdiscip. Rev. Nanomed. Nanobiotechnol. **9**, e1450 (2017)
7. T. Zhang, D. Xu, C.Y. Poon, X. Wang, F. Bolze, H.W. Li, M.S. Wong, Tuning the PKa of two-photon bis-chromophoric probes for ratiometric fluorescence imaging of acidic PH in lysosomes. Talanta **202** (2019)
8. S. Chu, X. Shi, Y. Tian, F. Gao, PH-responsive polymer nanomaterials for tumor therapy. Front. Oncol. **12** (2022)
9. A.S. Puranik, L.P. Pao, V.M. White, N.A. Peppas, Synthesis and characterization of PH-responsive nanoscale hydrogels for oral delivery of hydrophobic therapeutics. Eur. J. Pharm. Biopharm. **108**, 196 (2016)

10. J.J. Yan, X.Y. Wang, M.Z. Wang, D.H. Pan, R.L. Yang, Y.P. Xu, L.Z. Wang, M. Yang, Self-assembling nonconjugated poly(amide-imide) into thermoresponsive nanovesicles with unexpected red fluorescence for bioimaging, Biomacromolecules **20** (2019)
11. P. Sánchez-Moreno, J. de Vicente, S. Nardecchia, J.A. Marchal, H. Boulaiz, Thermo-sensitive nanomaterials: recent advance in synthesis and biomedical applications. Nanomaterials **8** (2018)
12. I. Jiménez-Amezcua, F.J. Carmona, I. Romero-García, M. Quirós, J.L. Cenis, A.A. Lozano-Pérez, C.R. Maldonado, E. Barea, Silk fibroin nanoparticles as biocompatible nanocarriers of a novel light-responsive CO-prodrug. Dalton Trans. **47**, 10434 (2018)
13. Y. Xing, B. Zeng, W. Yang, Light responsive hydrogels for controlled drug delivery. Front. Bioeng. Biotechnol. **10**, 1075670 (2022)
14. T. Kang, F. Li, S. Baik, W. Shao, D. Ling, T. Hyeon, Surface design of magnetic nanoparticles for stimuli-responsive cancer imaging and therapy. Biomaterials **136**, 98 (2017)
15. N. Hoshyar, S. Gray, H. Han, G. Bao, The effect of nanoparticle size on in vivo pharmacokinetics and cellular interaction. Nanomedicine (Lond) **11**, 673 (2016)
16. X. Han, K. Xu, O. Taratula, K. Farsad, Applications of nanoparticles in biomedical imaging. Nanoscale **11**, 799 (2019)
17. Y.W. Wang, N.P. Reder, S. Kang, A.K. Glaser, J.T.C. Liu, Multiplexed optical imaging of tumor-directed nanoparticles: a review of imaging systems and approaches. Nanotheranostics **1**, 369 (2017)
18. J. Lavaud, M. Henry, P. Gayet, A. Fertin, J. Vollaire, Y. Usson, J.L. Coll, V. Josserand, Noninvasive monitoring of liver metastasis development via combined multispectral photoacoustic imaging and fluorescence diffuse optical tomography. Int. J. Biol. Sci. **16**, 1616 (2020)
19. S. Sargazi, I. Fatima, M. Hassan Kiani, V. Mohammadzadeh, R. Arshad, M. Bilal, A. Rahdar, A.M. Díez-Pascual, R. Behzadmehr, Fluorescent-based nanosensors for selective detection of a wide range of biological macromolecules: a comprehensive review. Int. J. Biol. Macromol. **206**, 115 (2022)
20. Y. Wang, K. Zhou, G. Huang, C. Hensley, X. Huang, X. Ma, T. Zhao, B.D. Sumer, R. J. Deberardinis, J. Gao, A nanoparticle-based strategy for the imaging of a broad range of tumours by nonlinear amplification of microenvironment signals. Nat. Mater. **13** (2014).
21. Y. Wang, K. Zhou, G. Huang, C. Hensley, X. Huang, X. Ma, T. Zhao, B.D. Sumer, R.J. Deberardinis, J. Gao, A broad nanoparticle-based strategy for tumor imaging by nonlinear amplification of microenvironment signals. Nat. Mater. **13**, 204 (2014)
22. L. Xiao, E.S. Yeung, Optical imaging of individual plasmonic nanoparticles in biological samples **7**, 89 (2014). https://doi.org/10.1146/Annurev-Anchem-071213-020125
23. S. Gao, L. Zhang, G. Wang, K. Yang, M. Chen, R. Tian, Q. Ma, L. Zhu, Hybrid Graphene/ Au activatable theranostic agent for multimodalities imaging guided enhanced photothermal therapy. Biomaterials **79** (2016)
24. R.W. Coatney, Ultrasound imaging: principles and applications in rodent research. ILAR J. **42**, 233 (2001)
25. A. Carovac, F. Smajlovic, D. Junuzovic, Application of ultrasound in medicine. Acta Informatica Medica **19**, 168 (2011)
26. J.A. Navarro-Becerra, K.H. Song, P. Martinez, M.A. Borden, Microbubble size and dose effects on pharmacokinetics. ACS Biomater. Sci. Eng. **8** (2022)
27. P. Frinking, T. Segers, Y. Luan, F. Tranquart, Three decades of ultrasound contrast agents: a review of the past, present and future improvements. Ultrasound Med. Biol.
28. R. Xiong, R.X. Xu, C. Huang, S. De Smedt, K. Braeckmans, Stimuli-responsive nanobubbles for biomedical applications. Chem. Soc. Rev.
29. L. Duan, L. Yang, J. Jin, F. Yang, D. Liu, K. Hu, Q. Wang, Y. Yue, N. Gu, Micro/nano-bubble-assisted ultrasound to enhance the EPR effect and potential theranostic applications. Theranostics
30. C.M.C. Tempany, B.J. McNeil, Advances in biomedical imaging. JAMA **285**, 562 (2001)
31. S.M. Amini, S.M. Rezayat, R. Dinarvand, S. Kharrazi, M.R. Jaafari, Gold cluster encapsulated liposomes: theranostic agent with stimulus triggered release capability. Med. Oncol. **40**, 1 (2023)

32. A. Wang, L. Yin, L. He, H. Xia, F. Chen, M. Zhao, J. Ding, H. Shi, An acidic PH/reduction dual-stimuli responsive nanoprobe for enhanced CT imaging of tumours in Vivo. Nanoscale **10**, 20126 (2018)
33. S. Caspani, R. Magalhães, J.P. Araújo, C.T. Sousa, Magnetic nanomaterials as contrast agents for MRI. Materials **13** (2020)
34. Z. Li, R. Bai, J. Yi, H. Zhou, J. Xian, C. Chen, Designing smart iron oxide nanoparticles for MR imaging of tumors. Chem. Biomed. Imaging **2023**, 315 (2023)
35. Y. Yan, X. Sun, B. Shen, Contrast agents in dynamic contrast-enhanced magnetic resonance imaging. Oncotarget **8**, 43491 (2017)
36. M.E. Lorkowski, P.U. Atukorale, K.B. Ghaghada, E. Karathanasis, Stimuli-responsive iron oxide nano-theranostics: a versatile and powerful approach for cancer therapy. Adv. Healthc. Mater. **10**, e2001044 (2021)
37. A. Mohapatra, S. Uthaman, I.K. Park, External and internal stimuli-responsive metallic nanotherapeutics for enhanced anticancer therapy. Front. Mol. Biosci. **7**, 597634 (2021)
38. S.S. Chetty, S. Praneetha, A. Vadivel Murugan, K. Govarthanan, R.S. Verma, Human umbilical cord Wharton's jelly-derived mesenchymal stem cells labeled with Mn2+ and Gd3+ Co-doped CuInS2-ZnS nanocrystals for multimodality imaging in a tumor mice model. ACS Appl. Mater. Interfaces **12**, 3415 (2020)
39. D.E. Lee, H. Koo, I.C. Sun, J.H. Ryu, K. Kim, I.C. Kwon, Multifunctional nanoparticles for multimodal imaging and theragnosis. Chem. Soc. Rev. **41**, 2656 (2012)
40. S.S. Chetty, S. Praneetha, A. Vadivel Murugan, K. Govarthanan, R.S. Verma, Microwave-assisted synthesis of quasi-pyramidal CuInS2–ZnS nanocrystals for enhanced near-infrared targeted fluorescent imaging of subcutaneous melanoma. Adv. Biosyst. **3**, 1800127 (2019)
41. Y. Wang, H. Zhang, S. Xia, Conventional imaging techniques, in *Radiology of Infectious and Inflammatory Diseases*, vol. **2** (2022), p. 3
42. Q. Sun, F. He, C. Sun, X. Wang, C. Li, J. Xu, D. Yang, H. Bi, S. Gai, P. Yang, Honeycomb-satellite structured PH/H_2O_2-responsive degradable nanoplatform for efficient photodynamic therapy and multimodal imaging. ACS Appl. Mater. Interfaces **10**, 33901 (2018)
43. Y. Tang, H. Shi, D. Cheng, J. Zhang, Y. Lin, Y. Xu, X. Qian, W. Zhu, PH-Activatable tumor-targeting gold nanoprobe for near-infrared fluorescence/CT dual-modal imaging in vivo. Colloids Surf. B Biointerfaces **179**, 56 (2019)
44. Z. Zhao, H. Fan, G. Zhou, H. Bai, H. Liang, R. Wang, X. Zhang, W. Tan, Activatable fluorescence/MRI bimodal platform for tumor cell imaging via MnO_2 nanosheet-aptamer nanoprobe. J. Am. Chem. Soc. **136**, 11220 (2014)
45. R. Wei, X. Gong, H. Lin, K. Zhang, A. Li, K. Liu, H. Shan, X. Chen, J. Gao, Versatile octapod-shaped hollow porous manganese(II) oxide nanoplatform for real-time visualization of cargo delivery. Nano Lett. **19**, 5394 (2019)
46. Z. Dong, L. Feng, W. Zhu, X. Sun, M. Gao, H. Zhao, Y. Chao, Z. Liu, $CaCO_3$ nanoparticles as an ultra-sensitive tumor-PH-responsive nanoplatform enabling real-time drug release monitoring and cancer combination therapy. Biomaterials **110**, 60 (2016)
47. M. Bouché, M. Pühringer, A. Iturmendi, A. Amirshaghaghi, A. Tsourkas, I. Teasdale, D.P. Cormode, Activatable hybrid polyphosphazene-AuNP nanoprobe for ROS detection by bimodal PA/CT imaging. ACS Appl. Mater. Interfaces **11**, 28648 (2019)
48. J. Li, C. Liu, Y. Hu, C. Ji, S. Li, M. Yin, PH-responsive perylenediimide nanoparticles for cancer trimodality imaging and photothermal therapy. Theranostics **10**, 166 (2020)
49. Y. Wang, X. Hu, J. Weng, J. Li, Q. Fan, Y. Zhang, D. Ye, A photoacoustic probe for the imaging of tumor apoptosis by caspase-mediated macrocyclization and self-assembly. Angew. Chem. Int. Ed. **58**, 4886 (2019)

Chapter 12
Potential of MRI in Clinical Medicine

Adarsh Anil Kumar and C. Kesavadas

Abstract Magnetic resonance imaging (MRI) is a pivotal diagnostic tool in clinical medicine, providing detailed non-invasive imaging without the use of ionizing radiation. Its utility spans across numerous organs, leveraging various advanced MRI techniques to enhance diagnostic accuracy. In neuroimaging, MRI is indispensable for evaluating brain and spinal cord pathologies, including tumors, vascular anomalies, and degenerative diseases, utilizing techniques like diffusion tensor imaging (DTI) and functional MRI (fMRI) to assess neural connectivity and brain activity. Cardiac MRI (CMR) offers comprehensive evaluation of cardiac anatomy, function, and tissue characterization, aiding in the diagnosis of various pathologies including myocardial infarction, cardiomyopathies, and congenital heart disease. Abdominal MRI, including magnetic resonance cholangiopancreatography (MRCP), provides superior imaging of liver, biliary, and pancreatic diseases. Musculoskeletal MRI provides excellent soft tissue contrast, crucial for diagnosing joint, muscle, and bone pathologies. Whole-body MRI is emerging for metastatic disease detection and systemic conditions. Advanced techniques like MR spectroscopy, perfusion imaging, and quantitative MRI enhance tissue characterization and functional assessment, solidifying MRI's role as a cornerstone of modern diagnostic imaging.

12.1 Introduction

Magnetic resonance imaging (MRI) is a non-invasive technique that does not involve ionizing radiation, making it a valuable tool for examining the body's structure and function in both healthy and diseased states. It plays a crucial role in diagnosing a variety of diseases and monitoring treatment progress. MRI can produce detailed

A. A. Kumar
Department of Imaging Sciences and Interventional Radiology, SCTIMST, Thiruvananthapuram, Kerala, India

C. Kesavadas (✉)
Department of Radiology, SCTIMST, Thiruvananthapuram, Kerala, India
e-mail: kesav@sctimst.ac.in

N. Kalarikkal et al. (eds.), *Multimodal Biomedical Imaging Techniques*,
Biological and Medical Physics, Biomedical Engineering,
https://doi.org/10.1007/978-981-96-1124-9_12

3D images with high spatial resolution and excellent soft tissue contrast. Over time, technological advancements such as improved scanner hardware, stronger magnets, innovative pulse sequences, and enhanced radiofrequency coils have significantly boosted image quality and contrast. More recently, developments like quantitative MRI, new contrast agents, molecular imaging, and hybrid imaging methods have emerged, and ongoing research in these areas promises to deepen our understanding of physiology and pathology.

12.2 History of NMR and MRI

Nuclear magnetic resonance (NMR) was initially described in molecular beams by Isidor Rabi in 1938, a discovery for which he received the Nobel Prize in Physics. Felix Bloch and Edward Purcell expanded NMR's application to liquids and solids, earning them the Nobel Prize in 1952. Paul Lauterbur was the first to use NMR for imaging, creating an image of two water-filled test tubes. This imaging method was initially called "zeugmatography," but the term did not gain widespread use. Later, Peter Mansfield introduced methods for slice selection and echo-planar imaging (EPI). The discoveries of Lauterbur and Mansfield led to NMR being used to image the human body, and they were jointly awarded the Nobel Prize in Physiology or Medicine in 2003. However, it was Raymond Damadian, an American physician, who conducted the first full-body scan of a human in 1977 to diagnose cancer. He demonstrated that NMR could differentiate between healthy and cancerous tissues in animal models.

The term "nuclear" was dropped from NMR, and the technology was renamed magnetic resonance imaging (MRI), in order to avoid the negative connotations the word "nuclear" had with the general public. Since the early days of zeugmatography, MRI image quality has significantly improved, evolving from rudimentary images to the high-quality, high-resolution scans produced by modern MRI machines [1].

12.3 Basic Physics

The following are the four main components of an MR scanner (Fig. 12.1):

1. Magnet: May be resistive or superconducting (the latter being commonly used). Magnets have a field strength (B0) of typically 1.5–3.0 T. Superconducting magnets have coils that are cooled to near absolute zero with liquid helium.
2. Gradient coils lie in the x, y, and z planes of magnet. These are needed to spatially encode the MR signal during imaging.
3. Radiofrequency (RF) coil, which transmits the RF pulses needed to produce the MR signal
4. RF receiver coil that picks up the resonance signal created in the patient.

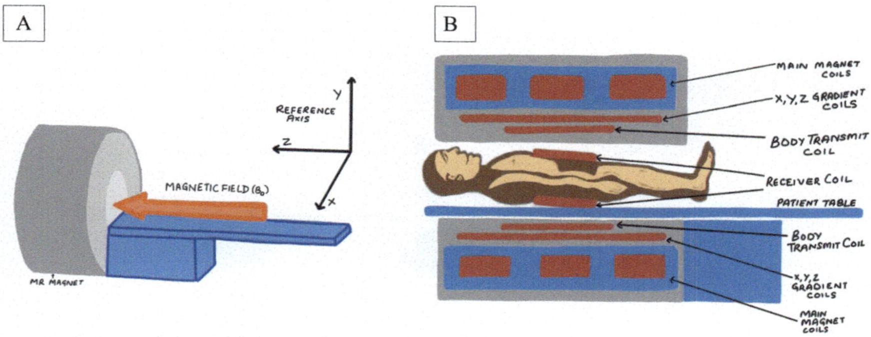

Fig. 12.1 (**A**) Pictorial illustration of a typical cylindrical magnet and the direction of magnetic field. By convention, the main magnetic field is oriented along the z axis. (**B**) Another pictorial representation demonstrating the main coils within the magnet

12.3.1 MR Signal Generation

Hydrogen (1H) nuclei in the body possess a nuclear spin, which creates a magnetic moment, much like a tiny bar magnet. When exposed to an external magnetic field, these spins will align either parallel or antiparallel to the field, with most aligning parallel (Fig. 12.2). In addition to this, the spins are at an angle to the main magnetic field and are precessing at a specific frequency, called the Larmor frequency (Fig. 12.3). This frequency depends on the type of nucleus and the strength of the magnetic field. A small net magnetization (M0) is generated in tissues, but it is quite weak compared to the external magnetic field. When the spins are aligned parallel to the magnetic field, little signal is produced. To detect a signal, an RF pulse from the RF coil is applied, which tips the magnetic field perpendicular to the main field. This RF pulse must match the Larmor frequency to achieve resonance (as illustrated in Figs. 12.4 and 12.5). The signal generated is then detected by the RF receiver coil. T1 relaxation, or spin–lattice relaxation (also known as longitudinal relaxation), occurs when the excited spins return to their preexcitation ground state. The time required for this is called the T1 relaxation time (T1), during which the excited nuclei release energy to surrounding nuclei as they return from a high-energy state to a lower-energy state.

T2 relaxation, also referred to as spin–spin relaxation or transverse relaxation, describes the gradual loss of phase coherence among spinning dipoles, which leads to a decay in magnetization in the transverse plane. Initially, after the RF pulse is applied, the spins are fully in phase, producing the maximum signal. After sometime, adjacent spins exchange energy and develop phase differences, causing the signal to decrease exponentially. T2 is defined as the time it takes for 63% of the signal to decay, leaving 37%. Inhomogeneities in the main magnetic field can cause spins to dephase more rapidly. The additional dephasing caused by these inhomogeneities is characterized by the relaxation time known as T2* (T2 star).

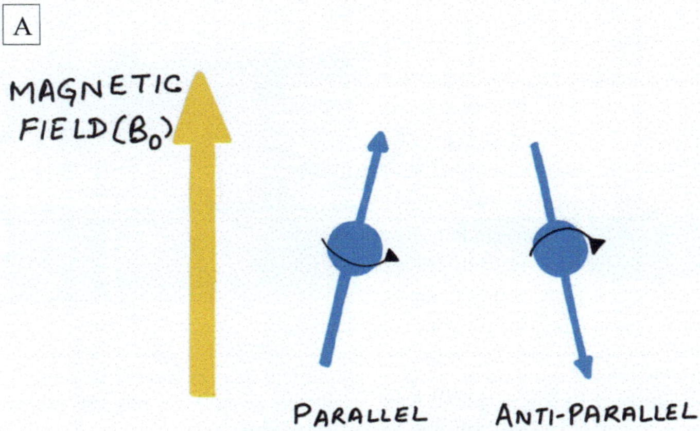

Fig. 12.2 (**A**) Illustration of the parallel and antiparallel orientation of magnetic moments after the protons have entered into an external magnetic field

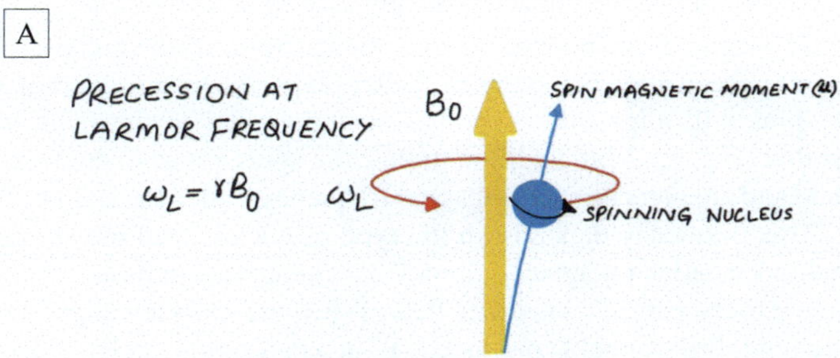

Fig. 12.3 Illustration of a spin processing at a particular frequency; γ represents the gyromagnetic ratio, B_0 represents the MR field strength in Tesla, and, ω_L represents the Larmor frequency

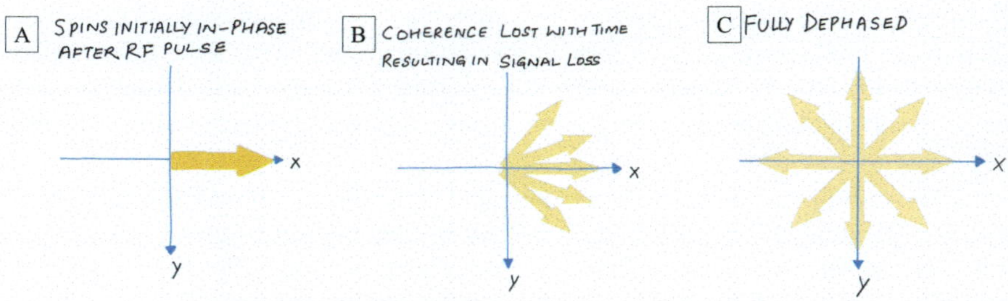

Fig. 12.4 Just after the initial 90 degree radiofrequency pulse, the spins are initially in phase (**A**), but after time the spins begin to diphase (**B**) and there is signal loss at a exponential rate. Finally the spins become fully dephased with no remaining signal (**C**)

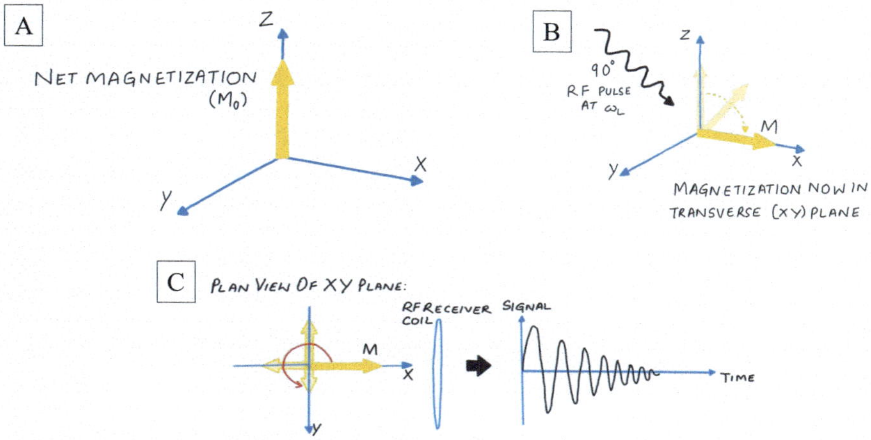

Fig. 12.5 Schematic diagram of creation of an MR signal. The net magnetization is along the z axis before excitation (**A**); After a 90 degree RF pulse at Larmor frequency, it is flipped into the xy plane (**B**); Initially the spins process in phase in the xy plane, around the z axis (**C**). Slowly they become out of phase, which causes decay in the signal

12.3.2 Imaging Sequences

There are two principal types of sequences: spin-echo pulse and gradient-echo sequences.

In spin echo sequence, an additional RF pulse is applied to create an echo. A 90-degree slice-selective pulse is followed by a 180-degree refocusing pulse, which compensates for magnetic field inhomogeneities, resulting in true T2 contrast. Although spin echo can generate T1-weighted images, it is less frequently used for this than gradient echo. After acquisition, the frequency data is processed using a Fourier transform to produce the final MRI image.

Gradient echo is a fast imaging technique that uses flip angles of less than 90 degrees and gradient fields to generate transverse magnetization. The application of a gradient helps recover the initial MR signal, but it does not correct for magnetic field inhomogeneities, leading to T2* contrast. Gradient echo is also capable of producing T1-weighted images [1].

12.3.3 MRI Contrast

Different types of images are generated by altering the sequence of RF pulses applied and collected. TR refers to the time between two consecutive pulse sequences applied to the same slice, while TE refers to the time between the delivery of the RF pulse and the reception of the signal. A short TE and TR produce a T1-weighted image, while longer TE and TR produce a T2-weighted image.

The most common MRI images are T1-weighted and T2-weighted. T1-weighted images are created with short TE and TR times, where the image contrast and brightness primarily depend on the T1 properties of the tissues. Similarly, T2-weighted images rely on the T2 properties of tissues for contrast and brightness. The appearance of tissues in MRI—whether bright or dark—depends on their T1 and T2 relaxation times. For example, cerebrospinal fluid (CSF) appears dark on T1-weighted images and bright on T2-weighted images. In Fluid Attenuated Inversion Recovery (FLAIR) sequences, an inversion pulse with a long inversion time is used to suppress the signal from CSF. Diffusion-weighted imaging, which examines the mobility of water in tissues, applies two additional gradients symmetrically on either side of a 180-degree pulse.

12.3.4 MR Contrast Agents

In addition to the intrinsic contrast mechanisms previously discussed, exogenous contrast agents are sometimes utilized in MR imaging. These agents are classified into two types: positive and negative. Positive contrast agents, such as gadolinium or manganese, enhance signal intensity and are typically paramagnetic metals. Gadolinium-based contrast agents are the most commonly used in clinical settings. Negative contrast agents, which decrease signal intensity, are based on superparamagnetic compounds like iron oxide. The alteration in signal intensity, whether an increase or decrease, is due to the magnetic interaction between the contrast agent and its surroundings. While some contrast agents are designed to target specific tissues, most are administered intravenously and spread throughout the body. Targeted agents are primarily used in molecular imaging [1].

12.3.5 Molecular MRI

This technique enables observing biological processes at both the cellular and molecular levels. To be more specific, contrast agents are used to modify tissue relaxation times so as to aid the identification and localization of molecular disease markers or therapeutic agents. Recent advancements in molecular contrast agents have occurred in both clinical and preclinical models, leading to earlier disease detection, drug discovery, and development. These developments have also contributed to biomedical research by providing insights into enzymatic and cellular processes involved in inflammation, angiogenesis, apoptosis, stem cell migration, and atherosclerosis [2]. In the field of medical diagnostics, this technique offers promise for enhancing early disease detection, improving treatment planning, monitoring treatment outcomes, and ultimately improving patient care.

12.4 Preclinical MRI

MRI is employed for preclinical imaging. Living animals are scanned for research purposes, such as drug development, studying physiological processes, and assessing treatment effects. Researchers can track changes at the organ, tissue, or cellular levels. Since it is non-invasive, MRI allows for longitudinal imaging of animal models. Advances in MR hardware, including higher-field scanners (greater than 4 T), have improved the signal-to-noise ratio (SNR) in preclinical imaging. Additionally, the use of small imaging coils that enhance signal reception has enabled high-resolution imaging in preclinical studies. Most of the techniques used in preclinical imaging are similar to those applied in routine clinical practice, which are elaborated below.

12.4.1 Magnetic Resonance Microscopy

Magnetic resonance microscopy (MRM) involves imaging at a voxel resolution of less than 100 μm and is typically used for ex-vivo tissue imaging. A key issue with MRM is that the MR signal obtained is directly related to the voxel size, meaning that the signal-to-noise ratio (SNR) is significantly lower. To achieve an acceptable SNR with high resolution, longer scan times are also required [1].

12.5 MRI: Clinical Applications

Magnetic resonance imaging (MRI) has become a valuable tool in both clinical applications and experimental research. It is highly effective in analyzing the anatomical, metabolic, functional, and molecular properties of tissues without the risks associated with radiation or invasiveness. These capabilities have led to its widespread use in diagnosing, characterizing, staging, and prognosticating outcomes of various diseases [3]. However, MRI is often viewed as an expensive procedure that contributes to the rising costs of healthcare. This makes it essential to highlight MRI's clinical effectiveness and its key role in non-invasive diagnostics. MRI's diagnostic reach spans multiple organ systems, including the central nervous system, neck, heart, lungs, liver, bones, joints, and breasts. As clinical applications expand, almost every organ can be imaged to some extent [4]. Initially, MRI was primarily used for anatomical assessment, but it has since evolved to focus on providing functional and metabolic information. Advanced MRI techniques, such as functional MRI, diffusion weighted imaging, diffusion tensor imaging, and MR perfusion, are now more commonly employed in clinical settings. There is also increasing research into using MRI to explore the diagnostic and therapeutic potential of nanotechnology. Recent technological advancements, like the development of highly sensitive receiving coils (particularly cryogenically cooled coils) and the use of stronger magnetic fields, have enabled microimaging with extremely fine spatial resolution [5]. Beyond diagnostics, MRI is becoming more integrated into therapeutic applications. The field of theranostics, which combines imaging and therapy, is gaining recognition due to its ability to monitor and control treatments in real-time for individual patients. MRI-guided therapy, for instance, allows precise and time-sensitive evaluations of disease progression and therapeutic responses. In 2020, the first human trial of MRI-guided nanoparticle theranostics demonstrated that brain tumor metastases could be visualized while simultaneously measuring local drug concentrations. This marked a significant milestone in MRI-guided nano-theranostics, showing its potential in precision medicine [6].

With this background of the plethora of uses of MRI in the diagnostic as well as therapeutic aspects of clinical medicine, let us now further explore the MR advances being used in clinical medicine and also further analyze in detail the clinical applications that MRI has in each of the various organ systems.

12.6 MR Advances in Clinical Medicine

12.6.1 MR Spectroscopy

In vivo nuclear magnetic resonance spectroscopy is referred to as MR spectroscopy (MRS). It was discovered in 1921 that certain magnetic nuclei, such as 1-H and 31-P, in a magnetic field can absorb radiofrequency energy at specific power levels. After absorbing this energy, these nuclei begin to resonate, with different atoms in a molecule resonating at distinct frequencies. This discovery formed the foundation of NMR [7]. On use in clinical settings, it became known as MR spectroscopy. MRS allows the analysis of small molecules within both intracellular and extracellular spaces. The spectra generated by MRS offers detailed information on metabolic composition and changes, enabling the assessment of metabolic abnormalities caused by various diseases and the evaluation of treatment efficacy [8]. The development of stronger magnetic fields, improved coils, and advanced radiofrequency pulse designs has increased MRS sensitivity, allowing it to be more widely used in routine clinical practice [9]. Various nuclei, including 1-hydrogen, 19-fluorine, and 31-phosphorus, can be utilized to obtain MR spectra, with 1-H MRS being the most common due to its high sensitivity, easy accessibility, and the abundance of hydrogen in metabolites [10]. MRS typically begins by acquiring conventional MR images to guide the selection of regions where the MRS spectrum needs to be assessed. In single voxel spectroscopy (SVS), a single voxel (volume of tissue) is selected within a tumor or a specific area where metabolism may be altered due to disease [11, 12]. Chemical shift imaging is another technique that detects and estimates metabolites across tissues, using large volumes divided into smaller voxels to capture all voxel spectra simultaneously (Fig. 12.6). Commonly used SVS techniques include point-resolved spectroscopy (PRESS) and stimulated echo acquisition mode (STEAM)

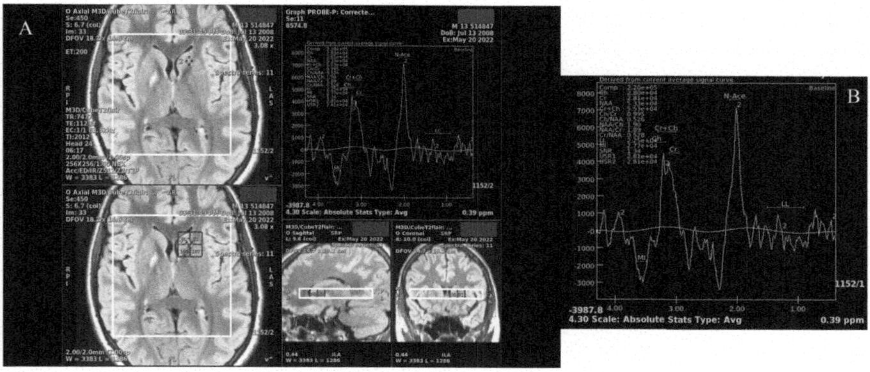

Fig. 12.6 Images of MR Spectroscopy (at intermediate TE: 144 msec) obtained at the level of capsuloganglionic region (**A**) and the corresponding spectrum (**B**)

[13, 14]. Although MRS is primarily used to study brain tissue, it is also applied in the detection, staging, and evaluation of response to treatment in various cancers, such as breast, prostate, and liver cancers. MRS also has non-cancerous applications, including the study of neural and psychiatric disorders, and the mapping of metabolites in non-neoplastic lesions of the prostate, breast, gastrointestinal, genitourinary systems, and liver [15].

12.6.2 Diffusion Weighted MRI

The human body consists of approximately 60–70% water. Diffusion weighted imaging (DWI) is a technique that creates contrast in images based on variations in the Brownian motion of water molecules caused by thermal energy. This method allows for the evaluation of molecular function and tissue microarchitecture. In DWI, signal contrast is quantified through apparent diffusion coefficient (ADC) maps, making it useful for assessing disease progression and monitoring treatment responses. When the medium is homogeneous, diffusion occurs randomly and isotropically, meaning water molecules have an equal probability of diffusing in all directions. However, in the human body, water is distributed between intracellular and extracellular compartments. In extracellular spaces, water molecules diffuse more freely, while intracellular molecules experience more "restricted diffusion." Different tissues have distinct diffusion properties, and pathological processes can alter water distribution between compartments. For instance, in high-grade malignancies or acutely infarcted brain tissue, the proportion of intracellular water increases (known as cytotoxic edema), leading to more restricted diffusion. DWI provides both qualitative and quantitative insights into diffusion properties and adds a functional dimension to MRI. In a DWI sequence, sensitization gradients are applied on either side of the 180° refocusing pulse (Fig. 12.7). The degree of diffusion weighting is controlled by a parameter called the "b value," measured in s/mm^2, which is proportional to the square of the gradient's amplitude and duration. Qualitative assessments are made on trace images, while quantitative evaluations use the ADC. Tissues with restricted diffusion appear bright on trace images and dark on ADC maps (Fig. 12.8). DWI has become an essential tool in imaging stroke, white matter disorders, and oncology. Recent advancements in diffusion imaging have broadened its applications, and it is believed that DWI has not yet reached its full potential. In the future, it is expected to help solve complex problems related to brain function [16].

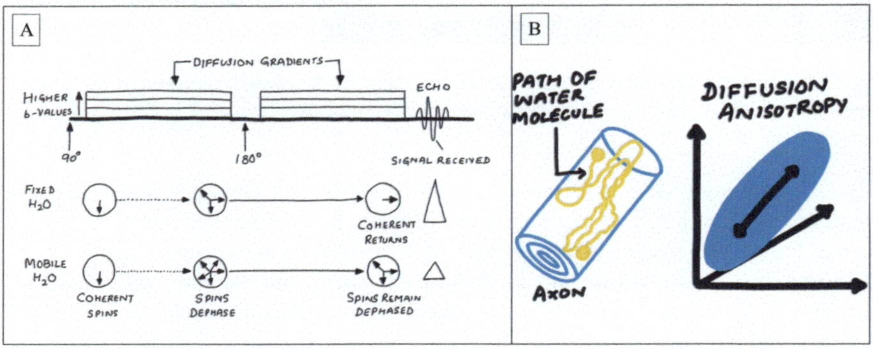

Fig. 12.7 (**A**) Illustration of the effect of diffusion gradients: Once diffusion gradients are applied, the signal intensity of fixed water remains unchanged, however, signal intensity of mobile water is reduced due to residual phase change. (**B**) Along the direction of axon, water diffusion is largely restricted; DTI uses this principle to identify the direction of principal anisotropy

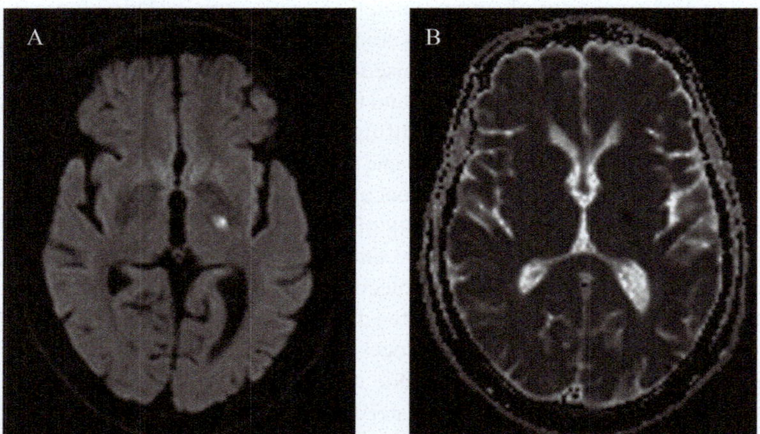

Fig. 12.8 Axial DWI (**A**) images and corresponding ADC map (**B**) demonstrating an ovoid focus of restricted diffusion in left anterolateral thalamus

12.6.3 Diffusion Tensor Imaging

Diffusion tensor imaging (DTI) is a technique used to analyze the directional properties of water molecule diffusion. It measures restricted water diffusion in tissues to create tractography images of neural pathways. The key information from DTI is described based on two parameters: diffusion anisotropy, which indicates the level of directionality, and the orientation of the axis along which water molecules preferentially move (Fig. 12.9). The diffusion tensor's six elements are calculated for each pixel using multivariate linear fitting. After diagonalization, three eigenvalues ($\lambda 1$, $\lambda 2$, $\lambda 3$) and three eigenvectors (v1, v2, v3) are generated. The anisotropy map calculates the fractional anisotropy (FA) value, which ranges from 0 (isotropic) to 1 (anisotropic) (Fig. 12.10). The eigenvector corresponding to the largest eigenvalue is used to determine fiber orientation. In the color map, red (R), green (G), and blue (B) represent right-left, anterior–posterior, and superior-inferior orientations, respectively. The color map employs a 24-bit color scheme, where each RGB color component has an 8-bit intensity level (0–255). To minimize orientation information in isotropic brain regions (areas without preferential orientation), the 24-bit color value is multiplied by the FA value. DTI matrices are evaluated using FA and apparent diffusion coefficient (ADC) values. These measurements are taken from the tumor core, peritumoral area, and corresponding normal white matter tracts in the contralateral hemisphere in cases of brain tumors. This helps on demonstrating the involvement of white matter tracts in the brain. MR tractography is particularly helpful for non-invasive visualization of important white matter fiber pathways. It also reveals the orientation and association of white matter tracts in cases of brain edema, as well as their displacement, infiltration, or disruption in brain tumors [17].

12.6.4 Functional MRI

Functional magnetic resonance imaging (fMRI) is a technique used to visualize dynamic changes in brain metabolism over time. These metabolic changes can occur due to task-induced cognitive state alterations or unregulated processes during the resting state. fMRI is popular because of its wide availability, non-invasive nature, and excellent spatial resolution [18]. Neural signaling processes in the brain require energy in the form of adenosine triphosphate (ATP). When a specific brain region is activated by a cognitive task (e.g., finger tapping), the energy demand increases locally, leading to an up-regulation of the cerebral metabolic rate of oxygen (CMRO2) in that region. As oxygen reserves near capillaries are temporarily depleted due to glycolysis and metabolite accumulation, chemical signals (like CO2, NO, H+) trigger a vasomotor response that dilates upstream arterial sphincters, increasing blood flow. This increased blood flow restores the necessary oxygen levels to the region, compensating for the transient deficit. Neural up-regulation initially causes an accumulation of deoxygenated hemoglobin [Hb] and a reduction in oxygenated hemoglobin

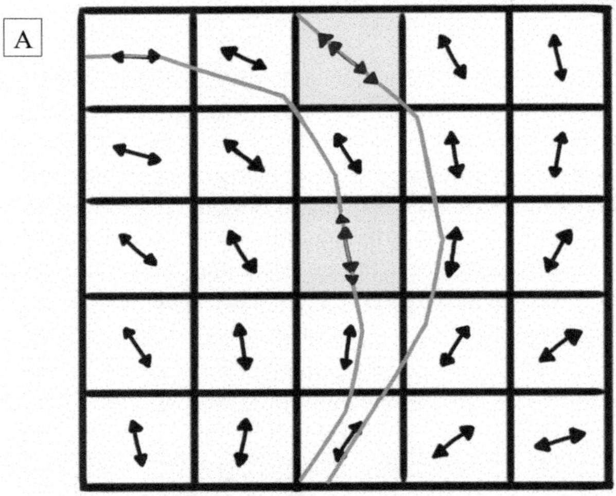

Fig. 12.9 (**A**) Two dimensional view of adjacent streamlines generated by voxel to voxel anisotropy data; Expected path of streamlines has been plotted based on the principal direction of diffusion

[HbO2] in both intra- and extravascular spaces. A vasodilatory response occurs within one to two seconds, leading to an increase in [HbO2] and a decrease in [Hb] compared to the resting state. This process is referred to as the "hemodynamic response to a neural event" [19]. The two main outcomes of increased neural activity detectable by MRI are the rise in local cerebral blood flow (CBF) and the changes in blood oxygenation levels, known as blood oxygen level dependent (BOLD) contrast [18]. BOLD fMRI contrast arises from changes in local magnetic fields around red blood cells, which are influenced by hemoglobin's oxygenation state. When hemoglobin is fully oxygenated, oxyhemoglobin is diamagnetic and indistinguishable from nearby brain tissue. However, fully deoxygenated hemoglobin has four unpaired electrons, making it highly paramagnetic. This paramagnetic property causes local magnetic gradients, which vary depending on hemoglobin concentration. These gradients affect the T2 and T2* relaxation times of intra- and extravascular blood through diffusion and intravoxel dephasing. A gradient-refocused echo (GRE) pulse sequence is used to capture the T2* and T2 effects, with T2* contrast being most prominent in venules at clinical magnetic field strengths of 1.5 and 3 T, while T2-weighted contrast dominates at higher field strengths [20, 21]. In fMRI task activation experiments (paradigms), stimuli such as visual or auditory cues are used to induce different cognitive states in the subject while MRI scans are continuously acquired (Fig. 12.11). A two-condition design typically includes an experimental and a control condition, testing whether signals differ between these two states. Block designs alternate between experimental and control conditions, with each block usually lasting several tens of seconds [22]. The collected fMRI data undergoes preprocessing, which involves: (1) time-slice correction to account for differences in acquisition time for each slice, (2) motion

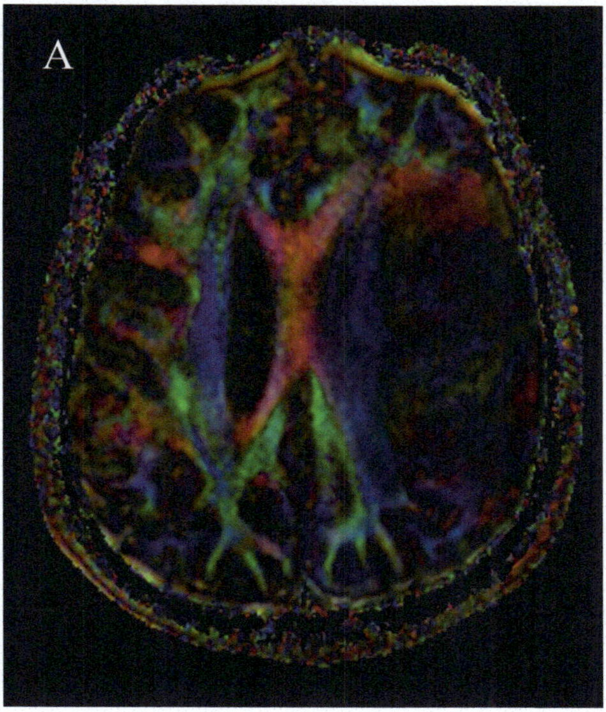

Fig. 12.10 Axial Fractional Anisotropy (FA) map obtained from Diffusion Tensor Imaging (DTI) demonstrating reduced FA values in left fronto-parietal region caused by a high grade glial neoplasm

correction to detect and adjust for head movements by aligning each volume in the time series, (3) correction for physiological noise from respiration and heartbeats, along with temporal filtering to remove irrelevant frequency components, (4) spatial smoothing to improve the signal-to-noise ratio (SNR) and normalize the noise distribution, and (5) pre-whitening to address autocorrelation in the time series [23]. After preprocessing, the fMRI data is analyzed to evaluate functional activity across various brain regions [18].

12.6.5 MR Perfusion

Perfusion is defined as steady-state delivery of blood to a specific region of tissue. It is used to assess various physiological parameters that influence the MR signal, such as blood volume, flow, and oxygenation. There are several noninvasive MRI methods to measure perfusion, with much of the research focused on brain perfusion imaging. Two main approaches are commonly used to assess cerebral perfusion.

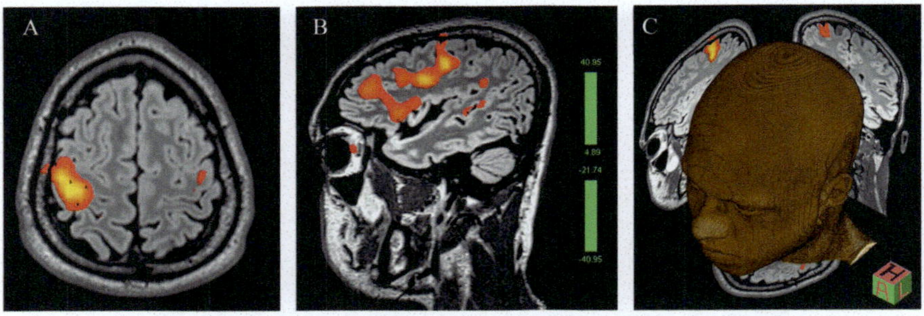

Fig. 12.11 Axial (**A**), Sagittal (**B**) and 3D (**C**) representative images of functional MRI demonstrating activation in right hand knob region while perform left finger tapping paradigm

The first involves an exogenous, intravascular, non-diffusible contrast agent (usually gadolinium-based) that leverages either the susceptibility effects of gadolinium (first-pass dynamic susceptibility contrast-enhanced or DSC MR perfusion) or its relaxivity effects on the signal echo (dynamic contrast-enhanced or DCE MR perfusion). The second approach uses an endogenous contrast agent, utilizing magnetically labeled arterial blood protons as tracers in arterial spin labeling (ASL) MR perfusion (Fig. 12.12). In contrast-enhanced perfusion MRI, a standard dose of 0.1 mmol/kg of body weight is used, and imaging is performed for at least 3 min after the contrast is administered. For DSC perfusion (which uses fast echo planar imaging), the bolus injection of the contrast should begin about 20 s after the DSC sequence starts. A minimum injection rate of 3 mL/s (with a range of 3–5 mL/s) is recommended to ensure a compact bolus arrival in the brain, followed by a saline flush at the same rate to push the bolus towards the heart. For DCE perfusion, a fast T1-weighted spoiled gradient echo sequence is used, with acquisition time depending on the parameters being measured. Acquisition typically lasts 3 min for Ktrans assessments and about 6 to 7 min for plasma volume and extravascular-extracellular space measurements. The injection begins 20 s after the DCE sequence starts, at a rate of 2–4 mL/s (according to the Tofts model), with an infusion lasting 30 s (Brix model). After the contrast medium injection, a saline injection of at least 10 mL (at the same rate) is required. Both DCE and DSC sequences can be performed in a single MRI protocol, with DCE done before DSC.

From DSC perfusion data, parametric maps of cerebral blood volume (CBV) and cerebral blood flow (CBF) can be generated (Fig. 12.13). DCE perfusion provides information on tissue kinetics, such as wash-in, plateau, and washout, offering insights into tissue properties at the microvascular level [24]. Using pharmacokinetic modeling, DCE MR perfusion data commonly yields metrics such as the transfer constant (Ktrans), the fractional volume of the extravascular-extracellular space (Ve), the rate constant (Kep, where Kep = Ktrans/Ve), and the fractional plasma volume (Vp) [25, 26].

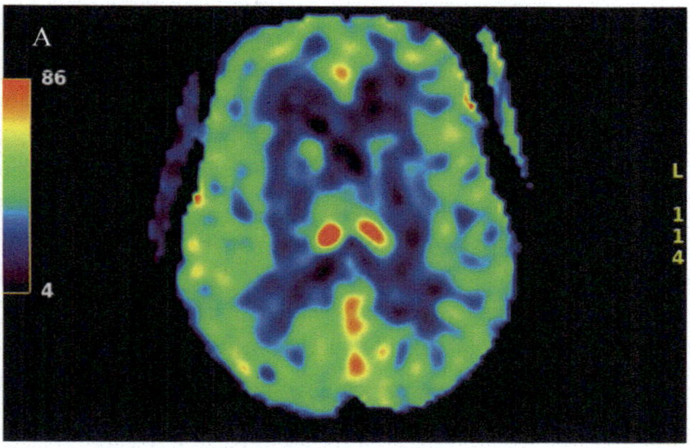

Fig. 12.12 Axial 3D ASL (**A**) image demonstrating normal cerebral parenchymal perfusion

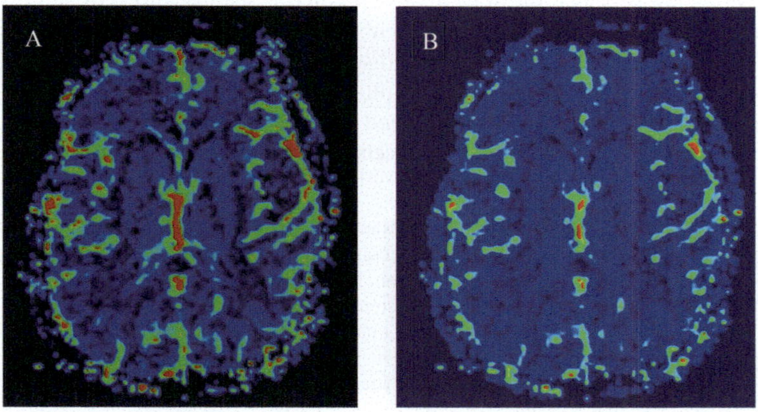

Fig. 12.13 Dynamic Susceptibility Contrast MR perfusion (DSC perfusion): rCBV (**A**) and rCBF (**B**) maps obtained from the same

12.6.6 Susceptibility Weighted MRI

Susceptibility weighted imaging (SWI) is an MRI technique that leverages differences in magnetic susceptibility of various substances, such as deoxygenated blood, blood products, iron, and calcium, to generate contrast. It is based on a high-resolution, 3D gradient-echo sequence with velocity compensation, using both magnitude and phase images, either separately or together. SWI is particularly useful in cases of trauma, acute stroke, vascular malformations, intracranial calcifications, cerebral microbleeds, neurodegenerative diseases, and brain tumors. With advanced MRI post-processing techniques, including quantitative susceptibility mapping, SWI allows for precise anatomical differentiation based on brain iron quantification [27].

Initially, phase imaging was primarily used for assessing large-vessel flow or in inversion recovery sequences, and diagnostic MRI largely relied on magnitude data. However, phase images hold valuable information regarding local susceptibility variations between tissues. This phase data can help measure iron levels and other substances that affect the local magnetic field. SWI provides insight into tissues that exhibit different magnetic susceptibilities compared to surrounding tissues, such as deoxygenated blood, hemosiderin, ferritin, and calcium [28]. SWI is particularly useful in evaluating diffuse axonal injury (DAI), which is often linked to small hemorrhages in deep and subcortical white matter. By cancelling signal intensity in magnitude images and applying additional suppression in SWI filtered-phase images, hypointense signals are produced in regions of acute and early subacute hemorrhage (Fig. 12.14). SWI can detect subarachnoid hemorrhage as blooming artifacts in the sulcal spaces. In the context of stroke, SWI provides crucial information, including: (1) identifying hemorrhagic components in infarcted areas, (2) highlighting hypoperfused regions to guide the need for perfusion-weighted imaging, (3) detecting intravascular thrombi, and (4) predicting the likelihood of hemorrhagic transformation before administering thrombolytic therapy. SWI can also detect iron deposition in conditions like Parkinson's disease, Alzheimer's disease, Huntington's disease, amyotrophic lateral sclerosis, multiple sclerosis (MS), Hallervorden-Spatz disease, and pantothenate kinase-associated neurodegeneration (PKAN). It is useful for identifying specific patterns of iron accumulation in these diseases [29].

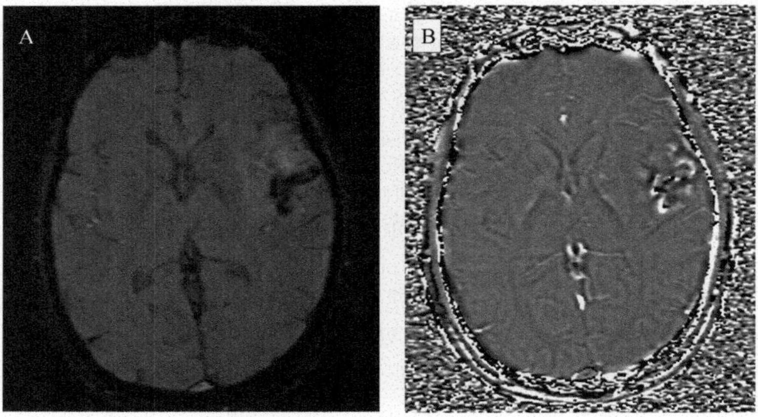

Fig. 12.14 Susceptibility Weighted Images (SWI): Magnitude and filtered phase images demonstrating intraparenchymal haemorrhage appearing dark on magnitude images (**A**) and bright on filtered phase images (**B**) {Acquired on a left handed system}

12.6.7 T1, T2, and T2 Mapping*

T1 mapping is a technique frequently used in cardiovascular magnetic resonance (CMR) imaging. In CMR, the T1 relaxation time of proton magnetization in myocardial tissue serves as an important biomarker for various pathological conditions. This is why T1 quantification has become a focus of attention, rather than just assessing its effect on image contrast. The technique has advanced to the point where pixel-level myocardial T1 mapping is now a routine part of CMR exams. When used with contrast agents, T1 mapping has been instrumental in investigating interstitial remodeling in both ischemic and nonischemic heart diseases [30]. Similarly, T2 mapping has also been incorporated into cardiac magnetic resonance imaging. Increased T2 relaxation time indicates elevated free water, higher myocardial water content, and is used as a marker for myocardial edema. The main advantages of quantitative T2 mapping are its accuracy in characterizing myocardial edema and its ability to detect reversible myocardial diseases early, without the use of contrast agents. It provides a reliable way to assess diffuse myocardial edema and can be used to diagnose, stage, and monitor myocardial injuries [31]. Cardiac T2* mapping is applied to detect iron accumulation within myocardial tissue in conditions like sickle cell disease, hereditary hemochromatosis, and β-thalassemia major. Advanced T2* measurement involves pixel-level quantification, which offers more precise spatial information about iron loading compared to region-of-interest (ROI) based measurements. Cardiac T2* mapping is also used for the early detection, monitoring, and treatment of iron overload [31]. T1, T2, and T2* mapping have also been employed in musculoskeletal imaging, with T2 mapping being the most commonly used [32]. T2 mapping typically involves acquiring multiple images with varying echo times (TEs), which generate signal intensities that follow a T2 relaxation curve. These T2 maps are useful for quantitatively assessing nearly all musculoskeletal tissues, but are particularly helpful in evaluating articular hyaline cartilage. In cartilage, the density and organization of the extracellular matrix are reflected as variations in T2 values, which can be represented using grayscale or color maps. This technique is valuable for detecting and quantifying early changes in water content and collagen composition, early cartilage degeneration, functional potential of cartilage, reparative tissue studies, and monitoring the effects of chondroprotective therapies [33].

12.6.8 MR Cartilage Imaging

MRI provides valuable insights into the composition and macromolecular structure of articular cartilage. It offers excellent tissue contrast, allowing for detailed assessment of early chondral-based lesions or pathologies. The availability of cartilage-specific sequences enables accurate predictions about the integrity of articular cartilage. These sequences help in evaluating cartilage morphology, thickness, volume, and any alterations in internal signal intensities [34].

Newer MRI techniques include cartilage mapping and delayed gadolinium-enhanced MR imaging of cartilage (dGEMRIC) sequences. Contrast agents, injected either intravenously or directly into the joint, are useful for assessing cartilage damage and glycosaminoglycan (GAG) content. Other compositional assessment methods include T1 mapping, T2 mapping, dGEMRIC, sodium imaging, and diffusion-weighted imaging. There are various versions of these techniques, depending on user preferences and the strengths and limitations of the available hardware platforms from different vendors [35].

Typically, MRI protocols for evaluating hyaline articular cartilage involve at least one cartilage-sensitive sequence in the sagittal plane. Additional acquisitions or 3D reformations in the axial and coronal planes are used to assess the patellar cartilage as well as the central weight-bearing areas of the femoral condyles and tibial plateau [36].

12.6.9 MR Angiogram

Magnetic resonance angiography (MRA) is an MRI technique used for vascular imaging, including both neurovascular and peripheral vascular applications. It allows detailed visualization of blood vessels without the need for invasive procedures. MRA works by manipulating magnetic fields and radiofrequency pulses to generate images of blood vessels. Several MRA techniques have been developed, including time-of-flight (TOF), phase-contrast (PC), contrast-enhanced (CE) MRA, and time-resolved MRA. TOF MRA creates high-resolution images of blood vessels by detecting the inflow of unsaturated blood. PC MRA measures the velocity of blood flow and provides information on flow dynamics. CE MRA uses contrast agents to highlight blood vessels, increasing sensitivity in detecting abnormalities. Time-resolved (4D) MRA involves capturing a series of images after intravenous contrast administration to track the flow of arterial contrast, a technique made possible by keyhole imaging [37]. MRA is widely used for evaluating cerebral, peripheral vascular, and cardiac systems. In neurovascular imaging, MRA aids in diagnosing conditions like arterial stenosis, aneurysms, and vascular malformations. Peripheral MRA is used for assessing peripheral arterial diseases and deep vein thrombosis. In cardiac imaging, MRA helps detect congenital heart diseases, valvular issues, and myocardial perfusion abnormalities [38]. The non-invasive nature of MRA makes it a safer alternative to conventional angiography, avoiding the risks associated with invasive procedures. However, MRA has some limitations, such as longer scan times, lower spatial resolution, and potential artifacts from patient movement or metallic implants. Recent advances are focusing on addressing these challenges through improvements in imaging protocols and hardware technology. Magnetic resonance angiography has revolutionized vascular imaging, providing detailed blood vessel visualization without invasive procedures, making it a vital tool for diagnosing various vascular diseases [37].

12.6.10 Non-contrast MR Angiogram

Non-contrast MR angiography (NC-MRA) is a rapidly advancing technique in vascular imaging that does not require contrast agents, making it a safer and more cost-effective option. It utilizes the inherent properties of blood and tissues to produce images without the use of contrast media. Several noncontrast MR angiography techniques exist, including time-of-flight (TOF) and phase-sensitive approaches. TOF MRA enhances vessel contrast by manipulating the longitudinal magnetization of unsaturated blood, which leads to signal loss in stationary tissues. Phase-sensitive techniques use phase differences between moving blood and stationary tissue for vessel visualization. TOF is the most widely used method, known for its excellent visualization of larger vessels (Fig. 12.15). Other techniques, such as balanced steady-state free precession (bSSFP) and phase-contrast, provide improved spatial resolution and flow quantification, respectively. Additional ones, like quiescent interval slice-selective (QISS) MRA and arterial spin labeling (ASL) MRA, are specifically useful for imaging arteries and evaluating perfusion respectively. In neurovascular imaging, NC-MRA helps assess intracranial arteries, detecting conditions such as stenosis, aneurysms, and arteriovenous malformations. Peripheral NC-MRA is beneficial for evaluating peripheral artery disease, deep vein thrombosis, and vascular anomalies [39]. Renal NC-MRA is used for diagnosing renal artery stenosis, while cardiac NC-MRA aids in evaluating congenital heart diseases and aortic abnormalities. It is also utilized in the assessment of abdominal vessels, including the mesenteric and hepatic arteries [40]. The absence of contrast agents in NC-MRA eliminates risks like allergic reactions and nephrotoxicity, along with the concerns of monitoring renal function and potential gadolinium deposition. It offers the advantages of shorter scan times, lower costs, and greater accessibility. However, some limitations include lower spatial resolution compared to contrast-enhanced techniques, reduced sensitivity to slow-flowing vessels, and artifacts caused by motion or vessel pulsatility. With ongoing improvements in imaging protocols and technology, NC-MRA has become a vital tool in diagnosing and managing vascular diseases [39].

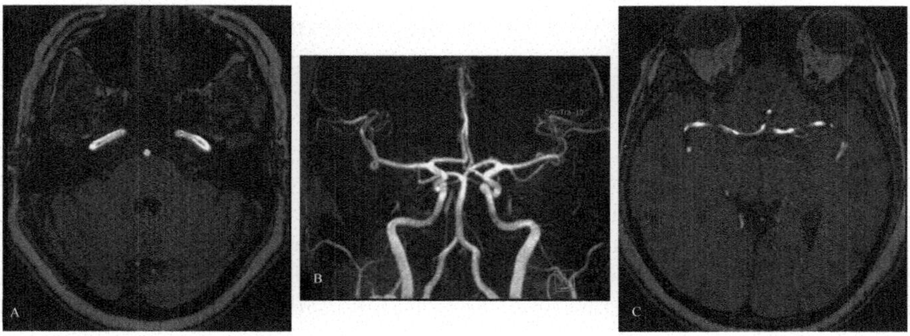

Fig. 12.15 Time of Flight (TOF) MR angiogram images: **A** and **C** representing the source images, and, **B** representing the MIP image

12.6.11 Kinematic MRI

Kinematic magnetic resonance imaging (MRI) is an advanced MRI technique that allows dynamic visualization of joint motion. By capturing a sequence of images in rapid succession it provides insights into the biomechanics and function of joints. Kinematic MRI combines the principles of conventional MRI with motion analysis. Specialized sequences and protocols are used to capture multiple images of a joint in different positions throughout a range of motion. By rapid acquisition of these images, kinematic MRI can create a dynamic representation of joint movement. This technique requires careful coordination between the patient, technologist, and radiologist to ensure accurate and reproducible motion. It has wide range of applications in assessing joint function and diagnosing musculoskeletal disorders. Common uses are in the evaluation of shoulder, knee, hip, spine, and wrist joints. In the shoulder, kinematic MRI can provide valuable information about instability, impingement, and labral tears during active movements such as internal rotation or abduction. In the knee, it aids in assessing ligamentous stability, patellar tracking, and meniscal abnormalities during flexion and extension. In the hip, kinematic MRI helps evaluate femoroacetabular impingement, labral tears, and hip instability during movements like flexion and internal rotation. The main advantage lies in the fact that it enables visualization of joint kinematics during functional movements, providing a more comprehensive understanding of joint mechanics and pathologies. It also allows for better patient-specific treatment planning and monitoring of treatment outcomes. Limitations are that it requires patient cooperation and active participation, which can be challenging for individuals with pain, limited mobility, or cognitive impairments. Additionally, complexity and time required for image acquisition and analysis can limit its availability and accessibility. Kinematic MRI has the potential to enhance understanding of joint mechanics, improve treatment planning, and ultimately contribute to better patient care in the field of musculoskeletal medicine [41].

12.7 Applications of MRI in Various Organs

12.7.1 Brain

Magnetic resonance imaging has revolutionized the field of neuroimaging. The non-invasive nature and ability to provide detailed structural and functional information make it very useful in evaluation of various neurological and neurosurgical conditions. It is widely employed for visualizing brain anatomy in great detail. High-resolution structural imaging techniques, such as T1-weighted and T2-weighted imaging, enable the identification of anatomical structures, including cerebral cortex, white matter tracts, ventricles, basal ganglia, and posterior fossa structures (Fig. 12.16). It further aids in the detection and characterization of pathologies such as developmental disorders, tumors, vascular malformations, and degenerative diseases. In addition, sequences like fluid-attenuated inversion recovery (FLAIR) and susceptibility-weighted imaging (SWI) enhance the visualization of subtle abnormalities such as white matter lesions and microbleeds [42]. Diffusion-weighted imaging (DWI) and diffusion tensor imaging (DTI) techniques allow the assessment of white matter microstructure and fiber connectivity. DWI provides information about random motion of water molecules, aiding in detection of acute ischemic stroke as well as white matter abnormalities. DTI enables visualization of major white matter tracts and the quantification of diffusion anisotropy, aiding in the study of various pathologies such as traumatic brain injury, multiple sclerosis, and neurodevelopmental disorders [43]. Perfusion-weighted imaging techniques, such as arterial spin labeling (ASL) and dynamic susceptibility contrast (DSC) imaging, help to assess cerebral blood flow (CBF) and vascular characteristics. It aids in the evaluation of cerebrovascular diseases, tumor vascularity, and assessment of ischemic penumbra in acute stroke. Perfusion imaging can also help in differentiating tumor recurrence from treatment-related effects, aiding in treatment planning and monitoring [44]. MRI has also improved our understanding of brain function through techniques such as functional magnetic resonance imaging (fMRI). By measuring changes in blood flow and oxygenation, fMRI allows mapping of brain activity during various cognitive tasks or in resting-state conditions. It is also useful in identifying language and motor areas preoperatively, localizing epileptic foci, and studying brain connectivity networks. Resting-state fMRI is used to investigate various intrinsic brain networks and connectivity disruptions in neurological disorders like Alzheimer's disease and schizophrenia [18]. Proton magnetic resonance spectroscopy (1H-MRS) is a technique that evaluates the chemical composition of brain tissue, providing insights into metabolic changes in neurological disorders. 1H-MRS can detect abnormalities in metabolite levels, such as N-acetylaspartate (NAA), choline (Cho), and creatine (Cr), aiding in the diagnosis and monitoring of brain tumors, neurodegenerative diseases, and epilepsy. Its applications encompass the visualization of brain anatomy, detection of pathological changes, functional mapping, and monitoring treatment response [45]. As MRI technology continues to advance, it holds great promise in furthering our understanding of neurological disorders and improving patient care through more precise diagnosis and tailored treatment approaches [42].

12.7.2 Neck

Magnetic resonance imaging helps to provide detailed and comprehensive assessments of various anatomical regions. In neck imaging, MRI plays a crucial role in evaluation of diverse structures, including the cervical spine, soft tissues, blood vessels, and lymph nodes. It is highly valuable in assessing the cervical spine, providing detailed information about the spinal cord, vertebral bodies, intervertebral discs, and surrounding soft tissues (Fig. 12.17). It also aids in diagnosis and

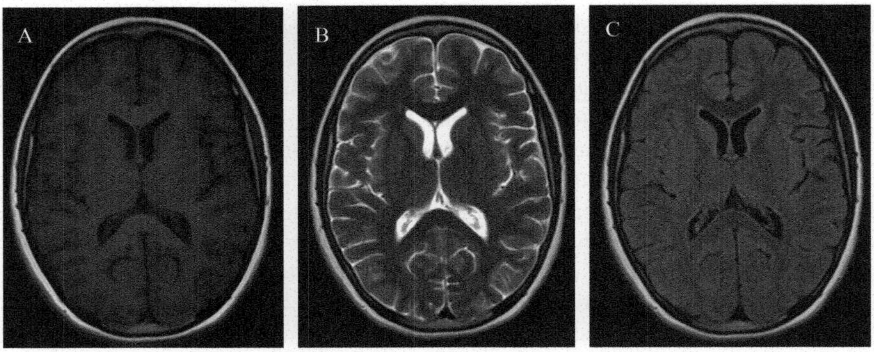

Fig. 12.16 Representative Axial T1W (**A**), T2W (**B**) and FLAIR (**C**) images of brain obtained using a 3T MRI

characterization of spinal conditions such as disc herniation, spinal stenosis, degenerative disc disease, spinal cord compression, and spinal tumors. Additionally, specialized sequences like T2-weighted imaging and fat-suppressed imaging can enhance the visualization of spinal cord lesions, nerve root impingements, and inflammation [46]. It is an essential tool for evaluating various soft tissue structures in the neck, including the muscles, thyroid gland, salivary glands, and paranasal sinuses. Detection and characterization of soft tissue masses, abscesses, inflammation, and tumors, is also another very important use of MRI in head and neck imaging. It can thus be used to differentiate between benign and malignant lesions, help in guiding biopsies, and assist in treatment planning. Dynamic contrast-enhanced sequences can be employed to assess the vascularity and enhancement patterns of tumors or inflammatory processes in the neck region. MRI also plays a significant role in evaluating vascular pathologies in the neck, including the carotid arteries and vertebral arteries [47].

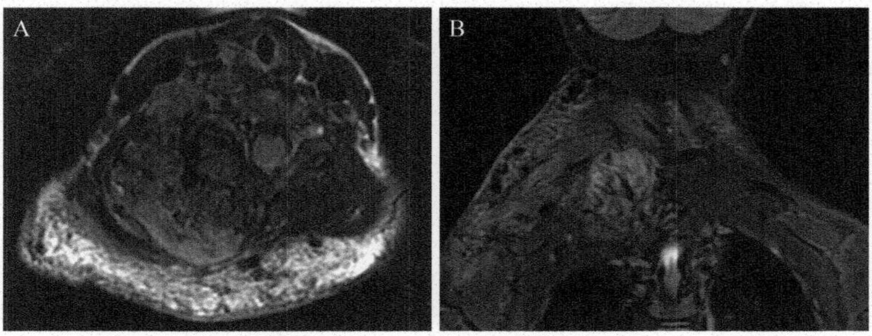

Fig. 12.17 Axial T1W (**A**) and Coronal STIR (**B**) images of MRI neck: Reveals a vascular malformation in the intermuscular plane of right side of neck demonstrating multiple flow voids within

12.7.3 Heart

Cardiac magnetic resonance imaging (MRI) is a useful imaging modality that has become an integral part of cardiovascular diagnostics. It offers detailed information about cardiac anatomy, function, and tissue characteristics, making it invaluable in the evaluation and management of various cardiovascular conditions. It allows precise visualization of the heart chambers, valves, and major blood vessels (Fig. 12.18). High-resolution cine MRI sequences capture dynamic images of the beating heart, enabling accurate assessment of cardiac wall motion, chamber volumes, and ejection fraction. These are vital in diagnosing and monitoring conditions such as cardiomyopathies, myocardial infarction, and heart failure. One of the key strengths of cardiac MRI is its ability to assess myocardial tissue characteristics. With specialized techniques such as T1 and T2 mapping, cardiac MRI can provide quantitative measurements of myocardial edema, inflammation, and fibrosis. This aids in the diagnosis and risk stratification of conditions such as infiltrative cardiomyopathies, myocarditis, and post-infarction scar tissue evaluation. It can also be used to evaluate myocardial perfusion and viability. By administering contrast agents during stress or rest, perfusion defects can be visualized, aiding in the detection of coronary artery disease. Moreover, delayed enhancement imaging allows the identification and quantification of myocardial scar tissue, providing valuable information for surgical planning, predicting recovery after revascularization, and assessing the viability of hibernating myocardium. Furthermore, cardiac MRI excels in assessing cardiovascular anatomy and function in congenital heart diseases. It enables detailed evaluation of complex cardiac structures, great vessels, and intracardiac shunts. By utilizing flow-sensitive sequences, cardiac MRI accurately measures blood flow and quantifies regurgitant or stenotic valves, contributing to preoperative planning and postoperative follow-up in congenital heart disease patients. In addition to its diagnostic capabilities, cardiac MRI plays a vital role in guiding interventional procedures and device implantation. Real-time imaging during cardiac catheterization procedures helps in guiding catheters and devices to the precise location, enhancing safety and accuracy. Cardiac MRI can also assist in the placement of pacemakers, defibrillators, and other cardiac devices by providing detailed anatomical information. With ongoing advancements in imaging technology and techniques, cardiac MRI continues to play a pivotal role in improving patient care and outcomes in the field of cardiology [48].

12.7.4 Lungs

While computed tomography (CT) and X-ray have traditionally been the primary imaging modalities for lung evaluation, magnetic resonance imaging (MRI) has emerged as a promising tool for lung imaging. MRI offers unique advantages, such as superior soft tissue contrast and absence of ionizing radiation, which make it an attractive alternative for lung evaluation. MRI has been employed in the evaluation of

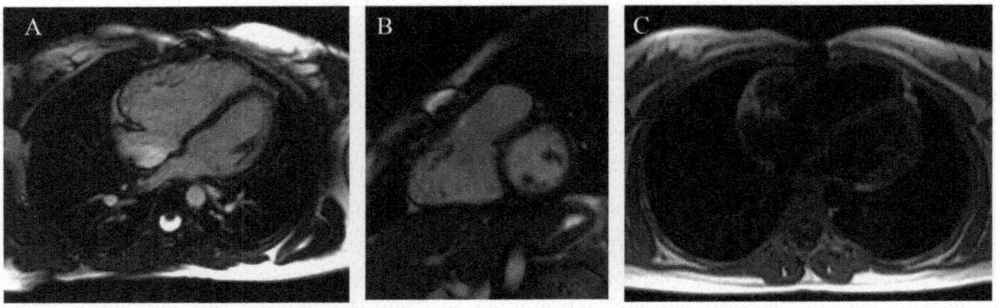

Fig. 12.18 Representative 4 chamber and short axis views of bright blood GRE sequences obtained in cardiac MRI are shown in **A** and **B** respectively; **C** demonstrates 4 chamber dark blood spin echo image

various lung pathologies, including bronchogenic carcinoma, pulmonary embolism, interstitial lung disease, and pulmonary infections. One of the key advantages of MRI is its ability to provide detailed anatomical information while simultaneously visualizing functional aspects of the lung tissue (Fig. 12.19). This capability is particularly useful in assessing lung ventilation, pulmonary perfusion, and tissue oxygenation. Additionally, it can aid in the characterization of lung lesions, differentiating between benign and malignant nodules, and guiding lung biopsies. MRI exploits the differences in tissue signal intensities to generate high-resolution images. This characteristic enables excellent visualization of lung structures, such as the pleura, mediastinum, and airway walls, facilitating the detection as well as characterization of abnormalities. It also allows imaging in multiple planes, providing a comprehensive assessment of lung pathologies from various perspectives. This feature is particularly valuable in evaluating complex lung diseases, such as interstitial lung disease, where a three-dimensional understanding of disease extent is crucial for accurate diagnosis and treatment planning. With the use of specialized techniques like dynamic contrast-enhanced MRI and oxygen-enhanced MRI, functional information such as perfusion, ventilation, and oxygenation can be obtained. This functional imaging capability enables a more comprehensive evaluation of lung diseases, aiding in treatment decisions and monitoring therapeutic response. Despite its many advantages, MRI has some limitations when it comes to lung imaging. One major challenge is the inherent low proton density and low signal-to-noise ratio in the lungs, leading to reduced image quality compared to CT. Additionally, lung motion during the respiratory cycle can cause image artifacts and affect diagnostic accuracy. As research and innovation continue, MRI is likely to play an increasingly significant role in the comprehensive evaluation of lung diseases, ultimately enhancing patient care and outcomes [49].

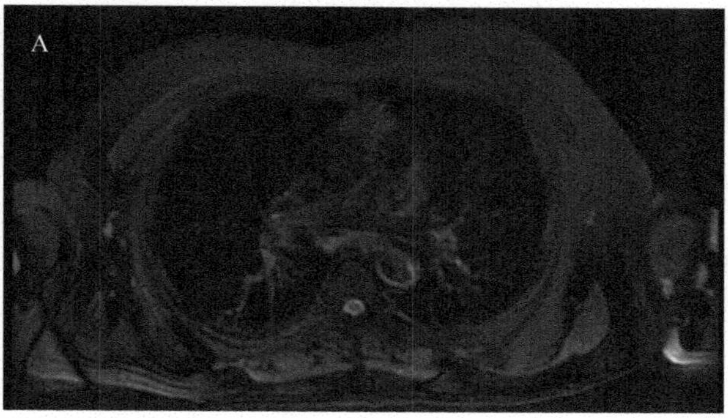

Fig. 12.19 Representative image of a normal free breathing axial radial T2W fat saturated MRI chest (**A**)

12.7.5 Liver

MRI has become an invaluable tool in diagnosis and management of liver diseases. With its ability to provide detailed anatomical and functional information without the use of ionizing radiation, MRI has gained significant popularity in liver imaging. It has a wide range of applications in liver imaging, including the evaluation of liver tumors, liver cirrhosis, liver fibrosis, liver metastases, and liver vascular abnormalities. There is the advantage of excellent soft tissue contrast and visualization of liver lesions with high sensitivity and specificity that it offers. It can aid in the characterization and differentiation of focal liver lesions, providing crucial information necessary for treatment planning and monitoring response to therapy. Additionally, it can also be used to assess hepatic perfusion, hepatic iron content, and biliary anatomy, contributing to a comprehensive evaluation of liver function. Superior soft tissue resolution of MRI is particularly useful in differentiating between liver lesions and adjacent healthy liver tissue, aiding in the diagnosis and staging of liver diseases. T1-weighted, T2-weighted, diffusion-weighted, and multiphasic contrast-enhanced sequences provide complementary information about liver lesions and surrounding tissues (Fig. 12.20). This improves the detection and characterization of liver abnormalities, enhancing diagnostic accuracy. MRI-based techniques such as magnetic resonance elastography (MRE) and diffusion-weighted imaging (DWI) can non-invasively assess liver fibrosis, eliminating the need for invasive liver biopsies. It also allows for dynamic imaging of the hepatobiliary system, providing valuable information on liver function. Techniques such as hepatobiliary phase imaging and MR cholangiopancreatography (MRCP) enable the visualization of the biliary system, aiding in the detection of biliary strictures, gallstones, and bile duct tumors. Quantitative MRI techniques, such as liver iron quantification and liver fat quantification, have gained prominence in assessing liver steatosis and iron overload, respectively. These techniques provide quantitative measurements, allowing for better monitoring of disease progression and treatment response. Advanced MRI techniques, such as

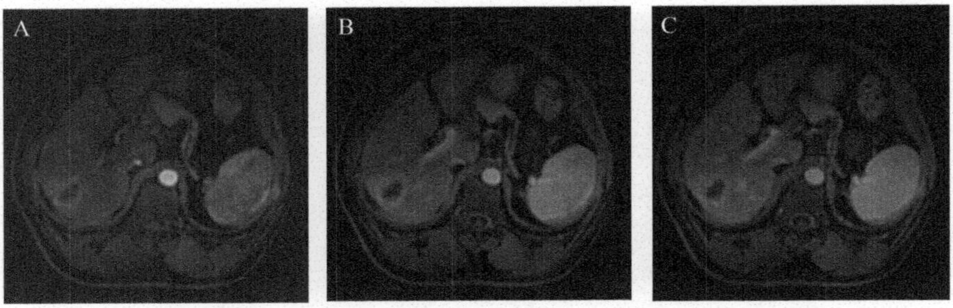

Fig. 12.20 Triple phase hepatic protocol CT: Arterial (**A**), Portal (**B**) and Venous (**C**) phase images demonstrating an arterial phase rim enhancing lesion in segment VI of liver demonstrating persistent enhancement on portal and venous phase images (LIRADS-3; Liver Imaging Reporting and Data System-3)

dynamic contrast-enhanced MRI (DCE-MRI) and arterial spin labeling (ASL), can assess liver perfusion and microvascular characteristics. These functional imaging modalities aid in evaluation of liver tumors and provide valuable insights into tumor angiogenesis. Its versatility in hepatobiliary imaging and recent advancements in technology have further enhanced its utility in the diagnosis, staging, and monitoring of various liver diseases [50].

12.7.6 Lower Limbs

MRI has become an integral tool in evaluation of lower limb pathologies, offering detailed anatomical and functional information. It has wide range of applications in lower limb imaging, including assessment of traumatic injuries, sports-related injuries, degenerative joint diseases, peripheral vascular diseases, and musculoskeletal tumors. It plays a crucial role in evaluating structures such as bones, muscles, tendons, ligaments, cartilage, and blood vessels in the lower extremities. MRI also aids in the detection and characterization of abnormalities, guides surgical planning, and facilitates post-treatment monitoring. Superior soft tissue contrast enables the identification and differentiation of pathologies such as muscle tears, ligamentous injuries, tendon disorders, and soft tissue tumors. The high spatial resolution of MRI enhances diagnostic accuracy, especially when evaluating complex anatomy and subtle abnormalities. Multiplanar imaging helps evaluate joint congruity, identify anatomical variants, and assess the extent of pathology, enabling appropriate therapeutic interventions. MRI is highly sensitive in detecting and characterizing soft tissue abnormalities in the lower limb, and it aids in the evaluation of muscle strains, tendonopathies, ligamentous injuries, and cartilage lesions. MRI can also assist in identifying the presence and extent of inflammatory processes, such as arthritis or cellulitis, providing crucial information for optimal treatment strategies. Magnetic resonance angiography (MRA) plays a significant role in assessing peripheral vascular diseases affecting the lower limb. MRA can non-invasively visualize the arterial and venous structures, aiding in the diagnosis of conditions like peripheral arterial disease, deep vein thrombosis, and vascular malformations. This information guides therapeutic decisions and helps plan interventions such as angioplasty or surgical revascularization [51]. Functional MRI techniques, such as diffusion-weighted imaging (DWI) and dynamic contrast-enhanced MRI (DCE-MRI), offer insights into tissue microstructure and perfusion, respectively. These techniques aid in the assessment of lower limb tumors, peripheral nerve disorders, and ischemic conditions, providing additional information beyond anatomical imaging. MR arthrography involves the injection of a contrast agent into a joint, enhancing visibility of intra-articular structures. This technique improves the diagnostic accuracy in evaluating conditions such as labral tears, cartilage lesions, and ligamentous injuries in the hip, knee, and ankle joints [52].

12.7.7 Joints

MRI has multiple applications in joint imaging, including assessment of traumatic injuries, degenerative conditions, inflammatory diseases, and neoplastic processes. It is used to evaluate joints such as the knee, shoulder, hip, ankle, wrist, and spine. MRI helps in the detection and characterization of ligamentous injuries, meniscal tears, cartilage defects, tendinopathies, synovial disorders, osteoarthritis, rheumatoid arthritis, and bone tumors. It provides information for treatment planning, monitoring disease progression, and assessing therapeutic response. Excellent soft tissue contrast enables the differentiation of various joint structures, such as bones, cartilage, tendons, ligaments, synovium, and surrounding soft tissues (Fig. 12.21). This capability allows for the detection of subtle abnormalities and assessment of pathological changes, facilitating accurate diagnosis and treatment decisions. Multiplanar imaging enhances diagnostic accuracy and aids in the identification of lesions that may be missed in other imaging modalities. Advanced MRI techniques, such as magnetic resonance arthrography and dynamic imaging, provide functional and dynamic information about joint structures. MR arthrography involves the injection of a contrast agent into the joint, enhancing the visualization of intra-articular structures. Dynamic imaging, such as cine MRI, can assess joint movement and stability. These techniques offer additional insights into joint biomechanics and help in the evaluation of conditions like instability, impingement, and adhesive capsulitis. Diffusion weighted imaging (DWI) measures the diffusion of water molecules within tissues and can be applied to joint imaging. It helps in the evaluation of inflammatory processes, tumor characterization, and the detection of bone marrow edema [53, 54].

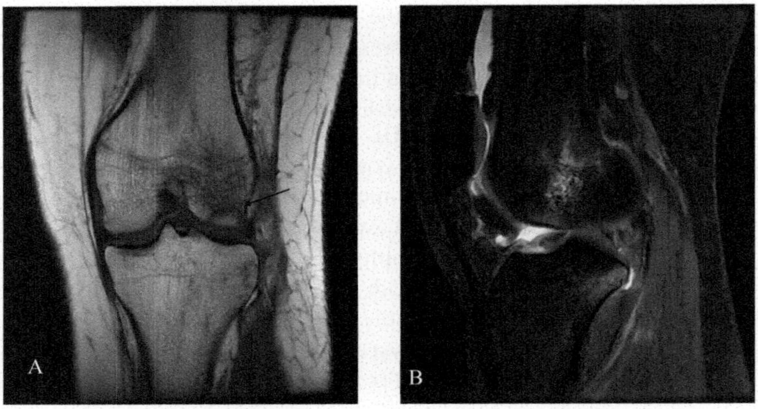

Fig. 12.21 Coronal T1W (**A**) and Sagittal PD FS (**B**) images demonstrating a fracture of the lateral femoral condyle (arrow in **A**) with surrounding marrow edematous changes and joint effusion

12.7.8 Bone Marrow

MRI is used in evaluation of various bone marrow disorders, including benign and malignant neoplastic lesions, infections, metabolic disorders, and bone marrow edema (Fig. 12.22). It is especially useful in the detection as well as characterization of bone marrow malignancies, such as multiple myeloma and lymphoma, and in the evaluation of bone marrow involvement in other malignancies, such as breast and prostate cancer. MRI can also provide valuable information about bone marrow infiltration by systemic diseases, such as sickle cell anemia and Gaucher's disease. In addition to conventional T1-weighted and T2-weighted sequences, others such as Short Tau inversion recovery sequences (STIR) and contrast-enhanced sequences are also very important in various bony pathologies. Advanced MRI techniques, such as diffusion-weighted imaging (DWI) and dynamic contrast-enhanced MRI (DCE-MRI), provide functional and dynamic information about bone marrow. DWI measures diffusion of water molecules within tissues and can be applied to bone marrow imaging to assess tissue viability, edema, and cellularity. DCE-MRI evaluates the perfusion of bone marrow and can help in the detection and characterization of infiltrative lesions. MR spectroscopy enables the assessment of biochemical changes within bone marrow tissues, providing insights into metabolic alterations associated with bone marrow disorders. It aids in the evaluation of conditions like leukemia, lymphoma, and metabolic bone diseases. MR angiography can be applied to bone marrow imaging to assess the vascularity of infiltrative lesions and to evaluate bone marrow perfusion in conditions like sickle cell anemia [55].

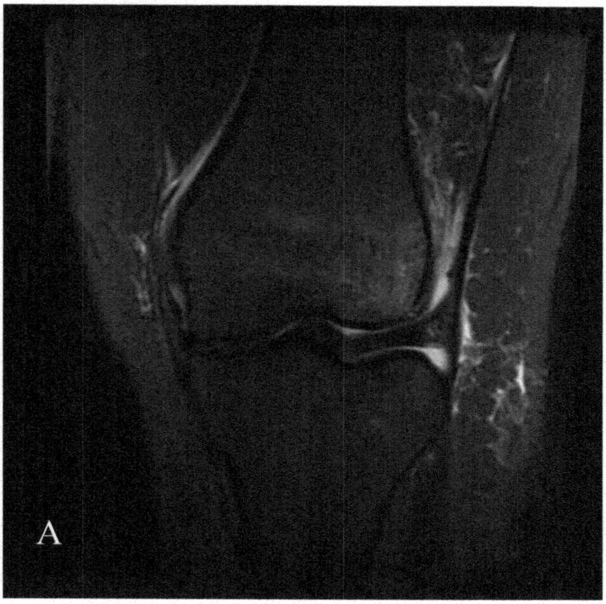

Fig. 12.22 Coronal STIR image (**A**) of knee joint reveals marrow edema in the distal femur predominantly in the lateral condyle, along with joint effusion and surrounding edematous changes

12.8 Conclusion

This article has detailed the role of MRI in current medical practice. Various MR techniques and their clinical applications have been discussed. MR imaging of various body organs has also been mentioned with examples of specific diseases. Before discussing this main topic of clinical MRI the history, physics of MR technology, preclinical MRI, molecular MRI & MR microscopy have been presented in short.

(Pictorial representations courtesy of: Dr. Tania Singh, Senior Resident, Department of Cardiothoracic and Vascular Anaesthesia, SCTIMST, Thiruvananthapuram).

References

1. J.O.S.H. Cleary, A.R. Guimarães, Magnetic Resonance Imaging, in *Pathobiology of Human Disease,* ed. by L.M. McManus, R.N. Mitchell (Academic Press, 2014), pp. 3987–4004
2. A. Hengerer, J. Grimm, Molecular magnetic resonance imaging. Biomed. Imaging Interv. J. **2**, e8 (2006)
3. J. F. Xian, M. Chen, Z. Y. Jin, Magnetic resonance imaging in clinical medicine: current status and potential future developments in China. Chin. Med. J. (Engl.) **128**, 569 (2015)
4. E. van Beek, C. Kuhl, Y. Anzai, P. Desmond, R. Ehman, Q. Gong, Value of MRI in medicine: more than just another test? J. Magn. Reson. Imaging **49**, 1626 (n.d.)

5. F. Bruno, V. Granata, F. Cobianchi Bellisari, F. Sgalambro, E. Tommasino, P. Palumbo, Advanced magnetic resonance imaging (MRI) techniques: technical principles and applications in nanomedicine. Cancers (Basel) **14**, 1626 (n.d.)
6. Z. Zhang, F.L. Zhou, G.L. Davies, G.R. Williams, Theranostics for MRI- guided therapy: recent developments. View **3** (2022)
7. N.F. Ramsay, Phys. Perspect., in Early History of Magnetic Resonance **1**(2), 123–135 (1999)
8. M. Van der Graaf, In vivo magnetic resonance spectroscopy: basic methodology and clinical applications. Eur. Biophys. J. **39**, 27 (2010)
9. R. Loffler, R. Sauter, H. Kolem, A. Haase, M. con Kienlin, Localized spectroscopy from anatomically matched compartments: improved sensitivity and localization for cardiac 31 P MRS in humans. J. Magn. Reson. **134**, 287 (1998)
10. S. Bluml, Magnetic resonance sepctroscopy: basics, in *MR Spectroscopy of Paediatric Brain Disorders* (Springer, New York, 2013), pp.11–23
11. T. Brown, B. Kincaid, K. Ugurbil, NMR chemical shift imaging in three dimensions. Proc. Natl. Acad. Sci. U. S. A. **79**, 3523 (1982)
12. A. Haase, J. Frahm, W. Hanicke, D. Matthaei, 1H NMR chemical shift selective (CHESS) imaging. Phys. Med. Biol. **30**, 341 (1985)
13. R. Ordidge, A. Connelly, J. Lohman, Image-selected in vivo spectroscopy (ISIS) a new technique for spatially selective NMR spectrosocpy. J. Magn. Reson. **66**, 283 (1969)
14. J. Frahm, H. Bruhn, W. Hanicke, K.D. Merboldt, E. Markakis, Localized proton NMR spectroscopy of brain tumors using short-echo time STEAM sequences. J. Comput. Assist. Tomogr. **15**, 915 (1991)
15. R. Faghihi, B. Zeinali-Rafsnajani, M.A. Mosleh-Shirazi, M. Saeedi-Moghadam, M. Lotfi, R. Jalli, Resonance spectroscopy and its clinical applications: a review. J. Med. Imaging Radiat. Sci. **48**, 233 (2017)
16. V. Baliyan, C.J. Das, R. Sharma, A.K. Gupta, Diffusion weighted imaging: technique and applications. World J. Radiol. **8**, 785 (2016)
17. S. Shaikh, A. Kumar, A. Bansal, Diffusion tensor imaging: an overview. Neurol. India **66**, 1603 (2018)
18. G.H. Glover, Overview of functional magnetic resonance imaging. Neurosurg. Clin. N. Am. **22**, 133 (2011)
19. R.B. Buxton, L.R. Frank, A model for the coupling between cerebral blood flow and oxygen metabolism during neural stimulation. J. Cereb. Blood Flow Metab. **17**, 64 (1997)
20. A. Shmuel, E. Yacoub, D. Chaimow, N.K. Logothetis, K. Ugurbil, Spatio-temporal point-spread function of functional MRI signal in human gray matter. Neuroimage **35**, 539 (2007)
21. K.R. Thulborn, J.C. Waterton, P.M. Matthews, G.K. Radda, Oxygenation dependence of the transverse relaxation time of water protons in whole blood at high field. Biochim. Biophys. Acta **714**, 265 (1982)
22. R.L. Buckner, P.A. Bandettini, K.M. O'Craven, R.L. Savoy, S.E. Petersen, M.E. Raichle, B.R. Rosen, Detection of cortical activation during averaged single trials of a cognitive task using functional magnetic resonance imaging. Proc. Natl. Acad. Sci. U. S. A. **93**, 14878 (1996)
23. S.A. Huettal, A.W. Song, G. McCarthy, *Functional Magnetic Resonance Imaging* (Sunderland: Sinauer Associates, Inc, 2004)
24. M. Essig et al., Perfusion MRI: the five most frequently asked technical questions (2013)
25. M.J. Paldino, D.P. Barboriak, Fundamentals of quantitative dynamic contrast-enhanced MR imaging. Magn. Reson. Imaging Clin. N. Am. **17**, 277 (2009)
26. P.S. Tofts et al., Estimating kinetic parameters from dynamic contrast-enhanced t 1-weighted MRI of a diffusable tracer: standardized quantities and symbols. J. Magn. Reson. Imaging **10**, 223 (1999)
27. A.M. Halefoglu, D.M. Yousem, Susceptibility weighted imaging: clinical applications and future directions. World J. Radiol. **10**, 30 (2018)
28. E.M. Haacke, S. Mittal, Z. Wu, J. Neelavalli, Y.-C.N. Cheng, Susceptibility-weighted imaging: technical aspects and clinical applications, part 1. Am. J. Neuroradiol. **30**, 19 (2009)

29. S. Mittal, Z. Wu, J. Neelavalli, E.M. Haacke, Susceptibility-weighted imaging: technical aspects and clinical applications, part 2. Am. J. Neuroradiol. **30**, 232 (2009)
30. A.J. Taylor, M. Salerno, R. Dharmakumar, M. Jerosch-Herold, T1 Mapping basic techniques and clinical applications. JACC Cardiovasc. Imaging **9**, 67 (2016)
31. A.T. O'Brien, K.E. Gil, J. Varghese, O.P. Simonetti, K.M. Zareba, T2 mapping in myocardial disease: a comprehensive review. J. Cardiov. Magn. Reson. **24**(1) (2022)
32. P. Triadyaksa, M. Oudkerk, P.E. Sijens, Cardiac T2* mapping: techniques and clinical applications. J. Magn. Reson. Imaging **52**, 1340 (2020)
33. R. De Mello, Y. Ma, Y. Ji, J. Du, E.Y. Chang, Quantitative MRI musculoskeletal techniques: an update **213**, 524 (2019). https://doi.org/10.2214/AJR.19.21143
34. J. Katta, Z. Jin, E. Ingham, J. Fisher, Biotribology of articular cartilage–a review of the recent advances. Med. Eng. Phys. **30**, 1349 (2008)
35. G.E. Gold, B.A. Hargreaves, S.B. Reeder, S.S. Vasanawala, C.F. Beaulieu, Controversies in protocol selection in the imaging of articular cartilage. Semin. Musculoskelet. Radiol. **9**, 161 (2005)
36. B.K. Paunipagar, D.D. Rasalkar, Imaging of articular cartilage. Indian J. Radiol. Imaging **24**, 237 (2014)
37. A.H. Kuo, P. Nagpal, B.B. Ghoshhajra, S.S. Hedgire, Vascular magnetic resonance angiography techniques. Cardiovasc. Diagn. Ther. **9**, S28 (2019)
38. E.K. Yucel, C.M. Anderson, R.R. Edelman, T.M. Grist, R.A. Baum, W.J. Manning, A. Culebras, W. Pearce, Magnetic resonance angiography. Circulation **100**, 2284 (1999)
39. R.R. Edelman, I. Koktzoglou, Non-contrast MR angiography: an update. J. Magn. Reson. Imaging **49**, 355 (2019)
40. A.J. Wheaton, M. Miyazaki, Non-contrast enhanced MR angiography: physical principles. J. Magn. Reson. Imaging **36**, 286 (2012)
41. F.G. Shellock, C.M. Powers, *Kinematic MRI of the Joints : Functional Anatomy, Kinesiology, and Clinical Application* (n.d.)
42. A.G. Van Der Kolk, J. Hendrikse, J.J.M. Zwanenburg, F. Visser, P.R. Luijten, Clinical applications of 7 T MRI in the brain. Eur. J. Radiol. **82**, 708 (2013)
43. T.A.G.M. Huisman, Diffusion-weighted and diffusion tensor imaging of the brain, made easy. Cancer Imaging **10**, S163 (2010)
44. B.E. McGehee, J.M. Pollock, J.A. Maldjian, Brain perfusion imaging: how does it work and what should i use? J. Magn. Reson. Imaging **36**, 1257 (2012)
45. G. Öz et al., Clinical proton MR spectroscopy in central nervous system disorders. Radiology **270**, 658 (2014)
46. A. Alghamdi, A. Alqahtani, Magnetic resonance imaging of the cervical spine: frequency of abnormal findings with relation to age. Medicines **8**, 77 (2021)
47. F.J. Wippold, Head and neck imaging: the role of CT and MRI. J. Magn. Reson. Imaging **25**, 453 (2007)
48. W.Y.I. Tseng, M.Y.M. Su, Y.H.E. Tseng, Introduction to cardiovascular magnetic resonance: technical principles and clinical applications. Acta Cardiol. Sin. **32**, 129 (2016)
49. H. Hatabu et al., Expanding applications of pulmonary MRI in the clinical evaluation of lung disorders: Fleischner society position paper. Radiology **297**, 286 (2020)
50. K.J. Fowler, J.J. Brown, V.R. Narra, Magnetic resonance imaging of focal liver lesions: approach to imaging diagnosis. Hepatology **54**, 2227 (2011)
51. T.J. Mosher, Musculoskeletal imaging at 3T: current techniques and future applications. Magn. Reson. Imaging Clin. N. Am. **14**, 63 (2006)
52. L. Shapiro, M. Harish, B. Hargreaves, E. Staroswiecki, G. Gold, Advances in musculoskeletal MRI—technical considerations. J. Magn. Reson. Imaging **36**, 775 (2012)
53. T.W. Hash, Magnetic resonance imaging of the knee. Sports Health **5**, 78 (2013)
54. C.J. Burke, H. Alizai, L.S. Beltran, R.R. Regatte, MRI of synovitis and joint fluid. J. Magn. Reson. Imaging **49**, 1512 (2019)
55. M.R. Nouh, A.F. Eid, Magnetic resonance imaging of the spinal marrow: basic understanding of the normal marrow pattern and its variant. World J. Radiol. **7**, 448 (2015)

Chapter 13
Progress and Prospects in Functional MRI (fMRI) Studies of Neural Networks

S. N. Kumar, A. Lenin Fred, H. Ajay Kumar, Parasuraman Padmanabhan, and I. Christina Jane

Abstract Functional magnetic resonance imaging (fMRI) has revolutionized the understanding of neural networks and brain function. It is a powerful non-invasive imaging technique that measures changes in blood flow and oxygenation levels in the brain, enabling the researchers to map neural activity and connectivity. While fMRI traditionally has lower temporal resolution compared to techniques like electroencephalography (EEG) or magnetoencephalography (MEG), recent advances such as multiband imaging have significantly improved its ability to capture rapid changes in neural activity. This is important for studying dynamic network interactions. fMRI has found a wide range of clinical applications, from pre-surgical planning in neurology to understanding the neural basis of mental health disorders. It has opened up new possibilities for personalized treatment approaches. Combining fMRI with other neuroimaging techniques like electroencephalography (EEG), magnetoencephalography (MEG), and positron emission tomography (PET) can provide a more comprehensive view of neural networks. This integration of data sources can enhance our understanding of brain function. Real-time fMRI combined with neurofeedback techniques may enable individuals to attain voluntary influence over their brain functions, potentially offering therapeutic benefits for a range of conditions.

S. N. Kumar (✉)
Department of Electrical and Electronics Engineering, Amal Jyothi College of Engineering, Koovappally, Kerala, India
e-mail: appu123kumar@gmail.com

A. L. Fred
Department of Computer Science and Engineering, Mar Ephraem College of Engineering and Technology, Marthandam, Tamil Nadu, India

H. A. Kumar · I. C. Jane
Department of Electronics and Communication Engineering , Mar Ephraem College of Engineering and Technology, Marthandam, Tamil Nadu, India

P. Padmanabhan
Nanyang Technological University, Singapore, Singapore

N. Kalarikkal et al. (eds.), *Multimodal Biomedical Imaging Techniques*,
Biological and Medical Physics, Biomedical Engineering,
https://doi.org/10.1007/978-981-96-1124-9_13

13.1 Introduction

Functional magnetic resonance imaging (fMRI) is a potent neuroimaging tool that enables the study of brain function by measuring alterations in blood flow and oxygenation levels. Over time, fMRI investigations into neural networks have seen significant advancements, yielding valuable insights into the organization and operation of the human brain. Technological progress has resulted in higher spatial and temporal resolution in fMRI research. This enables researchers to more accurately pinpoint brain activity and explore intricate functional networks [1]. Resting-state fMRI observes spontaneous brain activity when subjects are at rest. This method has uncovered intrinsic connectivity networks (ICNs) that persist even without a specific task. Studies of resting-state connectivity have identified distinct networks like the default mode network (DMN), implicated in self-referential processes and mind-wandering [2].

Task-based fMRI studies entail recording brain activity while subjects engage in specific cognitive tasks. These studies have furnished crucial understandings into the functional changes of the brain for tasks like language processing, motor control, and sensory perception [3]. Merging fMRI with other imaging techniques such as structural MRI, EEG, MEG, and PET provide a more comprehensive understanding of neural networks. This multimodal approach offers complementary information about brain structure, electrophysiological activity, and neurochemical processes [4, 5]. Unlike traditional fMRI studies that assume stable connectivity, recent research has focused on dynamic functional connectivity, examining how networks change over time. This approach can unveil the dynamic reconfiguration of neural networks during different tasks and states [6]. Machine learning methods have been applied to fMRI data for decoding brain states, predicting behavior, and classifying cognitive states. These approaches hold promise for identifying biomarkers in various neurological and psychiatric conditions [7, 8]. MRI has found practical applications in clinical settings for studying and diagnosing various neurological and psychiatric disorders. It has been employed to investigate conditions like Alzheimer's disease, schizophrenia, depression, and autism spectrum disorders [9].

fMRI-based neurofeedback enables individuals to learn to regulate their own brain activity, potentially offering therapeutic benefits for conditions like chronic pain, anxiety, and depression. Additionally, fMRI has been integrated into BCIs, enabling direct communication between the brain and external devices [10]. fMRI has been instrumental in studying how neural networks develop and change throughout the lifespan. It provides insights into how functional connectivity patterns evolve from childhood to adolescence and into old age, shedding light on the neural basis of cognitive and behavioral changes [11]. Large-scale collaborative projects and open-access data repositories have facilitated the sharing and analysis of fMRI data. This has led to more reproducible and robust findings, as well as the development of standardized analysis pipelines[12].

Functional magnetic resonance imaging (fMRI) is instrumental in diagnosing and comprehending various neurodisorders. It furnishes crucial insights into how different brain regions engage in specific tasks or when at rest [13]. fMRI aids in pinpointing specific functions within the brain, such as language processing, motor control, and sensory perception. This is especially critical in surgical planning for conditions like brain tumors or epilepsy, where preserving essential brain functions is paramount [13]. fMRI can evaluate cognitive functions like memory, attention, and executive functions. It helps in comprehending how these functions are impacted in neurodegenerative disorders like Alzheimer's disease. Distinctive patterns of functional connectivity irregularities across various brain regions may serve as indicators of neurodisorders. Resting-state fMRI, which observes spontaneous brain activity, is employed to detect alterations in connectivity patterns observed in conditions like schizophrenia, autism spectrum disorders, and depression. fMRI assists in differentiating between neurological conditions (e.g., stroke and multiple sclerosis) and psychiatric disorders (e.g., schizophrenia and bipolar disorder) by revealing distinct patterns of brain activation and connectivity associated with each [14]. fMRI can track changes in brain function in response to treatment. This is vital in assessing the effectiveness of interventions for disorders like depression, where alterations in brain activity are often evident. Investigation of Pain Perception: Studies using fMRI to explore pain perception have provided insights into how the brain processes and experiences pain [15]. This information is valuable in conditions involving chronic pain. Examination of Neurodevelopmental Disorders: fMRI is employed in the study of disorders like autism spectrum disorders and attention deficit hyperactivity disorder (ADHD), offering insights into the underlying neurobiology and potential biomarkers. fMRI can gauge the extent and impact of traumatic brain injuries (TBI) [16]. It identifies regions of the brain that may have been affected and provides information for rehabilitation planning. fMRI can be used alongside other techniques like EEG to pinpoint regions of abnormal brain activity in epilepsy. This information is crucial for surgical planning to remove or isolate the epileptic focus. fMRI proves as a precious tool for researchers investigating **the** underlying mechanisms of neurodisorders and for pharmaceutical companies engaged in drug development and testing [17]. Section 13.2 focus on application of FMRI in the diagnosis of various neurodisorders and conclusion is drawn in Sect. 13.3.

13.2 Applications of fMRI in the Diagnosis of Various Neuronal Diseases

13.2.1 fMRI in the Detection of Epilepsy

Functional magnetic resonance imaging (fMRI) is a valuable clinical instrument employed to delineate the brain's associations with cognitive functions, particularly in cases of epilepsy and language deterioration following neurosurgery. This chapter is designed to offer an easily comprehensible summary of fMRI, addressing aspects like epilepsy, cognitive evaluation, fMRI principles, statistical methods, and the connections between brain activity and behavior. The primary emphasis is on offering practical advice and pertinent resources and references in this field [18]. Accurately pinpointing the source of epileptic seizures in cases where traditional treatments are ineffective is a complex task. However, recent progress in neuroimaging methods such as non-Gaussian diffusion models and arterial spin labeling, coupled with the application of machine learning and network analysis, holds promise for enhancing the identification and comprehension of the lesions responsible for these seizures in focal epilepsy [19]. This paper provides an overview of the most common anomalies identified through neuroimaging in patients with epilepsy, including conditions such as hippocampal sclerosis, malformations of cortical development, and vascular malformations. The goal is to comprehend how these abnormalities relate to the progression, management, and outlook of epilepsy. Various neuroimaging techniques, including magnetic resonance imaging (MRI), fluorodeoxyglucose positron emission tomography (FDG-PET), quantitative image analysis, magnetic resonance spectroscopy (MRS), functional MRI (fMRI), diffusion tensor imaging (DTI), and fiber tract reconstruction, are employed to detect structural alterations and minor malformations. These techniques contribute to the assessment of surgical treatment options and the prediction of prognosis in epilepsy cases [20]. Neuroimaging is of paramount importance in the diagnosis, ongoing monitoring, and pre-surgical evaluation of pediatric epilepsy, offering both qualitative and quantitative insights into structural and functional irregularities. Cutting-edge MRI methods, such as diffusion-weighted imaging (DWI), functional MRI (fMRI), and diffusion tensor imaging (DTI), prove instrumental in detecting lesions, pinpointing eloquent cortical regions, and locating the source of epileptic seizures, facilitating tailored surgical strategies for each patient. Additionally, other techniques such as PET, SPECT, and electrical and magnetic source imaging can be employed for precise localization and intracranial EEG recording methods, enhancing the accuracy of diagnosis and treatment planning [21]. The different applications of FMRI are depicted in Fig. 13.1.

Neuroimaging methods, particularly magnetic resonance imaging (MRI), are continuously advancing in their ability to identify structural and functional aspects of the brain that are relevant to epilepsy. The integration of these techniques with machine learning is enhancing the detection of subtle epileptogenic lesions and the creation of dependable prognostic tools for epilepsy management [22]. The paper delves into the fundamental principles of evaluating neuropsychological results after epilepsy surgery and investigates how these principles can be adjusted and implemented in functional magnetic resonance imaging (fMRI) research designs. It highlights the existing disparity between sophisticated predictive methodologies and conventional outcome assessments in fMRI studies related to epilepsy surgery, underscoring the importance of optimizing clinical significance in these studies [23].

The aim of this research is to authenticate and validate a rapid analysis technique for assessing patients with pharmacoresistant focal epilepsy as part of their pre-surgical evaluation. The study demonstrates the clinical usefulness of "spike-and-slow-wave-activation-summary" (SSWAS) maps in pinpointing areas for surgical resection that are likely to lead to positive surgical outcomes. To achieve this, the researchers employed simultaneous EEG-fMRI at 3 T to generate subsecond

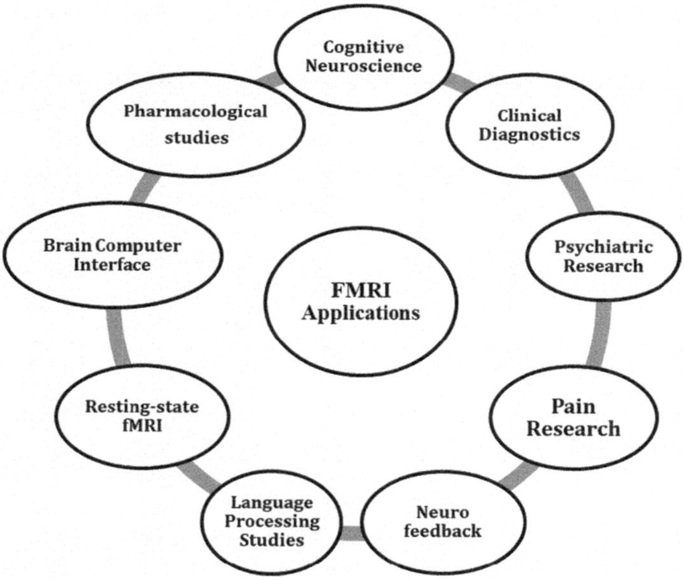

Fig. 13.1 Different applications of fMRI

fMRI activation maps that match the temporal resolution of the EEG data. These SSWAS maps were shown to aid surgeons in the selection of resection areas that have the potential for successful surgical outcomes. The study's results revealed that subsecond EEG-fMRI analysis can assist surgeons in identifying resection areas with a sensitivity of 83.3% and specificity of 75.0% for SSWAS maps [24]. The paper explores recent developments in neuroimaging methods to precisely pinpoint the origin of seizures in pediatric epilepsy. These advancements encompass various techniques, such as structural MRI, ultra-high-field imaging, diffusion tensor imaging, arterial spin-labeling perfusion MRI, simultaneous EEG-fMRI, and magnetoencephalography (MEG). By integrating these advanced imaging modalities with post-processing analyses and machine learning algorithms, it becomes possible to identify subtle structural anomalies, map white matter connections, localize epileptogenic sources, and recognize vital brain regions. This, in turn, helps reduce the risk of neurological deficits following surgical interventions [25].

The aim of the research was to evaluate the contribution of (fMRI) in the management of epilepsy and determine whether the laterality index (LI) derived from fMRI data could be a supplementary or alternative tool for assessing language lateralization and impairment, in place of preoperative neuropsychological testing. The results indicated that language lateralization can vary depending on the specific paradigm used, and the LI can be useful as an additional tool during preoperative assessment. However, it cannot entirely replace the need for traditional neuropsychological testing. Temporal lobe epilepsy (TLE) can result in cognitive side effects, particularly impacting language and memory functions. fMRI has been increasingly employed to noninvasively pinpoint the brain regions involved in language and memory processes, thereby assisting in surgical strategies for patients with TLE [26]. Memory deficits in epilepsy patients can often be specific to the type of material they are trying to remember and can be associated with the location of the brain lesion. Functional MRI (fMRI) can help identify the networks involved in memory and assess how surgical interventions impact these networks. In addition, fMRI is commonly used for the preoperative assessment of epilepsy surgery, particularly in cases where language functions may be affected. It allows clinicians to visualize language-related brain functions, which is crucial for predicting and minimizing postoperative complications related to language impairment [27].

13.2.1.1 Comparison of fMRI with Other Medical Imaging Modalities in the Detection of Epilepsy

Functional magnetic resonance imaging (fMRI) serves as a tool for studying epilepsy and can offer insights into the functional aspects of the brain. To better understand its role, it's helpful to compare fMRI with other common imaging methods used in epilepsy detection. EEG is a primary diagnostic tool for epilepsy, directly recording brain electrical activity and capturing seizure events. While EEG provides high temporal resolution, it lacks spatial information. FMRI, on the other hand, offers superior spatial resolution and can identify brain regions involved in functional processes, complementing EEG findings [28]. Standard MRI is excellent at identifying anatomical abnormalities and structural changes that can contribute to epilepsy. In contrast, fMRI is focused on functional brain activity and does not offer the same level of structural information as conventional MRI. PET scans can reveal areas of abnormal glucose metabolism in the brain, potentially indicating epileptogenic zones. FMRI measures changes in blood flow and oxygenation, providing additional information about functional brain activity. FMRI is non-invasive and safer, as it does not involve radioactive tracers like PET [29]. SPECT is often used for locating epileptogenic foci in ictal and interictal studies. FMRI is non-invasive and offers better spatial resolution than SPECT, making it useful for pinpointing the exact regions involved in epilepsy [30]. MEG measures magnetic fields produced by neural activity and excels in localizing the source of epileptic activity. FMRI provides complementary insights into functional brain activation and networks but lacks MEG's source localization capabilities [31].

DTI investigates white matter tracts and brain connectivity. In contrast, fMRI focuses on functional activation and connectivity, offering information about how different brain regions communicate during cognitive tasks [32]. While fMRI is a powerful tool for studying epilepsy and understanding its functional aspects, it does not replace other imaging techniques like EEG, structural MRI, PET, SPECT, MEG, or DTI [29]. Instead, it complements them by providing insights into functional brain networks and activity, contributing to a more comprehensive understanding of epilepsy and guiding treatment decisions.

13.2.2 fMRI in the Detection of Alzheimer's

The fMRI with multimodalities and different data analysis tools have improved the detection studying the Alzheimer's disease. The hippocampal activation in the medial temporal lobe (MTL) during the picture encoding and long-term memory was demonstrated with fMRI [33] [34]. The patients with Alzheimer's disease and other dementias have less functional signals in the MTL regions when performing learning task with fMRI [35]. The fMRI studies with task-related activation in AD patients, have irreversibly altered brain function for the performing the task [36]. From the resting state fMRI, compared the functional connectivity correlation of the brain with healthy controls and AD patients and concluded AD could potentially disrupt the correlation or anti-correlation effect observed in two inherently anti-correlated networks [37]. The spontaneous brain activity in AD have unbelievable interference in the neural signals at the spatial levels so it seems connectivity type biomarkers can identify the AD at early stages [38].

The fMRI studies in the AD patients has hypoactivation mainly located in visual, default, and ventral attention networks when compared to healthy patients [39]. Functional dynamic imaging biomarkers like task-related activation, task-related deactivation, and resting state activity with different fMRI studies are performed for detection of AD [40]. The Neurovascular dysfunctions related to Alzheimer's disease cerebral blood flow (CBF) dysregulation, cerebral blood volume (CBV), and cerebral metabolic rate of oxygen consumption (CMRO2) are potentially calculated with BOLD fMRI [41]. In the resting state human brain, fMRI with FDG-PET was used for the functional activation of brain was studied for baseline activity [42].The fMRI analysis with SDBOLD as biomarker of Alzheimer's disease can analyze the early disease progression and human coginiton to neurodegenerative diseases [43].

13.2.2.1 Comparison of fMRI with Other Medical Imaging Modalities in the Detection of Alzheimer's

Functional magnetic resonance imaging (fMRI) is used to study the Alzheimer's disease (AD), but it is frequently used alongside other imaging methods to offer a comprehensive view of the condition. While fMRI assesses brain activity by monitoring changes in blood flow and oxygenation, structural MRI produces detailed images of brain structure [44]. Structural MRI is useful for detecting atrophy, structural modifications, and anatomical irregularities in AD, whereas fMRI has the capability to find variations in functional connectivity and networks. By combining structural MRI with fMRI, one can gain insights into both the physical structure and functional aspects of the disease [45]. PET scans utilize radioactive tracers to observe changes in brain metabolism and the presence of amyloid plaques linked to AD. In contrast, fMRI is non-invasive and does not require radioactive tracers, making it a safer and more widely applicable choice. PET can provide quantifiable data about specific AD biomarkers, while fMRI offers information about changes in brain networks and functionality [46]. SPECT is another nuclear medicine method used to measure cerebral blood flow, aiding in AD diagnosis. Similar to PET, SPECT relies on radioactive tracers, whereas fMRI does not expose individuals to radiation. Combining SPECT with fMRI allows for the collection of both functional and perfusion data [47].

DTI, a type of MRI, assesses the diffusion of water molecules in the brain's white matter, offering insights into structural connections. In contrast, fMRI concentrates on functional connectivity and brain activation. Integrating DTI with fMRI provides a more comprehensive perspective on how structural white matter changes are related to functional variations in AD [48]. In summary, fMRI is advantageous in AD research have gained prominence due to its non-invasive nature, its capacity to examine functional connectivity, and its potential to detect early functional changes. However, it is frequently used in conjunction with other imaging modalities like structural MRI, PET, SPECT, and DTI collectively enhance our understanding of the intricate mechanisms behind Alzheimer's disease. Each imaging method offers distinct insights into the condition, and their combination provides a comprehensive assessment for diagnostic, monitoring, and research purposes.

13.2.3 fMRI in the Detection of Parkinson

FMRI is a multimodal approach for diagnosing Parkinson's disease by integrating data from functional magnetic resonance imaging (fMRI) and electroencephalograms (EEGs). Parameters related to EEG synchronization, EEG spectrum power, and resting-state functional connectivity were extracted from the data was put forward in [49]. A feature selection method was applied to identify the most informative features for Parkinson's disease (PD) diagnosis. Subsequently, a support vector machine (SVM) machine learning classifier was utilized to categorize the selected features. The approach achieved a classification accuracy of 93.5%, a notable improvement over earlier methods, when using either EEG or fMRI data alone. This underscores the effectiveness of the multimodal strategy in enhancing the accuracy of Parkinson's disease diagnosis.

The objective of the research presented in [50] was to create a predictive marker for motor development using resting-state fMRI (rs-fMRI) in recently diagnosed, medication-naive Parkinson's disease (PD) patients. The study included 27 healthy controls and 48 PD patients. The fractional amplitude of low-frequency fluctuations (fALFF) was computed for each patients. Using the least absolute shrinkage and selection operator (LASSO) regression model, two significant fALFF features were identified: fALFF in the putamen and fALFF in the precentral gyrus. The model, designed to differentiate between rapid and slow progressors based on the yearly increase in motor scores, achieved an (AUC) of 0.897. This suggests the potential of the identified fALFF features as a valuable marker for predicting motor development in recently diagnosed PD patients who have not yet initiated medication.

The study [51] investigated changes in the functional connectivity within the fear circuit of the brain in individuals with anxiety who also have Parkinson's disease (PD). The research employed seed-based functional connectivity analysis and rs-fMRI on a sample of thirty PD patients with anxiety and thirty without anxiety. The results revealed a reduction in the functional connectivity of the amygdala with the insula, prefrontal cortex, and hippocampal regions—key areas involved in the processing and regulation of fear. The findings suggest that there are changes in the neural circuitry associated with fear in individuals with Parkinson's disease and comorbid anxiety, although the specific relationship between alterations in the fear circuit and anxiety in Parkinson's disease remains unclear.

The primary objective of the study in [52] was to employ a sliding window approach to examine the temporal dynamics of Regional Homogeneity (ReHo) changes in individuals with Parkinson's disease (PD). rs-fMRI scans were acquired, involving 36 PD patients and 36 healthy controls. The ReHo value for each voxel was determined using a sliding window method. The temporal dynamics were assessed through a two-way repeated-measures analysis of variance (ANOVA). The findings revealed notable group x time interactions in the putamen, precentral gyrus, and supplementary motor region. Specifically, PD patients exhibited a decreasing trend in ReHo over time in these regions. This suggests that there are dynamic alterations in ReHo in certain brain regions in PD patients, highlighting the importance of investigating temporal changes for a more comprehensive understanding of the disease-related neural dynamics.

13.2.3.1 Comparison of fMRI with Other Medical Imaging Modalities in the Detection of Parkinson

Functional magnetic resonance imaging (fMRI) is a potent research tool for studying brain function in the context of Parkinson's disease. However, it is vital to note that fMRI is predominantly utilized for academic and scientific investigations to gain insights into the neural mechanisms associated with the disease. It is not typically employed for clinical diagnosis or routine medical imaging.

fMRI offers a deep understanding of brain function and connectivity, aiding researchers in unraveling the neural changes linked to Parkinson's disease. It is non-invasive and does not expose patients to ionizing radiation; provides high-resolution spatial mapping of brain activity; mainly accessible in research settings and not commonly used in clinical practice; and cannot directly visualize pathological brain changes. PET directly measures dopamine levels in the brain, a crucial aspect of Parkinson's disease. It is widely available for the diagnosis and evaluation of Parkinson's disease in clinical settings and it involves exposure to radiation due to the use of radioactive tracers. Also, it offers lower spatial resolution compared to fMRI [53].

SPECT assesses dopamine receptor function and cerebral blood flow, contributing to the diagnosis of Parkinson's disease; also involves exposure to ionizing radiation;

and provides limited spatial resolution. MRI supplies detailed structural information about the brain, aiding in the exclusion of other conditions that may mimic Parkinson's disease symptoms and is widely available and frequently utilized in clinical practice. It cannot directly measure functional changes or dopamine levels in the brain [30].

CT offers structural brain images but is less commonly used in the identification of Parkinson's disease. It involves exposure to ionizing radiation, similar to PET and SPECT. In summary, fMRI is a valuable instrument for scientific exploration of the functional aspects of Parkinson's disease and gaining insights into the underlying brain mechanisms. However, for clinical diagnosis and routine assessment, other imaging techniques such as PET and SPECT, which can directly measure dopamine levels and assess functional changes in the brain, are more commonly employed. Structural imaging techniques like MRI and CT are also crucial for ruling out other conditions and providing complementary information [54].

13.2.4 fMRI in the Detection of Schizophrenia

In [55], researchers developed a deep learning method for diagnosing schizophrenia using rs-fMRI data. They utilized a hybrid model that combined a long short-term memory (LSTM) recurrent network with a three-dimensional convolutional neural network (3DCNN) to capture both temporal patterns from functional connectivity (FC) characteristics and spatial information from activity maps. The approach was evaluated on the COBRE dataset, which included 98 subjects (58 with schizophrenia and 40 healthy controls). The results demonstrated a classification accuracy of 92.3%, surpassing previous methods and highlighting the potential of deep learning in schizophrenia diagnosis based on rsfMRI data.

A novel deep learning architecture named GRU-CNN was introduced for the multimodal diagnosis of schizophrenia in [56]. The GRU-CNN model combines convolutional neural networks (CNNs) and gated recurrent unit (GRU) networks to identify both spatial and temporal patterns in data. The CNN component focuses on extracting spatial information from low-level features in fMRI, MRI, or DTI data, while the GRU component evaluates functional connectivity measurements from fMRI data. The integration of these features is accomplished using a support vector machine (SVM) classifier to classify the data. The GRU-CNN architecture was evaluated on the COBRE dataset, which included fMRI, MRI, and DTI data from 98 subjects (58 diagnosed with schizophrenia, 40 healthy controls). The results demonstrated exceptional performance, with an area under the curve (AUC) of 99.7% and a remarkable classification accuracy of 99.50%, suggesting that the GRU-CNN architecture holds great promise as a highly effective method for diagnosing schizophrenia.

The objective of the analysis [57] was to employ whole-brain rs-fMRI for discriminating between individuals with schizophrenia and those without the disorder. Four distinct fMRI parameters—regional homogeneity, voxel-mirrored homotopic connection, fractional amplitude of low-frequency fluctuations, and amplitude of low-frequency fluctuations—were integrated. The proposed feature fusion method demonstrated superior performance compared to models based on individual features, achieving an average test accuracy of 98.71%. This underscores the importance of combining multiple fMRI parameters for a more dependable and precise approach to identifying schizophrenia.

The objective of the study [58] was to leverage deep learning methodologies for the classification of fMRI data, with the goal of enhancing the diagnostic accuracy of schizophrenia. The researchers utilized the VGG16 deep learning method to analyze fMRI data from 82 subjects, evenly split between 41 individuals with schizophrenia and 41 healthy controls. The algorithm surpassed conventional techniques, achieving an impressive classification accuracy of 84.3%. This highlights the transformative potential of deep learning techniques in revolutionizing the diagnosis of schizophrenia by analyzing fMRI data and mitigating the subjective nature of clinical assessments for the disorder.

In [59], a two-phase method was devised for the automated diagnosis of attention deficit hyperactivity disorder (ADHD) and schizophrenia based on rs-fMRI data. The first phase involved the use of convolutional autoencoder (CAE) models, which encode input data into a lower-dimensional representation and then reconstruct the original data. In the second phase, the extracted features were classified into schizophrenia (SZ), ADHD, or healthy control groups using Interval Type-2 Fuzzy Regression (IT2FR). The approach was tested on rs-fMRI data from 140 subjects in the ADHD-200 dataset. Demonstrating superior performance compared to previous techniques, the results showcased a high classification accuracy of 92.5% for schizophrenia diagnosis and 87.5% for ADHD diagnosis. This underscores the effectiveness of the novel method in accurately diagnosing schizophrenia and ADHD using rs-fMRI data.

13.2.4.1 Comparison of fMRI with Other Medical Imaging Modalities in the Detection of Schizophrenia

Functional magnetic resonance imaging (fMRI) is one of the imaging methods used for studying schizophrenia, but it's essential to compare it with other commonly used medical imaging techniques in detecting and researching this disorder. While fMRI focuses on brain activity, sMRI primarily offers information about brain structure. fMRI tracks dynamic changes during tasks, whereas sMRI provides static images. Combining fMRI and sMRI allows for a comprehensive understanding of schizophrenia by considering both functional and structural aspects [60]. PET employs radioactive tracers to measure brain activity, while fMRI relies on changes in blood flow. fMRI is non-invasive and does not expose patients to ionizing radiation, making it safer for repeat scans. PET may have higher sensitivity in detecting certain neurochemical abnormalities, but fMRI offers superior spatial resolution [61].

SPECT, like PET, is a nuclear medicine imaging method used to assess brain perfusion and function. In contrast to fMRI, SPECT involves radiation exposure. For localized brain function, fMRI's better spatial resolution is advantageous [62]. EEG and MEG directly measure electrical brain activity, providing high temporal resolution. In contrast, fMRI primarily examines the functional aspects of the brain and offers better spatial resolution. Combining fMRI with EEG or MEG can provide a comprehensive understanding of both the timing and location of brain activity changes in schizophrenia [63]. DTI is a specialized MRI technique used to study white matter structure and connectivity. While fMRI primarily focuses on brain function, DTI explores structural connections [48].

Combining fMRI and DTI can offer insights into both functional and structural changes in the brain related to schizophrenia. In summary, fMRI is a valuable tool for schizophrenia research due to its ability to provide good spatial resolution and capture dynamic changes in brain function. However, it is often used alongside other imaging methods, such as structural MRI, PET, SPECT, EEG, MEG, and DTI, collectively enhancing our understanding of the structural and functional changes associated with the disorder. The choice of imaging technique depends on the specific research objectives and the aspects of schizophrenia being investigated.

13.3 Conclusion

FMRI has made significant progress in mapping neural networks and understanding brain function. Ongoing research and emerging technologies promise even more insights into the organization, dynamics, and clinical relevance of these networks. The future of fMRI studies in neural networks looks promising and holds the potential for transformative discoveries in neuroscience. Advancements in fMRI are expected to help map specific cognitive functions to neural networks with increasing precision, shedding light on the neural basis of complex cognitive processes. Real-time fMRI combined with neurofeedback techniques may enable individuals to gain voluntary control over their brain activity, potentially offering therapeutic benefits for a range of conditions.

References

1. R.S. Menon, B.G. Goodyear, Spatial and temporal resolution in fMRI, in *Functional MRI: An Introduction to Methods* (2001)
2. Y. Luo, F. Kong, S. Qi, X. You, X. Huang, Resting-state functional connectivity of the default mode network associated with happiness. Soc. Cogn. Affect. Neurosci. **11**(3) (2016)
3. S.A.H. Batouli, M. Sisakhti, Some points to consider in a task-based fMRI study: a guideline for beginners. Front. Biomed. Technol. (2020)
4. Z. Liu, L. Ding, B. He, Integration of EEG/MEG with MRI and fMRI. IEEE Eng. Med. Biol. Mag. **25**(4) (2006)

5. B. He, Z. Liu, Multimodal functional neuroimaging: integrating functional MRI and EEG/MEG. IEEE Rev. Biomed. Eng. **1** (2008)
6. F.A. Espinoza, N.E. Anderson, V.M. Vergara, C.L. Harenski, J. Decety, S. Rachakonda, E. Damaraju et al., Resting-state fMRI dynamic functional network connectivity and associations with psychopathy traits. NeuroImage: Clin. **24**, 101970 (2019)
7. J. Li, P. Zhao, Deep learning applications in fMRI–a review work, in *Proceedings of the 2023 13th International Conference on Bioscience, Biochemistry and Bioinformatics* (2023), pp. 75–80
8. J. Peedicayil, Identification of biomarkers in neuropsychiatric disorders based on systems biology and epigenetics. Front. Genet. **10**, 469644 (2019)
9. M. Quaak, L. van de Mortel, R. Mani Thomas, G. van Wingen, Deep learning applications for the classification of psychiatric disorders using neuroimaging data: systematic review and meta-analysis. NeuroImage: Clin. **30**, 102584 (2021)
10. H. Marzbani, H. Reza Marateb, M. Mansourian, Neurofeedback: a comprehensive review on system design, methodology and clinical applications. Basic Clin. Neurosci. **7**(2) (2016)
11. M. López-Vicente, O. Agcaoglu, L. Pérez-Crespo, F. Estévez-López, J. María Heredia-Genestar, R.H. Mulder, J.C. Flournoy et al., Developmental changes in dynamic functional connectivity from childhood into adolescence. Front. Syst. Neurosci. **15**, 724805 (2021)
12. R.N. Boubela, K. Kalcher, W. Huf, C. Našel, E. Moser, Big data approaches for the analysis of large-scale fMRI data using apache spark and GPU processing: a demonstration on resting-state fMRI data from the human connectome project. Front. Neurosci. **9**, 165497 (2016)
13. R.L. Savoy, Functional magnetic resonance imaging (fMRI). Encycl. Neurosci. (1999)
14. E. Caantio, D. Chen, B. Biswal, A review of resting-state fMRI and its use to examine psychiatric disorders. Psychoradiology **1**(1) (2021)
15. D. Borsook, L.R. Becerra, Breaking down the barriers: fMRI applications in pain, analgesia and analgesics. Mol. Pain **2** (2006)
16. N.P. Ryan, C. Catroppa, S.C. Ward, K.O. Yeates, L. Crossley, M. Hollenkamp, S. Hearps, M.H. Beauchamp, V.A. Anderson, Association of neurostructural biomarkers with secondary attention-deficit/hyperactivity disorder (ADHD) symptom severity in children with traumatic brain injury: a prospective cohort study. Psychol. Med. **53**(11) (2023)
17. R.G. Wise, I. Tracey, The role of fMRI in drug discovery. J. Magn. Reson. Imaging Off. J. Int. Soc. Magn. Reson. Med. **23**(6) (2006)
18. C.F.A. Benjamin, K. Gkiatis, G.K. Matsopoulos, K. Garganis, Presurgical language fMRI in epilepsy: an introduction. Transl. Neurosci. Speech Lang. Disord. (2020)
19. D. Sone, Making the invisible visible: advanced neuroimaging techniques in focal epilepsy. Front. Neurosci. **15**, 699176 (2021)
20. B. Adamczyk, K. Węgrzyn, T. Wilczyński, J. Maciarz, N. Morawiec, M. Adamczyk-Sowa, The most common lesions detected by neuroimaging as causes of epilepsy. Medicina **57**(3) (2021)
21. A. De Vito, K. Mankad, S. Pujar, A. Chari, D. Ippolito, F. D'Arco, Narrative review of epilepsy: getting the most out of your neuroimaging. Transl. Pediatr. **10**(4), 1078 (2021)
22. N. Bernasconi, I Wang, Emerging trends in neuroimaging of epilepsy. Epilepsy Curr. **21**(2) (2021)
23. S. Baxendale, What are we really predicting with fMRI in epilepsy surgery? Epilepsy Behav. **145**, 109298 (2023)
24. Y. Ito, S. Maesawa, E. Bagarinao, Y. Okai, D. Nakatsubo, H. Yamamoto, H. Kidokoro et al., Subsecond EEG-fMRI analysis for presurgical evaluation in focal epilepsy. J. Neurosurg. **134**(4) (2020)
25. Y.L. Lee, Advanced neuroimaging techniques for evaluating pediatric epilepsy. Clin. Exp. Pediatr. **63**(3) (2020)
26. A. Prabhakar, N. Khandelwal, S. Vyas, V. Gupta, P.S. Kharbanda, M. Mohanty, Functional MRI in epilepsy–comparison of lateralization index and language scoring. Indian J. Radiol. Imaging **29**(2) (2019)

27. V. Schmidbauer, S. Bonelli, Recent developments in cognitive fMRI for temporal lobe epilepsy. Zeitschrift für Epileptologie **33**(1) (2020)
28. B. He, Z. Liu, Multimodal functional neuroimaging: integrating functional MRI and EEG/MEG. IEEE Rev. Biomed. Eng. **1** (2008)
29. R.N. Annavarapu, S. Kathi, V.K. Vadla, Non-invasive imaging modalities to study neurodegenerative diseases of aging brain. J. Chem. Neuroanat. **95** (2019)
30. A.O. Ceballos-Baumann, Functional imaging in Parkinson's disease: activation studies with PET, fMRI and SPECT. J. Neurol. **250**(Suppl 1) (2003)
31. E.L. Hall, S.E. Robson, P.G. Morris, M.J. Brookes, The relationship between MEG and fMRI. Neuroimage **102**(Pt 1), 80–91 (2014)
32. K. Zhang, B. Johnson, D. Pennell, W. Ray, W. Sebastianelli, S. Slobounov, Are functional deficits in concussed individuals consistent with white matter structural alterations: combined FMRI & DTI study. Exp. Brain Res. **204** (2010)
33. L.R. Squire, S. Zola-Morgan, The medial temporal lobe memory system. Science **253**, (5026) (1991)
34. C.E. Stem, S. Corkin, R.G. González, A.R. Guimaraes, J.R. Baker, P.J. Jennings, C.A. Carr, R.M. Sugiura, V. Vedantham, B.R. Rosen, The hippocampal formation participates in novel picture encoding: evidence from functional magnetic resonance imaging. Proc. Natl. Acad. Sci. **93**(16) (1996)
35. S.A.R.B. Rombouts, F. Barkhof, D.J. Veltman, W.C.M. Machielsen, M.P. Witter, M.A. Bierlaagh, R.H.C. Lazeron, J. Valk, P. Scheltens, Functional MR imaging in Alzheimer's disease during memory encoding. Am. J. Neuroradiol. **21**(10) (2000)
36. Y. Stern, J.R. Moeller, K.E. Anderson, B. Luber, N.R. Zubin, A. A. DiMauro, A. Park et al., Different brain networks mediate task performance in normal aging and AD: defining compensation. Neurology **55**(9) (2000)
37. K. Wang, M. Liang, L. Wang, L. Tian, X. Zhang, K. Li, T. Jiang, Altered functional connectivity in early Alzheimer's disease: a resting-state fMRI study. Hum. Brain Mapp. **28**(10) (2007)
38. C. Sorg, V. Riedl, R. Perneczky, A. Kurz, A.M. Wohlschlager, Impact of Alzheimer's disease on the functional connectivity of spontaneous brain activity. Curr. Alzheimer Res. **6**(6) (2009)
39. H.J. Li, X.H. Hou, H. Liu, C.L. Yue, Y. He, X.N. Zuo, Toward systems neuroscience in mild cognitive impairment and Alzheimer's disease: a meta-analysis of 75 fMRI studies. Hum. Brain Mapp. **36**(3) (2015)
40. D. Prvulovic, A.L.W. Bokde, F. Faltraco, H. Hampel, Functional magnetic resonance imaging as a dynamic candidate biomarker for Alzheimer's disease. Prog. Neurobiol. **95**(4) (2011).
41. A. Montagne, D.A. Nation, J. Pa, M.D. Sweeney, A.W. Toga, B.V. Zlokovic, Brain imaging of neurovascular dysfunction in Alzheimer's disease. Acta Neuropathol. **131** (2016)
42. T. Watabe, J. Hatazawa, Evaluation of functional connectivity in the brain using positron emission tomography: a mini-review. Front. Neurosci. **13**, 455975 (2019)
43. V. Scarapicchia, E.L. Mazerolle, J.D. Fisk, L.J. Ritchie, J.R. Gawryluk, Resting state BOLD variability in Alzheimer's disease: a marker of cognitive decline or cerebrovascular status?. Front. Aging Neurosci. 10 (2018)
44. P.M. Matthews, P. Jezzard, Functional magnetic resonance imaging. J. Neurol. Neurosurg. Psychiatry **75**(1) (2004)
45. C. Wu, F. Ferreira, M. Fox, N. Harel, J. Hattangadi-Gluth, A. Horn, S. Jbabdi et al., Clinical applications of magnetic resonance imaging based functional and structural connectivity. Neuroimage **244**, 118649 (2021)
46. A. Otte, U. Halsband, Brain imaging tools in neurosciences. J. Physiol. Paris **99**(4–6) (2006)
47. B.F. Hutton, M. Occhipinti, A. Kuehne, D. Máthé, N. Kovács, H. Waiczies, K. Erlandsson et al., Development of clinical simultaneous SPECT/MRI. Br. J. Radiol. **91**(1081), 20160690 (2018)
48. D. Zhu, T. Zhang, X. Jiang, X. Hu, H. Chen, N. Yang, J. Lv, J. Han, L. Guo, T. Liu, Fusing DTI and fMRI data: a survey of methods and applications. NeuroImage **102**(2014)
49. G.W. Jiji, A. Rajesh, M. MahaLakshmi, Diagnosis of Parkinson's disease using EEG and fMRI. Multimedia Tools Appl. **82**(10) (2023)

50. Y. Hou, L. Zhang, R. Ou, Q. Wei, X. Gu, K. Liu, J. Lin et al., Motor progression marker for newly diagnosed drug-naïve patients with Parkinson's disease: a resting-state functional MRI study. Hum. Brain Mapp. **44**(3) (2023)
51. K. Chen, L. Zhang, F. Wang, H. Mao, Q. Tang, G. Shi, Y. You, Q. Yuan, B. Chen, X. Fang, Altered functional connectivity within the brain fear circuit in Parkinson's disease with anxiety: a seed-based functional connectivity study. Heliyon **9**(5) (2023)
52. K. Li, Y. Tian, H. Chen, X. Ma, S. Li, C. Li, S. Wu, F. Liu, Y. Du, W. Su, Temporal dynamic alterations of regional homogeneity in Parkinson's disease: a resting-state fMRI study. Biomolecules **13**(6) (2023)
53. N.S.R. Bidesi, I. Vang Andersen, A.D. Windhorst, V. Shalgunov, M.M. Herth, The role of neuroimaging in Parkinson's disease. J. Neurochem. **159**(4) (2021)
54. T. Mortezazadeh, H. Seyedarabi, B. Mahmoudian, J.P. Islamian, Imaging modalities in differential diagnosis of Parkinson's disease: opportunities and challenges. Egypt. J. Radiol. Nucl. Med. **52** (2021)
55. M. Ghanbari, A.H. Pilevar, N. Bathaeian, Diagnosis of schizophrenia using brain resting-state fMRI with activity maps based on deep learning. Sign. Image Video Process. **17**(1) (2023)
56. B. Masoudi, S. Danishvar, Deep multi-modal schizophrenia disorder diagnosis via a GRU-CNN architecture (2022)
57. A.H. Algumaei, R.F. Algunaid, M.A. Rushdi, I.A. Yassine, Feature and decision-level fusion for schizophrenia detection based on resting-state fMRI data. Plos One **17**(5), e0265300 (2022)
58. J.C. Zheng, X.L. Wei, J.Y. Wang, H.S. Lin, H. Pan, Y. Shi, Diagnosis of schizophrenia based on deep learning using fMRI. Comput. Math. Methods Med. (2021)
59. A. Shoeibi, N. Ghassemi, M. Khodatars, P. Moridian, A. Khosravi, A. Zare, J.M. Gorriz, A. Hossein Chale-Chale, A. Khadem, U. Rajendra Acharya, Automatic diagnosis of schizophrenia and attention deficit hyperactivity disorder in rs-fMRI modality using convolutional autoencoder model and interval type-2 fuzzy regression. Cogn. Neurodyn. **17**(6) (2023)
60. R. Barnaly, M.R. Arbabshirani, E. Damaraju, M.S. Cetin, R. Miller, G.D. Pearlson, V.D. Calhoun, Classification of schizophrenia and bipolar patients using static and dynamic resting-state fMRI brain connectivity. Neuroimage **134** (2016)
61. V. George, The physiological processes underpinning PET and fMRI techniques with an emphasis on the temporal and spatial resolution of these methods. Psychol. Thought **6**(2) (2013)
62. B. Mohammed, M. Saqr, H. Hoffman, D. Patterson, S. Sharar, S. Minoshima, D.H. Lewis, Utility of SPECT functional neuroimaging of pain. Front. Psychiatry **12**, 705242 (2021)
63. M.S. Cetin, J.M. Houck, B. Rashid, O. Agacoglu, J.M. Stephen, J. Sui, J. Canive et al., Multimodal classification of schizophrenia patients with MEG and fMRI data using static and dynamic connectivity measures. Front. Neurosci. **10** (2016)

Chapter 14
Biomedical Imaging: Scope for Future Studies and Applications

Sukumaran Anil, Biba Vikas, Nebu George Thomas, and Vishnupriya K. Sweety

Abstract Biomedical imaging has revolutionized medical research and practice by enabling non-invasive visualization of anatomical structures, physiological processes, and molecular interactions within living systems. This chapter provides a comprehensive overview of the fundamental technologies and advanced applications of biomedical imaging in research. It covers the key modalities, including X-ray radiography, computed tomography (CT), magnetic resonance imaging (MRI), ultrasound, positron emission tomography (PET), single photon emission computed tomography (SPECT), and optical imaging techniques. The chapter highlights their unique capabilities in disease diagnosis, treatment monitoring, drug development, and understanding biological mechanisms. It also explores cutting-edge applications in oncology, neuroscience, and regenerative medicine. The integration of artificial intelligence, nanotechnology, and hybrid imaging systems is discussed as future directions that promise to enhance the power and precision of biomedical imaging. Furthermore, the chapter emphasizes the expanding role of molecular imaging in personalized medicine and theranostics. Despite the immense potential, challenges related to data management and ethical considerations are also addressed. Overall, this chapter underscores the critical role of biomedical imaging in driving scientific discoveries, improving patient care, and shaping the future of medical research.

S. Anil (✉)
Department of Dentistry, Hamad Medical Corporation, Oral Health Institute, Qatar University, Doha, Qatar
e-mail: drsanil@gmail.com; asukumaran1@hamad.qa

B. Vikas
Department of Medicine - Hematology & Medical Oncology, Ellis Fischel Cancer Center, University of Missouri, Columbia, MO 65211, USA

N. G. Thomas · V. K. Sweety
Pushpagiri Institute of Medical Sciences, Thiruvalla 689101, Kerala, India

N. Kalarikkal et al. (eds.), *Multimodal Biomedical Imaging Techniques*,
Biological and Medical Physics, Biomedical Engineering,
https://doi.org/10.1007/978-981-96-1124-9_14

14.1 Introduction

Modern biomedical research relies heavily on advanced imaging techniques to penetrate living organisms and visualize disease pathways, treatment responses, regenerative processes, and fundamental biological phenomena [1]. Researchers are leveraging biomedical imaging's unique capabilities for interrogation across scales, from model organisms to human subjects, while minimally perturbing native physiology. These non- or minimally invasive imaging modalities catalyze breakthrough discoveries across life science and medical research [2]. Biomedical imaging modalities foster paradigm shifts in research workflows across domains beyond typical clinical applications. Continued progress expanding the envelope of imaging resolutions, depth penetration, assay multiplexing, and computational quantitative analytics presages ever-broadening utility in revealing biological mechanisms, improving medical treatments, and furthering patient health [3]. Augmenting anatomical imaging with real-time functional, cellular, and even molecular level information clarifies connections across genes, molecular signals, cell behaviors, and tissue anomalies preceding and underlying whole-body outcomes [4].

Hybrid techniques combining anatomical, functional, and molecular imaging with external physical recorders, actuators, and stimulators will clarify physiological signaling networks driving outcomes in extraordinary detail [5]. Expanding imaging power, sensitivity, and throughput enough to transform population-wide screening and surveillance may move medicine progressively toward earlier disease interception and preventative health. Overall, biomedical imaging innovations continue to catalyze paradigm shifts in what realms of biology research can unravel in the twenty-first century—the impact is likely to only grow exponentially with technological and computational co-advances in the coming decades. With an increasing ability to peer within living systems, we march toward the promises of "images into understanding diseases, guiding treatments, developing drugs, and demystifying life and consciousness," as articulated by pioneering researchers [6].

14.1.1 Revolutionizing Cancer Research

Cancer research is one of the most prolific domains benefiting from biomedical imaging advances. Right from early detection and diagnosis to drug development and surgical planning, imaging is woven into nearly every facet of the cancer care continuum. Research geared toward revealing tumor genesis routes, treatment resistance mechanisms, prevention strategies, and more reveals key insights enabled by biological imaging approaches [7]. Intravital microscopy techniques, for instance, have opened new windows into metastatic spread, angiogenesis, and detailed tumor microenvironment dynamics by providing continuous visualization of cancer pathologies progressing in situ [8]. Hybrid positron emission tomography-magnetic resonance imaging (PET-MRI) systems allow the combining of gene expression reporters visible on PET with soft tissue anatomical MRI scans to clarify how genetic abnormalities drive phenotypes [9]. Bioluminescence platforms tracking tumor growth via implanted window chambers facilitate efficient in vivo screening of novel targeted and immune-based anti-cancer agents [10]. Such imaging advances clarify cancer complexity, improve treatments, and transform routine cancer research protocols. Theranostics approaches leveraging imaging for therapy optimization on a personalized level presage the next leap [11].

14.1.2 Uncovering Neurological Disease Mechanisms

Modern neuroscience research also increasingly depends on biomedical imaging techniques for basic science queries on brain function and unlocking pathological mechanisms driving neurological disorders for precision medicine applications [12]. For instance, manganese-enhanced MRI enables pinpointing specific brain circuit activity alterations underlying neuropsychiatric disease behaviors and mapping intricate neuronal connectivity in response to stimuli or during development [13, 14]. Photoacoustic microscopy further empowers imaging vasculature dynamics in living rodents and human brains, facilitating research on stroke mechanisms and the effects of vascular deficiencies in dementia [15]. Hybrid MEMS devices combining optical stimulation with electrical recordings are advancing retinal prosthetic development and optic nerve regeneration research [16]. Brain-wide clarification and microscopy techniques provide unprecedented views of the central nervous system cell architecture for gleaning neuroanatomical principles [17]. Overall, imaging platforms actively drive innovation in revealing normal brain function and pathological deviations underlying debilitating neurological conditions. The capacity to extract detailed structural and functional maps non-invasively or minimally invasively provides researchers unparalleled windows into this enigmatic organ in vivo across models and human patients.

14.1.3 Engineering Tissues Guided by Imaging Feedback

The interdisciplinary field of tissue engineering, encompassing areas like biomaterials, stem cells, and developmental biology, similarly relies heavily on imaging techniques. From assessing cell implantation, migration, and engraftment to monitoring scaffold integration, vascularization, and long-term viability of lab-grown tissues, imaging provides researchers with indispensable feedback [18, 19] . Fluorescent labels tracking transplanted cell population fates get quantified through in vivo imaging to optimize regenerative therapies and clarify endogenous repair mechanisms [20]. Raman spectroscopic imaging of collagen architecture informs efforts to recapitulate the native extracellular matrix complexity vital for lab-grown construct strength and bioactivity [21]. Pulse-echo ultrasound monitors cardiac tissue maturation, while POSM tracks graft stiffness [22]. Photoacoustic imaging visualizes real-time oxygenation changes in thick organoids to model developmental processes and disease events [23]. Overall, imaging enables pioneering breakthroughs in generating functional, biomimetic substitute tissues tailored to treat injuries and replace diseased organs.

14.2 Fundamental Technologies in Biomedical Imaging

14.2.1 X-ray Radiography

Since their serendipitous discovery by Wilhelm Röntgen in 1895, X-rays have revolutionized medical science and healthcare by enabling visualization of internal anatomical structures in living organisms for the first time. This capability gave birth to the new fields of medical imaging and radiology. Among the numerous imaging modalities later developed, plain X-ray radiography remains one of the most common and indispensable diagnostic tools in medicine today. Their continued utility stems from being relatively inexpensive, convenient, accessible, and able to clearly image bone and lung pathologies with reasonable resolution [24]. Beyond routine diagnostics, X-ray radiography also plays some critical and often underappreciated roles in empowering cutting-edge biomedical research, even in this high-tech era.

Enabling Function Investigation through Motion: A unique capability unlocked by X-ray imaging is the capacity to visualize internal organ system activities that involve motion or flow. Researchers leverage this to study the mechanics and patterns of breathing, swallowing, digestion, blood circulation, locomotion, and other physiological processes without surgery or other complex setups. Barium swallow studies using contrast X-rays facilitate analyzing abnormalities in the esophagus motility behind conditions like dysphasia through distortions detected in the barium bolus flow pattern [25]. X-rays pinpoint ideological and anatomical causes of respiratory conditions like pneumonia and atelectasis, appearing as opacity changes indicating fluid buildup or lung collapse [26]. Joint motions during activities get captured by X-rays to assess interventions for injuries, arthritis, or related conditions [27]. Real-time X-ray imaging, called video fluoroscopy, offers functional investigation capabilities surpassing MRI or ultrasound in critical organs.

High-throughput Experimentation Platform: Another domain seeing innovative usage of X-ray imaging is small animal research, especially for high-throughput experimentation. For instance, small animal computed tomography (CT) allows rapid whole-body or organ-specific 3D morphological screening of disease models, inheritance patterns, and therapeutic responses in living mice or rats [28]. Besides micro CT, planar micrography setups leverage X-ray snapshots to track tumor sizes, implant movements, tissue distortions, and contrast agent perfusion dynamics without sacrificing animals across sequential time points in the same cohort [29]. This further boosts experimental efficiency and reduces artifacts relative to traditional histopathology endpoints. Together, such X-ray imaging applications enhance research reproducibility while minimizing animal numbers conforming to "3R" (replacement, reduction, refinement) principles guiding ethical animal research [30].

Archival Records and Anchors for Big Data Integration: Another application of conventional radiography is the widespread clinical adoption and, hence, availability of large digital chest X-ray image databases across healthcare systems [31]. This big data acts as a crucial patient record, anchoring additional molecular, genomic, and sensor data accrued during medical experiments, trials, or routine care. Radiomic feature extraction from X-ray data further augments its utility for computational studies seeking to integrate diverse test results for predicting personalized disease risk or for AI-based decision support [32]. Despite resolution limitations, the ubiquitous presence of X-ray images, relative standardization, and hospital interpretability aid emerging data-intensive research directions.

14.2.2 Computed Tomography

Computed tomography (CT) serves indispensably in pushing the boundaries of biomedical research on multiple fronts—from enhancing fundamental scientific understanding of health and disease to accelerating therapeutic development and innovation. A cardinal contribution of CT towards advancing medical research is facilitating detailed 3D visualization and measurements of anatomical structures in situ. The digital slice images obtained from CT scans are reconstructed into precise 3D meshes, accurately modeling complex bone geometry, tumors, vasculature, organs, and other features within living subjects [33]. Researchers leverage these models to uncover morphological correlates and underpinnings behind cardiovascular disease, musculoskeletal disorders, cancers, and neurological conditions. Orthopedic mechanisms and interventions are optimized by studying muscle–tendon architecture and forces elucidated via CT images [34]. Even micro-CT systems provide cellular-level resolution 3D reconstructions illuminating processes like bone microdamage accumulation in osteoporosis. Overall, CT facilitates unprecedented 3D perspectives into anatomy–function links driving health and disease mysteriously otherwise [35].

Besides structural mapping, the utility of CT research lies in enabling quantitative imaging—non-invasively measuring density, composition, and metabolic activity clues within tissues indicative of underlying molecular changes. Dual-energy CT, for instance, quantifies iodine uptake dynamics to track renal functioning or blood flow [36]. Spectral CT details tissue constituents down to chemicals linking back to disease pathways based on X-ray attenuation signatures. PET-CT integrates CT contrast with positron emission tomography to correlate anatomy with molecular signals like glucose metabolism levels associated with cancers [37].

Small animal CT systems designed explicitly for scanning rodent models also prolifically advance various realms of research, from genetic studies to tailored cancer models, toxicology assays, and pharmacokinetics. Quick micro CT phenotypic screening of aberrations across rodent generations enables associating genetic loci with disease risks [38]. CT measures living animals' tumor development, metastases, and pharmacodynamics, replacing cumbersome histology.

14.2.3 Magnetic Resonance Imaging (MRI)

Magnetic resonance imaging (MRI) has evolved into one of the most versatile and indispensable modalities within the biomedical imaging toolkit today. Clinically, its unmatched soft tissue contrast and capacity to extract functional and molecular information without harmful ionizing radiation underpin MRI's dominance for neurological, musculoskeletal, and oncological imaging. But MRI's utility stretches far beyond hospitals into the foundations driving healthcare—academic and industrial research unlocking the secrets of life [39].

Mapping Brain Structure and Function: Neuroscience research arguably utilizes and benefits the most from unique MRI capabilities. Beyond anatomy mapping through T1 and T2 contrasts, diffusion MRI traces white matter connections between brain regions, clarifying wiring architecture altered in psychiatric illnesses like schizophrenia [40]. Functional MRI charts neural activity signals across the brain responding to motor tasks or visual/auditory stimuli to identify networks involved in sensation, cognition, and behavior [41]. Real-time MRI helps probe the dynamics of blood oxygen utilization and cerebrovascular flow regulation, revealing the mechanics behind conditions like dementia and ADHD. MRI empowers both human and animal studies to uncover this enigmatic organ's inner workings non-invasively [42].

Enabling Cell Tracking: MRI leveraged across stem cell, immunology, and regenerative medicine research involves tracking the migration and fate of transplanted therapeutic cells toward tissue repair [43]. Magnetically labeling cells before delivery enables following their journeys through the body to clarify critical mechanisms of homing, participation in wound healing, and integration with native tissues for months after transfers [44]. This revolutionizes evaluating the safety and efficacy of emerging cellular therapies from stroke to heart disease treatments.

Phenotyping Model Organisms: MRI also transforms capabilities for phenotyping transgenic models and human gene variants to discover new disease markers and otherwise tricky mechanisms. Small animal MRI facilities enable rapid in vivo screening of abnormalities across rodent cohorts with various genetic makeups, diets, or drug exposures, revealing morphological and functional phenotypes—from heart defects to kidney fibrosis to brain connectivity changes [45]. Powerful 7–10 Tesla scanners provide imaging details rivaling pathology—all while allowing continuous longitudinal tracking of disease progression or onset in the same animals [46].

Guiding Drug Discovery: MRI advances multiple facets of new drug development as well—from early toxicity screening to proving clinical efficacy for approval [47]. Testing candidate safety involves profiling tissue changes post-drug administration in animals using quantitative MRI maps of composition, perfusion, inflammation, and shape deformities exceedingly sensitive to insult well before visible symptoms arise. Later phase trials demonstrate therapeutic activity through MRI-based metrics like tumor shrinkage, neurotransmitter levels, and blood flows in patients relative to placebo cohorts. MRI is both de-risking and enabling new treatments [48].

14.2.4 Ultrasound Imaging

Medical ultrasound uses high-frequency sound waves to visualize soft tissues in living subjects safely, conveniently, and cheaply. From bulky cart-based machines to portable devices, ultrasound has secured routine roles across clinical specialties—from obstetrics to cardiology [49]. However, the profound value ultrasound increasingly unlocks across diverse medical research domains often goes underappreciated.

Enabling Molecular Imaging: Contrast-enhanced ultrasound via gas microbubbles functionalized with targeting ligands fundamentally upgrades ultrasonography into a molecular imaging modality. By detecting bindings to biomarkers of diseases like cancer angiogenesis or inflammation, microbubble contrast provides anatomical and specific molecular signals that are otherwise only possible via modalities like PET [50]. This allows for studying disease pathways, predicting therapeutic responses, and monitoring treatments longitudinally through non-invasive ultrasound.

Guiding Image-Guided Interventions: Another research domain leveraging ultrasound is the image-guided delivery of therapeutic payloads directly into anatomical targets. High-intensity focused ultrasound (HIFU) employs focused sound energy for thermal ablation of tissues, explored in cancer and neurological treatments [51]. Beyond therapy, microbubble cavitation enables targeted drug and gene delivery across tissue barriers under ultrasound guidance. Ultrasound visualization verifies accurate needle biopsies, placements for brachytherapy, and catheter deliveries into precise locations like tumors and the heart [52].

Enabling Functional Assessments: Unique among imaging modalities, ultrasound provides versatile options for non-invasively quantifying dynamic physiological functional parameters relevant for research—from blood flow via Doppler signals to tissue elasticity mapping and shear wave readings indicating organ health [53]. Cardiac and vessel wall motion studies clarify developmental defects in animal models, while transient elastography tracks liver fibrosis development and resolution from interventions over time in the same subjects. Ultrasound empowers robust functional phenotyping crucial for disease modeling and drug efficacy studies [54].

Ultrasound technology lends itself to miniaturization—from wearable skin patches monitoring wound healing to pill-sized ingestible assessing gastrointestinal disorders [55]. Microscale ultrasound transducers integrated into needles, catheters, surgical tools, and other instruments also transform in situ visualization capacities during research procedures and interventions [56].

14.2.5 Positron Emission Tomography (PET)

Among the cornucopia of biomedical imaging modalities, positron emission tomography (PET) occupies a unique niche-providing 3D visualization and quantifying physiological, biochemical, and pharmacological processes in living subjects by detecting signals from injected radioactive tracers targeted to molecules and pathways implicated in disease. PET imaging transforms the capacity to peer within the living body at the molecular level—an astonishing superpower propelling discovery across diverse medical research fronts [57]. PET molecular imaging profoundly empowers fundamental biomedical research by enabling non-invasive mapping of molecular processes driving health and disease in vivo.

Elucidating Neurological Conditions: The foremost medical domain harnessing PET is neuroscience research geared towards unraveling normal brain functioning and clarifying neurological conditions from Alzheimer's to mood disorders through mapping regional signaling aberrations [58]. PET tracers like fluorodeoxyglucose help correlate behavioral deficits with changes in neuronal glucose metabolism patterns, while ligands targeting receptors and transporters provide biochemical fingerprints for conditions like Parkinson's disease and schizophrenia [59]. PET is propelling new treatment discoveries by enabling such human and animal model studies to clarify pathways that drive neuropathology.

Guiding Targeted Drug Development: PET also contributes to targeted drug development for oncology and an expanding roster of inflammation, infection, and metabolism disorders by providing phenotypic readouts of efficacy, selectivity, and pharmacodynamics non-invasively in subjects. Research leverages new PET tracers of processes like glucose metabolism alongside novel proprietary radioactive ligands for candidate drug targets to map target expression, binding potential, and downstream functionality changes post-therapy across sequential scans [60]. PET provides crucial pharmacokinetic data to improve experimental therapies while minimizing failures in late costly stages.

Enabling Image-Guided Interventions: PET imaging is incorporated to guide emerging avenues like immunotherapies, regenerative treatments, and photodynamic interventions towards previously untreatable conditions. For instance, PET helps deliver engineered T cells precisely to tumors exhibiting specific surface signatures for maximal efficacy of cytotoxic actions while avoiding off-target effects [61]. Likewise, external radiation beams for cancer therapy and light stimulation for optogenetic neurologic treatments are targeted based on PET signal maps, demarcating sites that most need intervention [62].

14.2.6 Single Photon Emission Computed Tomography (SPECT)

Single photon emission computed tomography (SPECT) is a nuclear imaging technique that allows for functional and molecular imaging of tissues and organs. It injects a radioactive tracer into the patient, emitting gamma rays as it decays. These gamma rays are detected by a gamma camera that rotates around the patient, collecting projections from multiple angles. A computer then reconstructs these projections into cross-sectional tomographic images, similar to CT or MRI scans. However, unlike CT and MRI, which provide anatomical images, SPECT includes information on physiology and biochemistry [63].

SPECT has become an indispensable tool in medical research, allowing researchers to visualize, quantify, and track molecular processes in vivo. The technique is used in cancer, neurological disorders, cardiac disease, infection, inflammation, and more studies. SPECT tracers or radiopharmaceuticals can target specific biological processes or molecular pathways. For example, in oncology, 99mTc-MDP can image bone metastases by binding to hydroxyapatite in bone tissue [64]. Similarly, 131I-MIBG is used to image adrenal and neuroendocrine tumors. Other SPECT tracers include 99mTc-HMPAO to measure regional cerebral blood flow, 111In-octreotide for imaging somatostatin receptors, and 201TI chloride for myocardial perfusion imaging. The appropriate tracer can provide crucial information about disease stage, progression, and treatment efficacy [65].

A significant advantage of SPECT over other imaging modalities is its extreme sensitivity and ability to detect nano to picomolar levels of radiotracer concentrations. This allows earlier detection of molecular processes compared to techniques like PET or MRI. SPECT systems also tend to be more widely available and affordable. The radiation exposure from most SPECT scans is tolerable, equivalent to background radiation over nine months to 3 years. Furthermore, SPECT images have excellent contrast and resolution, clearly delineating tissues and structures [66].

In neurology, SPECT is particularly useful in imaging blood flow, dopamine transport, and neurodegeneration. Differences in regional cerebral blood flow (rCBF) may indicate areas of neuronal dysfunction in conditions such as epilepsy, dementia, and movement disorders. 123I-FP-CIT SPECT helps differentiate Parkinsonian syndromes by imaging loss of dopamine transporter binding [67]. And 131I-MIBG uptake is reduced in Lewy Body dementia. SPECT enables understanding the relationship between molecular processes, physiology, and disease progression.

Within oncology, SPECT continues to play a significant role despite the growth of PET techniques. Using tumor-specific radiotracers, SPECT provides non-invasive characterization of primary tumors and metastases. This assists in diagnosis, determining tumor stage, planning treatment, and evaluating response [68]. SPECT also guides radionuclide therapy where radioactive materials directly damage tumors, as in 131I therapy for thyroid cancer and neuroendocrine tumors. SPECT-CT fusion improves anatomic localization and image quality.

Cardiovascular SPECT scanning assesses myocardial blood flow, viability, and function. Differences in tracer uptake indicate under-perfused regions and allow quantification of ischemia. Viability studies identify hibernating myocardium that may recover after revascularization [69]. Gated SPECT measures left ventricular volumes and ejection fraction. These capabilities lead to better risk stratification and clinical management in coronary artery disease. Research applications for SPECT are expanding with the development of novel targeted radiotracers. The ability to link radioisotopes to peptides, antibodies, nanoparticles, and drugs enables non-invasive quantification and tracking of molecular binding in vivo. SPECT provides insight into pharmacokinetics and drug biodistribution. Agents that target amyloid-beta plaques may improve diagnosis and monitoring of Alzheimer's disease. SPECT also guides radionuclide therapy dosimetry to optimize delivery while minimizing toxicity. Ongoing innovation in SPECT hardware and software also promises to advance research usage. Dedicated cardiac cameras offer improved diagnostic capabilities. Hybrid SPECT/CT devices provide better anatomic correlation and correction for photon attenuation [70]. Software solutions enable more patient-specific dosimetry and three-dimensional displays. Collimators with motion tracking capability can image freely moving structures like the heart.

14.2.7 Optical Imaging Techniques

Optical imaging techniques utilize light and the properties of photon propagation through tissue to probe structure and function non-invasively. These methods provide essential capabilities for medical research beyond traditional imaging modalities like MRI, CT, and ultrasound. Optical imaging offers specific advantages in real-time response, cost-effectiveness, sensitivity, and ability to guide interventions. Methods like optical coherence tomography (OCT), diffuse optical tomography (DOT), fluorescence imaging, and endoscopy have become indispensable tools for clinical and pre-clinical research across many fields [71]. OCT uses near-infrared light to capture micrometer-resolution images showing tissue morphology in real time. It works by measuring the time delay and magnitude of backscattered light. OCT is analogous to ultrasound imaging, except using light instead of sound waves. Essential areas of medical research using OCT include ophthalmology, cardiology, dermatology, and oncology [72]. In ophthalmology, OCT allows detailed imaging of retinal abnormalities and optic nerve head structure—critical for managing diseases like glaucoma and macular degeneration. Intravascular OCT enables minute evaluation of coronary artery plaque to guide stent placement. OCT assists in early cancer detection by improving the visualization of tissue microarchitecture [73].

DOT uses multiple external light sources and sensors to determine light absorption and scatter within larger tissue volumes. Computer reconstruction generates spatial maps showing oxygen saturation and hemoglobin concentration. DOT has major research applications in breast cancer, providing functional information on angiogenesis and tumor response to therapy [74]. It is also applied in brain imaging of cerebral oxygen metabolism related to neuronal activity. DOT offers bedside monitoring capabilities that complement other modalities like MRI. Fluorescence imaging encompasses a range of microscopic and macroscopic techniques utilizing fluorescent dye reporters. It allows sensitive, specific, and real-time tracking of molecular processes in vivo—critical in cancer targeting, surgical guidance, and drug development. Researchers can non-invasively monitor protease activity around tumors using activatable fluorescent probes. Intraoperative imaging assists surgeons in differentiating cancerous and normal tissues. Fluorescence labeling of pharmaceuticals elucidates biodistribution and pharmacokinetics. Fluorescence is also widely used in microscopy, cell culture assays, and molecular biology [75].

Other emerging optical techniques show increasing utility for medical research. Photoacoustic tomography combines pulsed laser excitation with acoustic detection of ultrasonic emission. This offers deeper tissue penetration for anatomical, functional, and molecular imaging compared to pure optical modalities [76]. Raman spectroscopy leverages inelastic scattering of light to characterize tissue biochemical composition. Nonlinear microscopy and super-resolution techniques like STED enable visualization below the diffraction limit. These methods provide unprecedented views of tissue microenvironments, molecular interactions, and nanoscale biomechanics. Optical reporters and instrument designs continue improving. For example, researchers have created fluorophores that switch on via enzymatic reactions or changes in redox potential, local heating, ion concentrations, membrane voltage, and protein conformation. Multispectral optoacoustic tomography can image multiple fluorescent probes simultaneously. New endoscopic probes enable microscopic imaging during clinical procedures. Fiber optic sensors facilitate internal body sensing of pressure, temperature, and analytes. Dedicated imaging systems integrate multiple modes like OCT, laser scanning, and fluorescence microscopy [77].

The benefits of optical imaging in medical research are ongoing. Optical modalities are inexpensive, versatile, and easily combined to provide structural, functional, and biochemical information. They enable the microscopic examination of living cells and tissues in real time. Imaging can guide surgery, biopsies, and other interventions [78]. Optical reporters advance understanding of myriad biological processes. Optical imaging sheds light on pathogenesis while paving the way for highly targeted diagnostics and therapy.

14.3 Advanced Applications in Biomedical Imaging in Research

14.3.1 Disease Diagnosis and Progress Monitoring

Advanced medical imaging allows physicians to diagnose diseases earlier and more accurately than traditional approaches. Modalities like MRI, PET, and SPECT scanning visually represent biochemical and functional changes in organs and tissues, revealing abnormalities at very early stages [79]. These sensitive techniques are also routinely applied to evaluate disease progression, helping assess ongoing impacts on tissue and organs. For example, functional MRI scans that show progressive loss of neuron activity combined with shrinking brain matter can confirm the worsening of neurodegenerative disorders like Alzheimer's. Tracking subtle changes through quantitative molecular imaging biomarkers supports precise disease staging while also determining the efficacy and timing of experimental and established therapeutic interventions [4].

14.3.2 Drug Development and Pharmacokinetics

Nuclear imaging methods, especially PET, have become indispensable in contemporary drug development pipelines and pharmacokinetic research. They allow unparalleled visualization and characterization of medication pharmacodynamics within target tissues or whole organisms by labeling drug compounds with radioactive tracers [60]. The distribution, binding kinetics, and dissociation can be monitored over time, quantifying absorption, bioavailability, and excretion. PET imaging visually maps dosages and receptor occupancy levels, identifying optimal doses in early-phase trials. By non-invasively demonstrating engagements with biological targets, ineffective compounds are eliminated faster while gathering data on human pharmacokinetic responses [80]. Molecular imaging pharmacometric technologies strongly support economical drug discovery, validation, and approval processes.

14.3.3 Gene Therapy and Molecular Imaging

As a critical tool in gene therapy development, multi-modality molecular imaging enables noninvasive spatiotemporal assessment of gene transfer vectors and expression in living subjects. The migration, delivery efficiency, and gene expression level over short and long terms can be examined using reporter genes and probes. Researchers can compare the performance of delivery platforms like adeno-associated viruses and lentiviruses across cell and animal models in inducing stable target protein generation. Similarly, the effects of RNA interference therapies utilizing methods like RNA antisense or siRNA can be monitored by measuring downstream molecular events [20]. These approaches provide rich longitudinal biomarker data on therapeutic gene manipulation efficacy and persistence, greatly benefiting the advancement of human clinical trials. They promise more effective gene treatments for cancer, neurological, cardiovascular, and genetic diseases.

14.3.4 Neuroimaging and Understanding Brain Functions

Neuroimaging techniques like fMRI have been crucial in furthering insights into structural and functional brain connectivity related to cognitive, behavioral, and emotional processes. Using blood oxygenation signals, fMRI detects neural activity changes while subjects are exposed to sensory, motor, or emotional stimuli or engaged in different tasks. This allows mapping activated brain regions to specific mental processes [81]. Diffusion MRI and tractography methods demonstrate white matter integrity and anatomical connectivity by measuring water diffusion and constructing wiring diagrams of fiber bundles between brain regions. With conditions like autism, schizophrenia, and ADD/ADHD linked to atypical connectivity, these powerful tools clarify networks and pathways disturbances behind psychiatric illnesses. Neuroimaging platforms enable unprecedented understanding of human brain functional organization in health and disease [82].

14.3.5 Oncology: Tumor Detection and Monitoring

In clinical cancer management, integrated anatomical and functional imaging using systems like combined PET/CT and PET/MRI scanners enable enhanced sensitivity in diagnosis, staging, and therapy planning. PET visualizes increased glucose metabolism and discriminates malignant lesions, while CT offers anatomical detail. Adding MRI improves soft tissue contrast, showing structural and vascular changes. By registering molecular, functional, and morphology, small lesions and metastasis missed on CT/MRI alone can be detected. Tracking standardized uptake values from PET alongside tumor shape throughout treatment also gives a reliable assessment of therapeutic response [83]. Using multiple imaging biomarkers fulfills precision oncology's goal of personalized therapy based on molecular profile. Newer approaches like photoacoustic imaging, exploiting both light and sound, promise to improve detection specificity further through better spectroscopic distinction of tumor tissue.

14.3.6 Cardiovascular Imaging

Cardiac imaging techniques have become the forefront of managing heart disease—a leading global cause of morbidity and mortality. Echocardiogram using ultrasound waves furnishes real-time moving images of heart chambers, valves, and vessels to evaluate anatomy and pumping activity. Stress echos examine function under induced hypertension. MRI offers 3D dynamic views, quantifying chamber volumes, muscle viability, and perfusion deficits, while PET maps decreased tissue glucose uptake, indicating coronary blockages [84]. CT angiography visualizes stenosis and plaque formations in arteries to gauge stroke risk and guide stent placements. SPECT scanning assesses cell death and ischemia by tracing radioactive tracers [66]. Applying multiple modalities provides comprehensive structural, functional, and metabolic perspectives for precise phenotyping to deliver optimized, personalized cardiovascular care, translating to better outcomes.

14.4 Future Directions and Emerging Technologies

The future of biomedical imaging in research is bright, with emerging technologies promising even greater insights. The integration of artificial intelligence in image analysis is set to revolutionize how we interpret imaging data, making it faster and more accurate. Nanotechnology is another frontier with the potential to dramatically enhance imaging resolution and sensitivity. Hybrid imaging systems, combining the strengths of different modalities, are already showing promise in clinical trials. Developments in molecular imaging continue to advance our understanding of disease processes at the molecular level, while personalized medicine stands to benefit significantly from tailored imaging strategies.

14.4.1 Artificial Intelligence in Image Analysis:

Artificial intelligence and machine learning technologies are poised to significantly assist medical research through advanced analysis of medical imaging data. Medical imaging techniques like X-rays, CT scans, MRI, and ultrasound generate massive volumes of complex visual information. While this data holds valuable insights, it can overwhelm human analysis capabilities. AI image recognition algorithms promise to unlock deeper understandings from medical scans to further research and improve patient care. Deep learning systems can be trained on vast databases of medical images to identify visual patterns imperceptible to the human eye. For example, AI can analyze CT scans for early signs of cancer or other pathological indications before they manifest physically [85]. This capacity for early and accurate detection through pattern recognition can aid disease diagnosis and monitoring. Further, AI image analysis can extract quantitative data on anatomical structures and physiological processes. This data fuels research on the underlying mechanisms of health and disease.

An essential application is using AI to prescreen medical images as an automated "second set of eyes" to surface cases needing urgent expert review. This helps direct clinical resources more efficiently so radiologists can focus on the most critical cases. Workforce shortages in radiology make such AI assistance crucial for handling growing imaging workloads. This will accelerate research that relies on extracting insights from medical images. Overall, AI image recognition stands to uncover new medical knowledge and optimize healthcare delivery, making it an invaluable partner for twenty-first-century medical research and practice [86]. The integration of AI in analyzing and interpreting medical images promises to transform biomedical research and practice by unlocking unrealized value from imaging data. This will ultimately lead to better-informed, more accurate, and earlier disease insights to improve diagnostic, prognostic, and therapeutic practices.

14.4.2 Nanotechnology and Imaging:

Nanotechnology is revolutionizing medical imaging by engineering nanomaterials that interface with biological systems at the molecular level. Scientists are developing customized nanoparticles to act as contrast agents and molecular probes to improve the resolution and specificity of medical imaging techniques. A significant focus is enhancing cancer imaging using nanoparticles engineered to identify and interact with cancer biomarkers. For example, researchers have designed gold nanoparticles to target and illuminate epidermal growth factor receptors overexpressed in specific cancers [87]. This enables more precise visualization of tumors via techniques like MRI and CT scans. Similarly, nanoparticles carrying radionuclides can spotlight cancer locations via PET and SPECT scanning when they accumulate there.

Beyond cancer, scientists are imaginatively applying multifunctional nanoparticles to reveal deeper biological insights. Magnetic nanoparticles induce contrast in MRI scans, while photoluminescent ones work for fluorescence imaging. Combined, these nanoparticles enable merging MRI and fluorescence imaging to provide anatomical, functional, and molecular information simultaneously. Nanoparticle engineering is also making ultrasound imaging dramatically more sensitive—a technique using nanoparticle-stabilized microbubbles has improved ultrasound imaging 100-fold. The unique properties of materials at the nanoscale present game-changing opportunities for medical imaging. Personalized, targeted nanoparticles demonstrate immense potential for spotting cancer earlier, visualizing cellular activities, and combining imaging modalities. With their high tunability, multifunctionality, and biocompatibility, we are likely seeing only the first glimmers of how nanomaterials can enhance medical imaging and research.

14.4.3 Hybrid Imaging Systems

Medical imaging techniques like X-ray, MRI, and CT scans have become essential for disease diagnosis and guiding treatment decisions. Each modality provides distinct information- X-rays capture bone, MRI offers soft tissue contrast, and CT reveals detailed body structures. However, relying on a single imaging method can miss important clinical indications that are only visible via a different technique. This has driven the development of hybrid imaging systems integrating multiple scanning modalities into a unified device. A significant focus has been combining anatomical, functional, and metabolic imaging in one examination. PET/CT pioneered this hybrid approach by fusing CT scans that depict anatomy with PET data that conveys body function through radioactive tracers. This improves the localization of abnormalities spotted on PET by mapping them onto CT images. PET/MRI aims to enhance this capability further—MRI offers superior soft tissue visualization and avoids the radiation exposure of CT. Consequently, PET/MRI provides the highest sensitivity metabolic PET imaging aligned with exquisite MRI anatomical images [88].

Further, MRI/radiation therapy systems enable real-time imaging of organs during radiation treatment. This allows finely focused dose delivery to be adjusted as the patient breathes. Fusing ultrasound with pre-acquired MRI or CT images assists minimally invasive procedures using live ultrasound footage. Overall, concurrent visualization via multiple imaging modalities empowers more targeted, adaptive, and insightful care. Image fusion systems merging X-ray, CT, PET, and MRI together could be on the horizon [89]. This would provide unprecedented views of integrated anatomical, functional, and metabolic information to reveal deeper biological relationships in health and disease. Hybrid medical imaging promises to drive better decision-making through consolidated diagnostic perspectives.

14.4.4 Developments in Molecular Imaging

Molecular imaging is a fast-evolving discipline within medical imaging that enables visualization and quantification of cellular function and molecular processes. By using targeted imaging agents as probes, molecular imaging allows noninvasive monitoring of disease biomarkers and pathways in vivo, ushering in new possibilities for understanding and precision management of disease. Recent advances are expanding the capabilities and applications of molecular imaging in crucial ways. Novel multifunctional nanoparticles and radiotracers enable detailed cellular-level views of dynamic biological interactions. From animal studies to humans, these targeted contrast agents exhibit the potential for earlier and specific detection of diseases like cancer-based on molecular changes [90]. Hybrid imaging systems like PET/MRI propel molecular imaging by merging susceptible molecular data with detailed anatomical images. Further, combining optical imaging techniques with MRI and CT enables the investigation of molecular dynamics simultaneously across vast fields of view and at micro-scales. Molecular imaging also advances personalized medicine through radiogenomics—linking imaging phenotypic signatures to genomic patterns. This can predict tumor response to therapies, enabling adaptive treatment regimens tailored to individuals. Overall, expansions in multimodal targeted imaging agents, hybrid imaging hardware, and linkage with genomic markers drive molecular imaging to the forefront of biomedical research and precision medicine. By peering deep within living systems, molecular imaging promises to provide powerful tools for combating disease.

14.5 Conclusion

Biomedical imaging has become an indispensable tool in medical research and clinical practice, allowing unprecedented insights into the structure, function, and molecular processes within living systems. From X-rays to advanced modalities like PET, MRI, and optical imaging, these techniques enable non-invasive visualization across scales—from whole-body to cellular and molecular levels. As this chapter highlights, biomedical imaging is crucial across various research domains. In oncology, it allows early detection, staging, and monitoring of tumors. Neuroimaging unravels the complexities of brain function and disorders. Cardiovascular imaging guides the management of heart diseases. Molecular imaging and theranostics are paving the way for personalized medicine. Furthermore, emerging technologies like artificial intelligence, nanotechnology, and hybrid imaging systems promise to enhance the capabilities and impact of biomedical imaging even further. However, the rapid growth of imaging technologies also presents challenges. The vast amounts of data generated require advanced computational methods for efficient management and analysis. Ensuring data privacy and security is critical when dealing with sensitive medical information. Moreover, the responsible use of imaging in research and clinical settings necessitates addressing ethical considerations, such as balancing the benefits against potential risks and ensuring equitable access. In conclusion, biomedical imaging has transformed the landscape of medical research and practice. It continues to evolve, pushing the boundaries of what we can visualize, quantify, and understand about the human body in health and disease.

References

1. H.S. Lahoti, S.D. Jogdand, Cureus **14**, e28923 (2022)
2. Q. Huang, A. Garrett, S. Bose, S. Blocker, A.C. Rios, H. Clevers, X. Shen, Cell Stem Cell **28**, 603 (2021)
3. D.P. Popescu, L.P. Choo-Smith, C. Flueraru, Y. Mao, S. Chang, J. Disano, S. Sherif, M.G. Sowa, Biophys. Rev. **3**, 155 (2011)
4. F.Y. Chiu, Y. Yen, Biomark Res. **11**, 35 (2023)

5. J. Cal-Gonzalez, I. Rausch, L.K. Shiyam Sundar, M.L. Lassen, O. Muzik, E. Moser, L. Papp, T. Beyer, Front. Phys. **6** (2018)
6. M. Flores, G. Glusman, K. Brogaard, N.D. Price, L. Hood, Per. Med. **10**, 565 (2013)
7. A. Pulumati, A. Pulumati, B.S. Dwarakanath, A. Verma, R.V.L. Papineni, Cancer Rep. (Hoboken) **6**, e1764 (2023)
8. E.M. Gabriel, D.T. Fisher, S. Evans, K. Takabe, J.J. Skitzki, Oncotarget **9**, 20165 (2018)
9. B.J. Pichler, H.F. Wehrl, A. Kolb, M.S. Judenhofer, Semin. Nucl. Med. **38**, 199 (2008)
10. D.M. Close, T. Xu, G.S. Sayler, S. Ripp, Sensors (Basel) **11**, 180 (2011)
11. L.B. Solnes, R.A. Werner, K.M. Jones, M.S. Sadaghiani, C.R. Bailey, C. Lapa, M.G. Pomper, S.P. Rowe, J. Nucl. Med. **61**, 311 (2020)
12. C. Yen, C.L. Lin, M.C. Chiang, Life (Basel) **13**, 1472 (2023)
13. J.M. Malheiros, F.F. Paiva, B.M. Longo, C. Hamani, L. Covolan, Front. Neurol. **6**, 161 (2015)
14. T.W. Uselman, C.S. Medina, H.B. Gray, R.E. Jacobs, E.L. Bearer, NMR Biomed. **35**, e4675 (2022)
15. X. Yang, Y.H. Chen, F. Xia, M. Sawan, Photoacoustics **23**, 100287 (2021)
16. Z. Ramezani, K. J. Seo, H. Fang, J. Micromech. Microeng. **31** (2021)
17. H.R. Ueda, H.U. Dodt, P. Osten, M.N. Economo, J. Chandrashekar, P.J. Keller, Neuron **106**, 369 (2020)
18. R. Bardini, S. Di Carlo, Comput. Struct. Biotechnol. J.**23**, 601 (2024)
19. J.R. Richards, J.P. McGahan, Radiology **283**, 30 (2017)
20. C. Ashmore-Harris, M. Iafrate, A. Saleem, G.O. Fruhwirth, Mol. Ther. **28**, 1392 (2020)
21. M.S. Bergholt, A. Serio, M.B. Albro, Front. Bioeng. Biotechnol. **7**, 303 (2019)
22. O. Pedreira, C. Papadacci, L. Augeul, J. Loufouat, M. Lo-Grasso, M. Tanter, R. Ferrera, M. Pernot, EBioMedicine **83**, 104201 (2022)
23. S.V. Bodea, G.G. Westmeyer, Front. Neurosci. **15**, 655247 (2021)
24. D. Strickland, A.N. Stranges, Med. Secoli. **3**, 207 (1991)
25. U. Debi, M. Sharma, L. Singh, A. Sinha, Indian J. Radiol. Imaging **29**, 141 (2019)
26. L.A. Rousan, E. Elobeid, M. Karrar, Y. Khader, BMC Pulm. Med. **20**, 245 (2020)
27. X. Wang, W. M. Oo, J. M. Linklater, Rheumatology (Oxford) **57**, iv51 (2018)
28. C.T. Badea, M. Drangova, D.W. Holdsworth, G.A. Johnson, Phys. Med. Biol. **53**, R319 (2008)
29. S. Das, K. Dholam, S. Gurav, K. Bendale, A. Ingle, B. Mohanty, P. Chaudhari, J.R. Bellare, Sci. Data **9**, 348 (2022)
30. M. Obrecht, S. Zurbruegg, N. Accart, C. Lambert, A. Doelemeyer, B. Ledermann, N. Beckmann, Front. Pharmacol. **14**, 1177421 (2023)
31. G. Irmici, M. Ce, E. Caloro, N. Khenkina, G. Della Pepa, V. Ascenti, C. Martinenghi, S. Papa, G. Oliva, M. Cellina, Diagnostics (Basel) **13** (2023)
32. B. Kocak, E.S. Durmaz, E. Ates, O. Kilickesmez, Diagn. Interv. Radiol. **25**, 485 (2019)
33. S.A.A. El-Gendy, B.M. Kamal, M.A.M. Alsafy, BMC Zool. **7**, 49 (2022)
34. D. Özer Kaya, in *Comparative Kinesiology of the Human Body*, ed. by S. Angin, I.E. Şimşek (Academic Press, 2020), p. 115
35. H. Yu, S. Wang, Y. Fan, G. Wang, J. Li, C. Liu, Z. Li, J. Sun, Front. Phys. **10** (2022)
36. X. Wang, D. Liu, X. Zeng, S. Jiang, L. Li, T. Yu, J. Zhang, Quant. Imaging Med. Surg. **11**, 1946 (2021)
37. A. Zhu, D. Lee, H. Shim, Semin. Oncol. **38**, 55 (2011)
38. D.P. Clark, C.T. Badea, Phys. Med. **88**, 175 (2021)
39. E.J.R. van Beek et al., J. Magn. Reson. Imaging **49**, e14 (2019)
40. F. Zhang, A. Daducci, Y. He, S. Schiavi, C. Seguin, R.E. Smith, C.H. Yeh, T. Zhao, L.J. O'Donnell, Neuroimage **249**, 118870 (2022)
41. J.E. Chen, G.H. Glover, Neuropsychol. Rev. **25**, 289 (2015)
42. K.S. Nayak, Y. Lim, A.E. Campbell-Washburn, J. Steeden, J. Magn. Reson. Imaging **55**, 81 (2022)
43. H. Nejadnik, J. Tseng, H. Daldrup-Link, Wiley Interdiscip. Rev. Nanomed. Nanobiotechnol. **11**, e1552 (2019)
44. X. Li et al., J. Nanobiotechnol. **18**, 113 (2020)

45. H. Li, A. Abaei, P. Metze, S. Just, Q. Lu, V. Rasche, Front. Phys. **8** (2020)
46. D.A. Feinberg et al., Nat. Methods **20**, 2048 (2023)
47. Y.X. Wang, M. Deng, J. Thorac. Dis.**2**, 245 (2010)
48. Y. Shi, W. Wu, BMC Med. **21**, 372 (2023)
49. X. Qian, R. Wodnicki, H. Kang, J. Zhang, H. Tchelepi, Q. Zhou, IEEE Trans. Ultrason. Ferroelectr. Freq. Control **67**, 2230 (2020)
50. G. Kose, M. Darguzyte, F. Kiessling, Nanomaterials (Basel) **10**, 1935 (2020)
51. V.S. Bachu, J. Kedda, I. Suk, J.J. Green, B. Tyler, Ann. Biomed. Eng. **49**, 1975 (2021)
52. W. Tzu-Yin, K.E. Wilson, S. Machtaler, J.K. Willmann, Curr. Pharm. Biotechnol. **14**, 743 (2013)
53. G. Cloutier, F. Destrempes, F. Yu, A. Tang, Insights Imaging **12**, 127 (2021)
54. N. Nagdyman et al., Cardiovasc. Diagn. Ther. **9**, S198 (2019)
55. S. Patel, F. Ershad, M. Zhao, R. R. Isseroff, B. Duan, Y. Zhou, Y. Wang, C. Yu, Soft Sci. **2** (2022)
56. A. Kumar, K.M. Kempski Leadingham, M.J. Kerensky, S. Sankar, N.V. Thakor, A. Manbachi, Front. Med. Technol. **5**, 1238129 (2023)
57. S. Katal, L.S. Eibschutz, B. Saboury, A. Gholamrezanezhad, A. Alavi, Diagnostics (Basel) **12**, 426 (2022)
58. Q. Qiao, Z. Qu, S. Tian, H. Cao, Y. Zhang, C. Sun, L. Jia, W. Wang, Neuropsychiatr. Dis. Treat. **18**, 2181 (2022)
59. C. Bouter, P. Henniges, T.N. Franke, C. Irwin, C.O. Sahlmann, M.E. Sichler, N. Beindorff, T.A. Bayer, Y. Bouter, Front. Aging Neurosci. **10**, 425 (2018)
60. S.G. Nerella, P. Singh, T. Sanam, C.S. Digwal, Front. Med. (Lausanne) **9**, 812270 (2022)
61. Z. Liu, Z. Zhou, Q. Dang, H. Xu, J. Lv, H. Li, X. Han, Theranostics **12**, 6273 (2022)
62. K. Koka, A. Verma, B.S. Dwarakanath, R.V.L. Papineni, Cancer Manag. Res. **14**, 1421 (2022)
63. S. Dorbala et al., J. Nucl. Cardiol. **25**, 1784 (2018)
64. V. Frantellizzi, M. Ricci, A. Cimini, L. Filippi, M. Conte, M.S. De Feo, G. De Vincentis, Appl. Sci. **13**, 1890 (2023)
65. W. Fang, S. Liu, Curr. Radiopharm. **12**, 171 (2019)
66. G. Crisan, N. S. Moldovean-Cioroianu, D. G. Timaru, G. Andries, C. Cainap, V. Chis, Int. J. Mol. Sci. **23** (2022)
67. B. Ballanger, T. van Eimeren, A.P. Strafella, in *Textbook of Stereotactic and Functional Neurosurgery*. ed. by A.M. Lozano, P.L. Gildenberg, R.R. Tasker (Springer, Berlin, Heidelberg, 2009), p.307
68. E. Ouvrard, A. Kaseb, N. Poterszman, C. Porot, F. Somme, A. Imperiale, Front. Med. (Lausanne) **10**, 1320574 (2023)
69. A.I. Loffler, C.M. Kramer, Interv. Cardiol. Clin. **7**, 355 (2018)
70. J.S. Lee, G. Kovalski, T. Sharir, D.S. Lee, J. Nucl. Cardiol. **26**, 543 (2019)
71. B. E. Bouma et al., Nat. Rev. Methods Primers **2** (2022)
72. J.G. Fujimoto, C. Pitris, S.A. Boppart, M.E. Brezinski, Neoplasia **2**, 9 (2000)
73. J. Wang, Y. Xu, S.A. Boppart, J. Biomed. Opt. **22**, 1 (2017)
74. Q. Zhu, S. Poplack, Eur. J. Radiol. **129**, 109067 (2020)
75. Y. Yang, H. Hong, Y. Zhang, W. Cai, Cancer Growth Metastasis **2**, 13 (2009)
76. J. Yao, L.V. Wang, Contrast Media Mol. Imaging **6**, 332 (2011)
77. M. Kaur, P.M. Lane, C. Menon, Appl. Sci. **10**, 6865 (2020)
78. D.L. Farkas, Molecules **26** (2021)
79. S. Hussain, I. Mubeen, N. Ullah, S. Shah, B.A. Khan, M. Zahoor, R. Ullah, F.A. Khan, M.A. Sultan, Biomed. Res. Int. **2022**, 5164970 (2022)
80. L.K. Vora, A.D. Gholap, K. Jetha, R.R.S. Thakur, H.K. Solanki, V.P. Chavda, Pharmaceutics **15** (2023)
81. C. Yen, C.L. Lin, M.C. Chiang, Life (Basel) **13** (2023)
82. L.Z. Chen, A.J. Holmes, X.N. Zuo, Q. Dong, Psychoradiology **1**, 272 (2021)
83. S.L. Rice, K.P. Friedman, PET Clin. **11**, 387 (2016)
84. W. Jin, L. Wang, T. Zhu, Y. Ma, C. Yu, F. Zhang, BMC Cardiovasc. Disord. **22**, 218 (2022)

85. L. Pinto-Coelho, Bioengineering (Basel) **10** (2023)
86. C.J. Cai, S. Winter, D. Steiner, L. Wilcox, M. Terry, Proc. ACM Hum. Comput. Interact. **3**, 1 (2019)
87. M. Chehelgerdi et al., Mol. Cancer **22**, 169 (2023)
88. L. Papp, C.P. Spielvogel, I. Rausch, M. Hacker, T. Beyer, Front. Phys. **6** (2018)
89. J. Ng, F. Gregucci, R.T. Pennell, H. Nagar, E.B. Golden, J.P.S. Knisely, N.J. Sanfilippo, S.C. Formenti, Front. Oncol. **13**, 1117874 (2023)
90. G. Liu, X. Yang, X. Zhou, Biosensors (Basel) **13** (2023)

Index

N. Kalarikkal et al. (eds.), *Multimodal Biomedical Imaging Techniques*, Biological and Medical Physics, Biomedical Engineering, https://doi.org/10.1007/978-981-96-1124-9

The manufacturer's authorised representative in the EU is Springer Nature Customer Service Centre GmbH, Europaplatz 3, 69115 Heidelberg, Germany. If you have any concerns regarding our products, please contact ProductSafety@springernature.com

Printed and bound by CPI Group (UK) Ltd, Croydon, CR0 4YY
30/09/2026
02225299-0005